Industrial Refrigeration

MECHANICAL ENGINEERING

A Series of Textbooks and Reference Books

Editor
L. L. Faulkner
Columbus Division, Battelle Memorial Institute
and Department of Mechanical Engineering
The Ohio State University
Columbus, Ohio

Vol. 78 Industrial Refrigeration: Principles, Design and Applications
P. C. Koelet

For information about additional volumes in this series, please contact the publisher:

Promotion Department
Marcel Dekker, Inc.
270 Madison Avenue
New York, NY 10016

800–228–1160

Industrial Refrigeration

Principles, Design and Applications

P. C. Koelet

with T. B. Gray

Marcel Dekker, Inc. New York • Basel • Hong Kong

First published in Great Britain 1992 by
THE MACMILLAN PRESS LTD
Houndmills, Basingstoke, Hampshire RG21 2XS
and London
Companies and representatives
throughout the world

ISBN 0-8247-8731-5

Published in the United States,
its dependencies, and Canada by:
MARCEL DEKKER, INC.
270 Madison Avenue
New York, NY 10016

Library of Congress Cataloging-in-Publication Data applied for

Printed in Hong Kong

Contents

Preface

Refrigeration is an exciting application of science and technology. It involves a range of subjects such as thermodynamics, mechanics, electrical technology, civil engineering, hydraulics, aerodynamics, chemistry and, last but not least, food technology. However, it must never be forgotten that refrigeration is not an end in itself but the means to an end – in most cases, to help condition man's environment and satisfy many of his nutritional needs. During the past 100 years, refrigeration has become indispensable to our society. Modern food processing, storage and transportation, for instance, would be impossible without the help of refrigeration.

I am convinced that all over the world engineers and technicians would like to learn more about the subject, and for this reason I have written this book in English. It is intended for those who, either as contractors or consultants, are involved in the design of refrigeration plants, and for those who are responsible for the purchasing or maintenance of these plants.

Industrial applications are generally larger than commercial applications and normally require a qualified operator on duty. Industrial refrigeration plants are found in the food processing, plastic, metallurgical and chemical industries, and on board ship. Such systems are based on a centralized design and normally use open-type compressors with high efficiency, long lifetime and power consumption above 30 kW.

This book deals not only with the installation and its components, but also with the products that are processed or stored in the refrigeration installation; it also considers the building in which the system operates.

The first chapters contain some useful and relevant thermodynamic background information. Later, a description of refrigeration and freezing systems is provided, as well as the calculation methods necessary to determine the refrigeration capacity and selection of the different components for the plant. This book is also about the application of refrigeration and therefore provides the reader with background information about the composition of and the optimum storage conditions for food products. Finally, it deals with the economic aspects, maintenance procedures, fault finding and repair.

Teachers and students in universities, polytechnics and technical colleges or others who would simply like to study refrigeration should find this book very useful.

I wish to thank everyone who was helpful in creating this book, especially my wife Mrs Maria Teresa Koelet-Paranhos de Oliveira for the extensive and difficult word processing and also Mrs Ilonka Perl-Chatterjee who, together with my wife, spent many hours editing the text.

I should also like to express my grateful thanks to Tom Gray for his painstaking work in helping to produce the final draft; his constructive comments have made a significant contribution throughout. Nonetheless, I remain ultimately responsible for any errors that may remain.

Last but not least I would like to acknowledge the assistance of all the individuals and companies quoted in the References at the end of the book.

PIETER C. KOELET
Meise, Belgium, 1991

Symbols and Units

Symbols

A m^2	Surface
a m^2/s	Temperature conductivity coefficient
c, c_p, c_v kJ/(kg K)	Specific heat
D m^2/s	Diffusion coefficient
E kJ	Energy
E kJ/s (= kW)	Energy flow
F	Force
H kJ	Enthalpy
h kJ/kg	Specific enthalpy
J $g/m^2/m/s$	Water vapour diffusion coefficient
k	Isentropic exponent
k $W/(m^2 K)$	Heat transmission coefficient
L kJ/kg	Specific evaporation enthalpy
l kJ/kg	Specific melting enthalpy
M kJ/kmol	Relative molecular mass
m kg	Mass
$\dot{m}$ kg/s	Mass flow
n	Exponent of polytrop
n 1/s; 1/min	Rotation number
n kmol	Molecular quantity
P kW	Power
p Pa = N/m^2; bar	Pressure
Q kJ	Heat load
Q_o kW	Refrigeration flow
$\dot{Q}_o$ kJ/s (= kW)	Heat flow
q kJ/kg	Specific heat load
q_o kJ/kg	Specific refrigerating load
q_o, v kJ/m^3	Volumetric refrigerant load

R kJ/(kg K)	Gas constant
R kJ/(kmol K)	Universal or absolute gas constant
S kJ/K	Entropy
s kJ/(kg K)	Specific entropy
T K	Absolute temperature
T_c K	Condensation temperature
T_e K	Evaporation temperature
T_o K	Room temperature
T_p K	Product temperature
t °C	Temperature
t h	Time
U m/s	Velocity
U kJ	Internal energy
V m^3	Volume
V' m^3/s; m^3/h	Volume flow
v m^3/kg	Specific volume
W kJ; kW h	Work
α W/(m^2 K)	Heat-transfer coefficient
β kg/(Pa/m^2/s)	Matter transfer coefficient
δ m	Thickness
ϵ kg/kg; %	Mass proportion
ϵ_0	Cold factor
η	Efficiency coefficient
η Pa s = N s/m^2	Dynamic viscosity
λ	Filling ratio
λ W/(m K)	Thermal conductivity coefficient
ν m^2/s	Kinematic viscosity
π	Pressure ratio
ρ kg/m^3	Density

International Units

The unit in which heat was expressed in the past was the calorie. 1 kcal of heat was the quantity necessary to raise the temperature of 1 kg of water by 1 degree. This unit was used until the 1970s in continental Europe. Often the British Thermal Unit (Btu) was used in English-speaking countries, and 'tons of refrigeration' in the USA. There is now an internationally agreed system of units called the SI (*Système International*). In this system a quantity of heat is expressed in joules (1 calorie = 4.1868 joules). The capacity of an installation is no longer expressed in calories/h, or in Btu, but in kW. The International System defines the unit of temperature as the kelvin, rather than °C or °F (1°C is equivalent to 1 kelvin, 0°C is 273.15 kelvin).

The SI or International System is based on six basic units plus two supplementary units. These are:

Unit of length	= metre (symbol m)
Unit of mass	= kilogram (symbol kg)
Unit of time	= second (symbol s)
Unit of electric current	= ampere (symbol A)
Unit of temperature	= kelvin (symbol K)
Unit of luminous intensity	= candela (symbol cd)
Unit of plane angle	= radian (symbol rad)
Unit of solid angle	= steradian (symbol sr)

The system uses the same units for energy, no matter what kind of energy is being considered. 150 years ago Newton propounded his laws of motion: his 'second law' is often expressed by force is equal to mass multiplied by acceleration, or $F = m \times a$. In honour of his achievement the new unit of force is called the newton (symbol N). It is the force required to accelerate a body of 1 kg mass at 1 m/s^2. Previously, the unit of force was the pound or the kg:

> 1 kg is the force that a body of 1 kg weight is subjected to by the pull of gravity at a certain point in Paris
>
> $1 \text{ kg force} = \text{kg mass} \times 9.807/s^2 = 9.807 \text{ N}$

The old system was confusing in as much as force and mass were both expressed in kg. Also in an age of space exploration, units based on gravitational force at a certain point of a certain planet are not logical.

Pressure = force/m^2. The unit of pressure is now 1 N/m^2, also called 1 Pascal (1 Pa). As this unit is small, for liquids and gases we use the unit 1 bar = 10^5 Pa = 10^5 N/m^2. 1 kg/cm^2 in the old metric system = 9.807×10^4 N/m^2 = 0.9807×10^5 Pa.

The unit of energy is derived from work = force × distance. Thus the unit of energy is 1 N × 1 m = 1 N m. This unit is now called 1 joule (1 J). 1 joule is the work done by a force of 1 N when it moves 1 m in the direction of the force. The old 1 kg (force) = 9.807 N.

$$1 \text{ kg m} = 1 \text{ m} \times (1 \text{ kg force} \times 9.807 \text{ N/kg force})$$

$$1 \text{ kg m} = 9.807 \text{ N m} = 9.807 \text{ J}$$

1 kcal was 427 kg m, which is 427 × 9.807 J or 4,187 J. Related to time, the old system used the unit of kcal/h. In the SI system, the unit of effort is the energy required to do work of 1 N m = 1 J in 1 second. The J/s is however called the watt (symbol W). We think J/s but say ‘W’.

$$1 \text{ kcal/h} = \frac{4187}{3600} \text{ J/s} = 1.163 \text{ J/s or } 1.163 \text{ W}$$

$$1 \text{ kcal/kg} = 4187 \text{ J/kg} = 4.187 \text{ kJ/kg}$$

Specific heat is now expressed in J/(kg K). Remember also that it is no longer permissible to use the metric unit of horse power (HP) but rather W or kW (1 HP was 75 kg m or 0.736 kW).

> Kinematic viscosity ν: 1 centistoke = 1 m m^2/s
>
> Dynamic viscosity η: 1 centipoise = 1 m Pa s

Speeds of revolution are no longer expressed as revolutions per minute (rpm):

$$1 \text{ rpm} = \pi/30 \text{ rad/s}$$

$$1 \text{ rad/s} = 30/\pi \text{ rpm}$$

Notwithstanding this, however, rpm are still used in this book. Experience has shown that few people in refrigeration technology use the new unit. In general, however, new units are being used routinely in many countries by refrigeration engineers.

1

Principles of Refrigeration

1.1 History

When discussing refrigeration we generally talk about relatively modern techniques, forgetting that in even prehistoric times people were storing food in caves with wet walls in an attempt to preserve it by taking advantage of the natural cooling effect. The inhabitants of the island of Crete in the Mediterranean were aware even around 2000 BC that a low temperature was of great importance in the preservation of food. Throughout the Minoan culture, specially constructed cellars were made for natural ice collected in winter, and this was used to preserve foodstuffs during the hot season. Records show that Alexander the Great, in about 300 BC, served his soldiers snow-cooled drinks to keep up their morale; in AD 755 Khalif Madhi operated a 'refrigerated transport' system from the Lebanon across the desert to Mecca, using snow as the refrigerant; and in AD 1040 the Sultans of Cairo used snow, transported daily from Syria, in their kitchens. The Arabs have known, from ancient times, how to keep water cool by storing it in earthern jars. Part of the water evaporates, thereby cooling the remainder.

From very early times until the beginning of the 20th Century people harvested natural ice during the winter and stored it in ice houses, cellars and later in ice stores for a year or more. The United States in the mid 19th Century, for example, had an important trade in natural ice mainly centred around the Hudson River and Maine. In Europe, at the same time, natural ice blocks from Norway were in great demand.

From 1805 to the end of the 19th Century, sailing vessels transported natural ice from North America to many of the warmer countries such as the West Indies, Europe, and even India and Australia (see figure 1.1). In 1872, 225 000 tonnes of natural ice were shipped to these places.

It was around this time, when mechanical refrigeration came of age, that most of the compressor types and the present-day cycles were conceived. It

Figure 1.1 *19th Century refrigeration 'engineers'*

could be said that mechanical refrigeration was first born when William Cullen, a Scot, succeeded in 1755 in making ice by evaporating ether under low pressure. In 1810 Sir John Lesley manufactured the first successful ice-making machine working on a similar principle. A milestone in the history of this development is 1834 when Jacob Perkins, an American, was granted British patent number 6662 on a vapour compression machine – nowadays the most widely used system in refrigeration. The specification is worth requoting: "What I claim is an arrangement whereby I am enabled to use volatile fluids for the purpose of producing the cooling or freezing of fluids, and yet at the same time constantly condensing such volatile fluids, and bringing them again and again into operation without waste."

Unfortunately, little evidence survives of Perkins taking the design further. James Harrison, a Scot, who emigrated to Australia in 1837, is credited with the invention of a successful hand-operated machine in the early 1850s, and Alexander Twinning was producing a tonne of ice a day in 1856 in Cleveland, Ohio.

While it was Carl von Linde who provided refrigeration with much of its thermodynamic background theory, many other scientists, British, German, French, American and Dutch, have contributed to the development of refrigeration: men such as Carré, Black, Faraday, Carnot, Joule, Mayer, Clausius, Thompson, Thomson (Lord Kelvin), Helmholtz and Kamerlingh Onnes. See also figures 1.2 and 1.3. For a complete survey of the history of refrigeration, reference should be made to Thevenot's *Histoire du Froid Artificiel.*

Figure 1.2 *One of the first early 20th Century refrigeration compressors*

Figure 1.3 *A modern refrigeration double-screw compressor*

1.2 Relationship between Pressure and Temperature

An important concept in refrigeration is the relationship between the pressure and temperature of a saturated vapour and the liquid associated with it. Before embarking on a detailed explanation, it is important to discuss atmospheric pressure and the pressure region below atmospheric pressure, commonly known as 'vacuum'.

It should help to follow a test originally carried out by an Italian physicist, Torricelli (1608–47). He took a glass tube sealed at one end, inverted it and completely filled it with mercury then, closing the open end, he turned the tube over so that the open end was at the bottom and immersed it into a bath of mercury. The end was then opened. Torricelli found that the liquid in the tube adjusted itself to balance the atmospheric pressure by falling down the tube until the appropriate height was attained. This was found to be 76 cm under normal conditions. The empty space above the mercury exerts no pressure and is a perfect or absolute vacuum. it is commonly referred to as the Torricellian vacuum (see figure 1.4a). This experiment can be repeated with water, but in this case we find that the height of water necessary to balance the atmospheric pressure is over 10 m.

When a vapour such as ether is injected into the space above the mercury, it is found that the level falls as the pressure above the mercury increases. However, if ether vapour is injected further, there comes a point at which liquid ether condenses and the pressure ceases to rise. At this point, when the small amount of liquid is in equilibrium with its vapour at a value of the pressure known as its saturated vapour pressure, we say that the space contains 'saturated' vapour and liquid (see figure 1.4b).

The pressure that we note, p_1, under these conditions is the saturation pressure and is affected only by the liquid and vapour temperature, which in this case is the ambient temperature t_1. When we increase the temperature of the ether in the space to t_2 it is found that the mercury level falls, indicating an increase in the saturation pressure to p_2 as shown on figure 1.4c. In fact some of the liquid will have evaporated in this process; however, so long as some liquid remains, the same pressure p_2 will always be indicated.

When we lower the temperature of the ambient to t_3 it is found that some of the liquid condenses and the vapour pressure falls to p_3 as shown in figure 1.4d. The relationships between the pressure and the temperature of ether and all practical working fluids have been determined and published in tables or charts. In the case of those fluids used as refrigerants, the tables provide additional information. See tables 3.10, 3.11, 3.12 and 3.13 in chapter 3.

Pressure gauges as used on compressors (usually indicating in bar, kilopascal or psi) are in fact showing the amount of pressure above

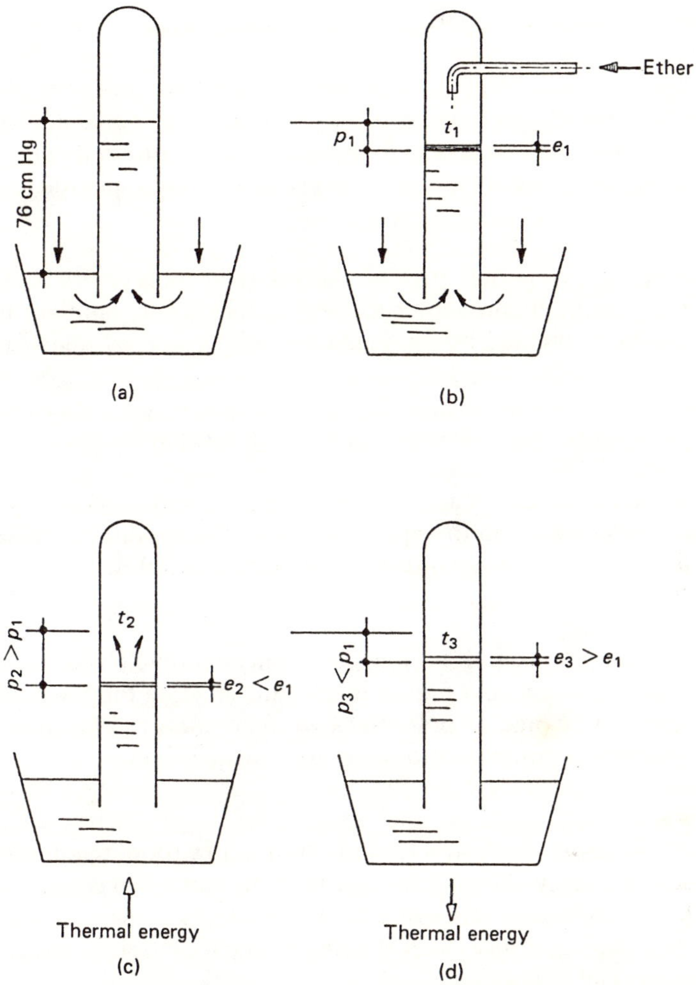

Figure 1.4 *The tube of Torricelli*

atmospheric pressure. If p_g is the gauge pressure, p_a the absolute pressure and p_{at} the atmospheric pressure, then

$$p_a - p_{at} = p_g$$

1.3 Energy

In order to understand what really happens during compression and expansion, it is necessary to recall a little about energy. There are various different forms of energy – nuclear, electrical, chemical, mechanical, hydraulic and thermal. Refrigeration is mainly concerned with thermal energy.

A recent concept in the field of thermodynamics has been the division of energy into two sorts – available energy or 'exergy' and unavailable energy or 'anergy'. Available energy is that portion of the internal energy that is available for useful work. It can be expressed by the equation

$$E_q/Q_t = 1 - T_a/T$$

Unavailable energy is that part of the internal energy which cannot be accessed by a heat engine under the restrictions of the ambient temperature. It can be expressed by the equation

$$Q_t = E_q + A_q$$

Here E_q = exergy, A_q = anergy, Q_t = total internal energy, T = system temperature and T_a = ambient temperature.

It is sufficient at this stage to remember that the amount of energy available is dependent on the application and the conditions prevailing at any particular time, as shown in the following examples.

1. *Hydraulic energy*

Stationary water at ground level posseses only anergy. To creat exergy we have to take the water to a higher level, thus creating the level difference that produces a change in potential hydraulic energy. The difference in level is created by pumping the water to a higher level.

2. *Thermal energy*

If the thermal energy to heat a room is at or below room temperature, it is in the form of anergy. Thus we cannot heat the room. To create exergy we must take the thermal energy to a higher level. By using a heat pump we can increase the difference in level between the room temperature and the available thermal energy.

1.3.1 Internal Energy

The total amount of energy that a system contains is the sum of its internal energy and external work.

Internal energy is mainly kinetic energy derived from the movement of molecules. It can be imagined that, when molecules move faster and become more excited, the temperature of the system increases. If, in contrast, the molecules lose kinetic energy, they slow down and the temperature decreases.

This part of the internal energy is called the 'sensible heat', and we can measure its value by measuring the temperature increase. The thermal energy, or sensible heat, needed to increase (or decrease) one unit of mass by one degree in temperature is called the *specific heat capacity* of the substance being heated (or cooled). It is the degree of intensity of the vibration of the molecules.

The value is different for all substances, and depends on whether the substance is in the solid, liquid or vapour state, as the distances between and kinetic energies of the molecules differ in all these cases.

Another part of the internal energy is the so-called *latent energy*, which is used to change the state of a substance. For instance in changing a substance from the liquid state to the vapour state, the increase in energy will result in an increase in vibration and in the distances between the molecules. When we add thermal energy to ice, the vibration of the molecules increases until they break away from their fixed positions and begin to move around separately or in groups. At a certain speed of vibration each molecule moves freely and the matter has then changed from the solid to the liquid state by increasing the latent energy.

The amount of latent energy needed to convert one mass unit of a substance from one state to another is called the *latent heat capacity* of the substance. This value is also different for all substances, for the same reason as explained for specific heat capacity. Similarly, its value depends on whether the change of state is from solid to liquid or from liquid to vapour.

Specific heat is denoted by c, latent heat of fusion by L_F and latent heat of vaporization by L_V.

1.3.2 Internal Energy in relation to Compression and Expansion

If the kinetic energy of the molecules increases, more collisions take place and stronger collisions occur with the surroundings, that is with the boundaries of the system. The molecule bombardment is heavier, or in other words the pressure on the boundaries, or the surroundings, increases.

Suppose we add heat (thermal energy) to a gas in a cylinder. The gas will try to expand in order to increase its volume, because the molecules become more excited, and their kinetic energy increases. If there is a piston on one side of the cylinder, that piston will move. We see the result of addition of thermal energy as work done on the piston. The added energy is used to increase the internal energy, which eventually results in external work. So the external thermal energy $dQ = dU + dW$. Here U stands for internal energy and W for work. Internally a lot of things have happened which we could not see. The molecules have received an increase in kinetic energy from another part of the internal energy – the potential energy of the molecules, the atoms and parts of atoms. This gain in internal energy, however, was converted into work.

Suppose molecules are thrown and pushed against the piston, but are subsequently thrown back with decreased energy. The energy 'stolen' from the potential forces is the force of attraction, which has decreased because of the increased distance between the molecules. If there had been expansion without the addition of thermal energy to compensate for the loss of internal energy, then the temperature would of course decrease.

Now let us look at the opposite action. There is an outside force pushing the piston. External work is now compressing the gas in the cylinder, or we could say the gas is delivering negative work, $-W$.

The kinetic energy is converted partly into forces of attraction, because the molecules come closer together; there is at the same time a gain in kinetic energy, because of the increased repulsive forces between the parts. Because of the dominance of the outside work, the result is always an increase in kinetic energy – the temperature of the system increases.

The internal energy depends on pressure, temperature and volume.

The exernal work depends on the ambient conditions, that is the 'resistance' of the surroundings.

1.4 Compression

Compression can be an isothermal, adiabatic or a polytropic process. If the process is isothermal the temperature of the gas will not change. This is a process with a high efficiency, but unfortunately it cannot be realized in practice. For a process to be isothermal it would need to be very slow, one that allows a decrease in internal energy or increase in internal energy at expansion.

An adiabatic process is a process without an exchange of thermal energy; in other words, the process would have to be completely insulated both in the case of expansion and compression.

The practical process lies somewhere between these two and is called the polytropic process. When thermal energy is added to a system the total amount of internal energy increases and so the possibility of delivering external work increases, as we have already seen. When such a system is a piston and gas-filled cylinder then using the relationship that work is force multiplied by the distance through which it travels, the work is calculated in this case from force, given by the pressure p multiplied by the area of the piston A, multiplied by the displacement of the piston $\mathrm{d}S$. The expression for work thus derived is $pA\mathrm{d}S$. This can also be written as $p\mathrm{d}V$.

The process of compression can also be imagined as the addition of an infinite number of very small movements of the piston, decreasing the volume from V_1 to V_2. When the process takes place between p_1 and p_2 then the total amount of work involved is the integral of $V\mathrm{d}p$ between p_1 and p_2.

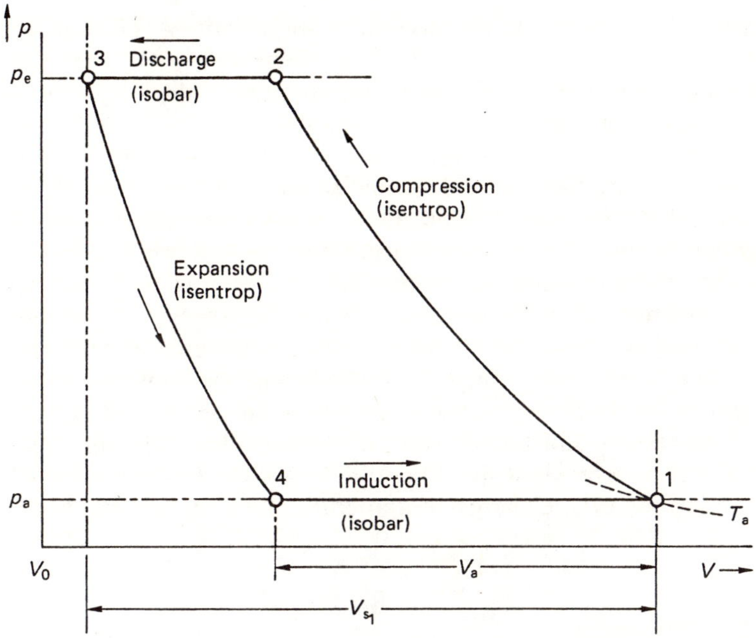

Figure 1.5 *An indicator diagram for an ideal reciprocating compressor*

Referring to the p/V diagram in figure 4.9 in which the process follows a curve with the equation pV^k = constant we see that, if the step dp is very small, the product Vdp can be considered as a rectangle with length V and height dp; so the total surface of the figure between the lines of the process represents the total work needed, or the integral of Vdp. (Theoretically, we consider pV^k = constant to be the equation of an adiabatic and reversible, or in other words an isentropic, process.) See figure 1.5.

The quantity of added energy is transformed into U, internal energy, and $\int V\mathrm{d}p$, the negative external work of the gas.

We leave here for the time being the subject of the theoretical aspects of the process taking place in a compressor.

1.5 Expansion

In order to bring a fluid at a high pressure down to a low pressure we can use several methods. The use of an expansion cylinder to do useful work by displacing a piston, so utilizing the potential external work, is unfortunately an irreversible process. If the process is carried out very slowly, we can achieve a reversible process in which the internal energy will not change. In the ideal case there should be laminar stream flow, with no friction or turbulent agitation of the molecules occurring. The kinetic energy should not decrease because there are no violent collisions with the piston.

With the expansion cylinder being an impractical device, the throttling valve in its various forms is now in common use. The throttling process involves no change in the total amount of energy; expressed more accurately, the total enthalpy does not change. The meaning of the term enthalpy will be explained later. In practical terms the internal kinetic energy will decrease as the temperature decreases. Some energy seems to disappear; this of course is impossible – the energy has been used to change some of the liquid into vapour and in the process some is converted into flow work or displacement energy. Note that to create gas or vapour in this situation some latent energy L_V is used. When we generate a fraction x of this so-called flash gas, we use xL_V units of energy. So far we have only considered vapour and liquid. Solid matter also has a certain amount of internal energy, a fact that should not be forgotten. Internal energy can be compared to the elastic energy contained in a steel spring or rubber band.

1.6 Enthalpy

Enthalpy is a calculated composite energy term defined as the sum of the flow work. Expressed mathematically this is

$$h = u + pv$$

where h = the specific enthalpy in joule/kg
u = the internal energy in joule/kg
p = the absolute pressure in pascal
v = the specific volume in m^3/kg.

Since refrigerants are usually in the liquid state or in the vapour region where the gas laws do not apply, the concept of enthalpy is widely used in refrigeration to evaluate energy changes. Refrigeration cycles are usually plotted on pressure/enthalpy or Mollier coordinates.

One point of difficulty in this work is the number of different datum points from which internal energy or enthalpy is measured. We measure the reference datum for enthalpy from different temperatures for different substances, and in certain cases we find that different sources use different datum points for the same substance. Steam tables and tables for dry and moist air use the liquid state at 273K as zero, as did early refrigeration tables; as lower temperatures were achieved the liquid state at 233K was adopted for refrigerants. For food products, enthalpy is zero at 253K. More recently some designers of Mollier charts for refrigerants have chosen the enthalpy at 273K to be 100 kJ/kg while others have chosen 200 kJ/kg. This serves as a reminder that energy is a relative and not an absolute value.

The enthalpy scale also starts at different temperature levels for different purposes, and even for the same purpose. When performing calculations using charts or tables, it is essential not to mix values from different sources.

An enthalpy change during a process is denoted by dh, for it is the enthalpy change which is needed in thermal calculations. For water this means the difference of thermal energy necessary to bring the material from 273K to TK: Q = specific heat ($c_1 \times dT_1$) + L_V (increase of internal energy to convert liquid to vapour) + the difference in thermal energy of the vapour between saturation point and the actual final vapour temperature ($c_2 \times dT_2$). For air it is a little more complicated because air contains a certain quantity of water vapour, so it is a mixture of two substances. If there is X kg/kg water vapour in the air, we can say $h = (c_1 \times T)$ + latent heat $L_V \times X$ + heat needed to raise the temperature of the water vapour (specific heat $c_2 \times X \times T_2$) kJ/kg, where $c_1 = 1.005$ kJ/kg K, $c_2 = 1.84$ kJ/kg K and $L_V = 2491$ kJ/kg.

In the next section you will see how 'cold' is generated; in other words, how the processes involved in a refrigeration system interrelate to achieve this end.

1.7 Principles of the Refrigeration Process

Refrigeration utilizes the fact that a relatively large amount of thermal energy is required to convert a liquid into a vapour. This thermal energy is extracted from the substance to be cooled by arranging for the liquid to boil at a temperature below that of the substance to be cooled. Try it yourself: put some eau de Cologne on your hand, and you will feel your hand cooling as the eau de Cologne evaporates.

Another test is to use an aerosol appliance, for instance a deodorant in a spray tube. Besides the deodorant, there is a propellant gas in the tube – a refrigerant. If you push the nozzle and keep your hand in front of the aperture, you will feel the cooling effect of the expansion of the gas. In fact the refrigerant (until recently a CFC) is released from the high pressure inside the tube to the lower pressure of the atmosphere. The conclusion is that not just evaporation but also expansion give a cooling effect.

Note that in most countries CFCs have already been replaced in aerosols by alternative propellants, as a consequence of the Montreal protocol which ultimately plans to eliminate the emission of all CFCs into the atmosphere.

In a refrigeration system the designer arranges for the working fluid to be liquid at its boiling point at such a temperature that it readily evaporates by absorbing thermal energy from the surroundings.

Let us examine the processes when solids, liquids and vapours change state. When we heat water it boils under atmospheric pressure at 100°C or

373K. At that temperature and pressure the water changes from liquid into vapour. In the liquid state the molecules are moving slowly, they are close together, and they have a relatively small amount of internal energy. When we add thermal energy, they start to move faster and become more widely spaced, while those at the surface have a tendency to escape. If the pressure on the water surface permits and the molecules have enough energy to overcome the cohesive forces that originally kept them in the liquid, some of them will break through the surface. This is because the forces here are weaker than those inside the liquid where they are surrounded and under the influence of the binding forces on all sides, and so molecules arriving at the surface can escape into the space above it.

Boiling is an isothermal process in that the temperature remains constant. Under these conditions, in order to add thermal energy, it is necessary that there should be a source of thermal energy available at a temperature above 100°C, for example a gas flame or an electric heater.

When water is boiled in a completely closed vessel, such as a pressure cooker, the pressure rises and the boiling temperature increases. Under these conditions the molecules will have more difficulty in escaping because of the higher pressure in the space above the liquid. The result of this is that the boiling process can only take place at a higher temperature. This also means that more energy must be added to the molecules. Again we see the relationship between saturation pressure and temperature – that the boiling temperature of a liquid depends on the pressure of the vapour above it.

The temperature at which a fluid changes from the liquid to the vapour phase and vice versa is called the *saturation temperature*. The liquid and vapour are then said to be in equilibrium with each other. The liquid under these conditions is called the *saturated liquid* and the vapour the *saturated vapour*. At any pressure where this condition can exist, the saturation temperature is the maximum temperature that the liquid can have and the minimum that the vapour can have. When the pressure over a liquid is lowered, then the saturation temperature falls correspondingly; when thermal energy is added to the fluid, the temperature rises and the pressure this time rises correspondingly. In other words, the point of equilibrium is displaced.

When selecting a substance to be used in a cooling or possibly a freezing process, a liquid that boils at atmospheric pressure at 100°C is not suitable, unless it is made to boil at a very low pressure, which is not very practical in refrigeration systems. The solution is to select a fluid with a suitably low boiling point at a reasonable pressure. Such fluids when used in refrigeration systems are called refrigerants; typical examples are ammonia and certain of the chlorofluorocarbon compounds (CFCs).

We know now that 'cooling' can be produced by evaporating a refrigerant. This is achieved in practice using a piece of equipment known as an evaporator, which is normally a battery of tubes into which refrigerant is injected or pumped. To create a continuous process, the vapour must be transformed again into liquid ready to continue the cycle. Now recall the vessel containing boiling water. When a glass plate is placed near the outlet of the vessel and a low temperature is maintained by blowing cold air onto it or by cooling it with the help of cold water, vapour condenses on the glass plate as shown in figure 1.6. When thermal energy is removed, the molecules move closer to each other and vibrate less violently, and so pass from the vapour to the liquid state. The water dripping from the plate can be returned once more into the vessel, where it can re-evaporate and we now have a closed circuit process.

Thus all we have to do to condense the vapour is to bring it into contact with a surface at a lower temperature. If we were to use a liquid with a low boiling point to achieve a low temperature, for instance in the range −10°C to −30°C, we would need another refrigeration installation to cool the vapour in order to achieve this condensation. This is obviously too complicated. One solution to this problem is to arrange for the condensation to take place at a higher temperature, using the ambient air or water from natural sources to remove the thermal energy. This results in a correspondingly higher pressure being required for the process. To achieve the high pressure we use a compressor, which has a dual task: it maintains the pressure low enough in the evaporator for the boiling to take place at a suitably low temperature; and at the same time it produces a pressure high enough to produce condensation at the appropriate high temperature.

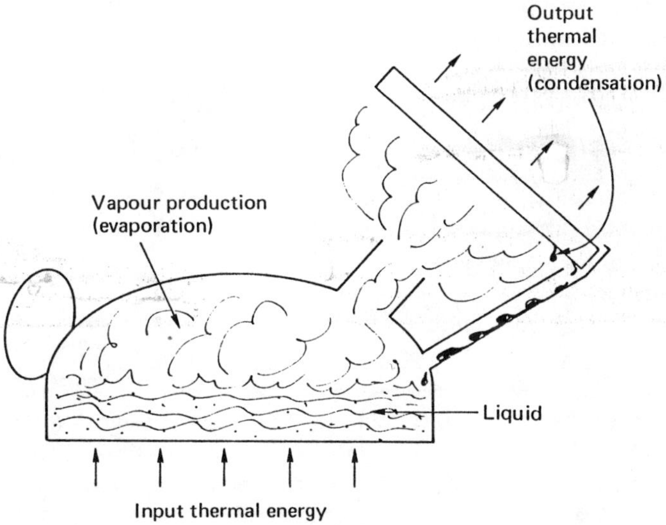

Figure 1.6 *Simplified representation of the refrigeration cycle*

The energy added to the vapour by the motor powering the compressor is the main energy requirement to drive the refrigeration installation. In order to use the minimum of energy we try to arrange condensation at the lowest practical pressure using, as mentioned above, air or preferably, if available, cold water. Unfortunately, local circumstances may dictate which has to be used.

Incidentally, every vapour has a temperature above which it is impossible to achieve condensation; at that critical point there is no distinction between vapour and liquid. In addition, compressors are limited as to their maximum operating pressure, corresponding approximately to temperatures of +55°C. For heat pumps, special compressors allowing higher operating temperatures have been developed.

After compression, the temperature of the vapour is higher than the temperature corresponding to the saturation pressure – it is overheated or superheated. The internal energy is increased and some of this results in an increase in the temperature. We have to extract this superheat in the condenser before condensation can start.

Superheat also occurs in the cycle at the end of the evaporation process. As the vapour leaves the evaporator additional thermal energy results in an increase in temperature, the so-called ‘suction superheat’.

On the other hand, once the vapour is liquefied and we continue to subtract thermal energy, then the temperature falls and the liquid is then said to be subcooled.

To compete this section we will examine the diagram of a typical refrigeration circuit, as shown in figure 1.7. The air of the coldstore is

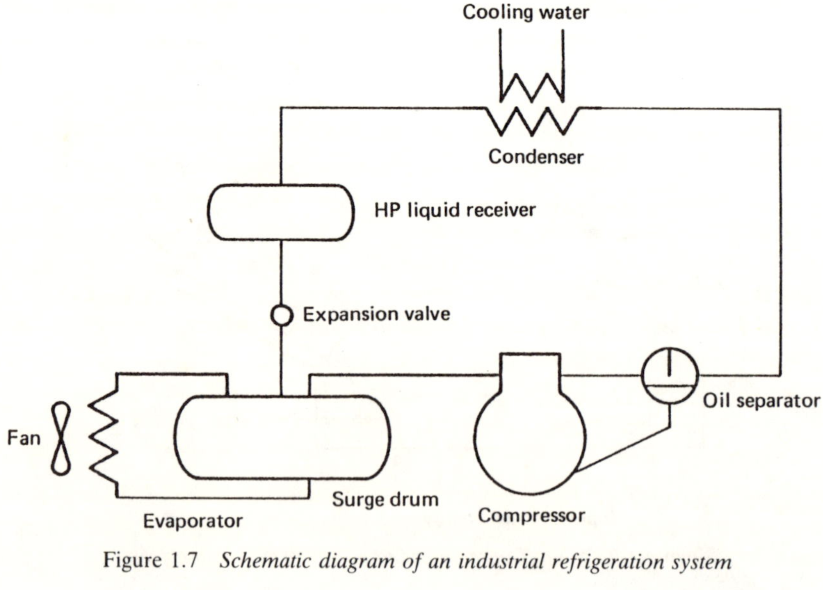

Figure 1.7 *Schematic diagram of an industrial refrigeration system*

blown over the evaporator and gives up its thermal energy to the liquid refrigerant inside. At a low pressure and temperature the liquid becomes superheated vapour and is pumped into the condenser by the compressor. Remember that the compressor produces the low pressure in the evaporator while at the same time elevating the vapour to the condenser pressure.

Cold air or water flowing over the condenser removes the thermal energy from the superheated vapour, converts it into liquid and lastly subcools it. The liquid next flows to the receiver, from high to low pressure, and finally back to the evaporator again ready to restart the cycle.

1.8 The Mollier Diagram or Chart for Refrigerants

Mollier proposed that the thermodynamic characteristics of refrigerants could be usefully displayed in diagrammatic form. He constructed the diagram with the pressure p, normally logarithmical, on the vertical axis and the enthalpy h on the horizontal base. Thus lines of constant pressure (or isobars) are horizontal straight lines, and lines of constant enthalpy (or isenthalps) are vertical straight lines.

See figures 1.8 and 1.9 for commercial Mollier diagrams of two typical refrigerants. Note how the saturation line divides the diagram into three main parts. To the left of the saturated liquid line is the subcooled liquid region, to the right of the saturated vapour line the superheated vapour region. In the middle, under the saturation line, the region of mixture of vapour and liquid is situated. Furthermore, lines of specific volume, lines of constant entropy s and lines of constant temperature are also shown on the diagrams. These parameters are also available in tables, in detail for saturated conditions but in less detail for superheated conditions.

The concept of entropy is not discussed in this book. It is sufficient to state that the theoretical vapour compression process in the refrigeration cycle normally follows the line of constant entropy, or the isentrop s = constant, quite closely. However, in reality entropy increases during each irreversible thermodynamic process.

One mathematical definition of the entropy change from point 1 to point 2 can be expressed as the integral of the change in thermal energy divided by the absolute temperature:

$$s_2 - s_1 = \int \mathrm{d}Q/T$$

Expressed in another way, entropy is a measurement of the degree of chaos of a system. A thermal energy system is more chaotic than a mechanical energy system.

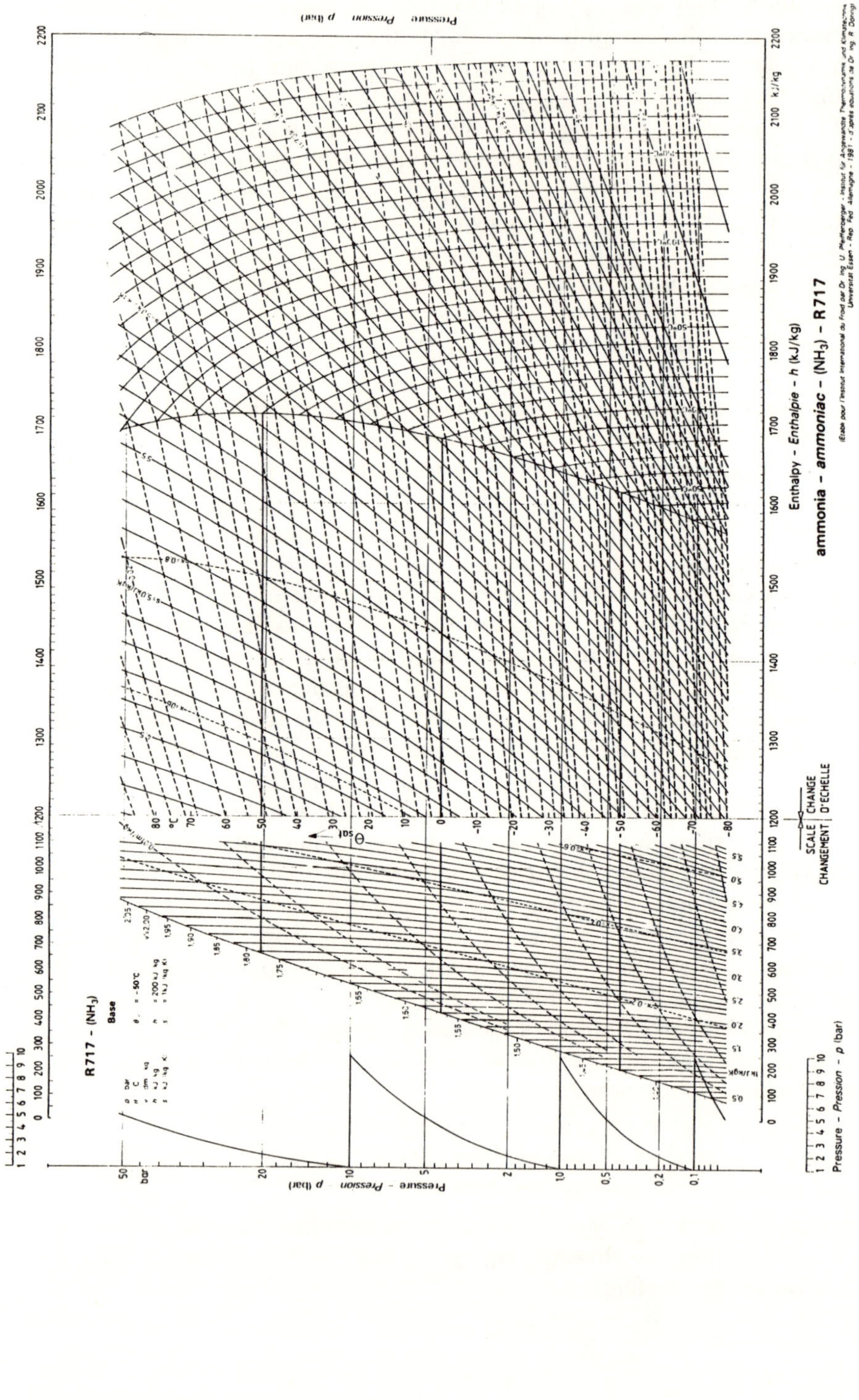

Figure 1.8 *Mollier diagram for R717*

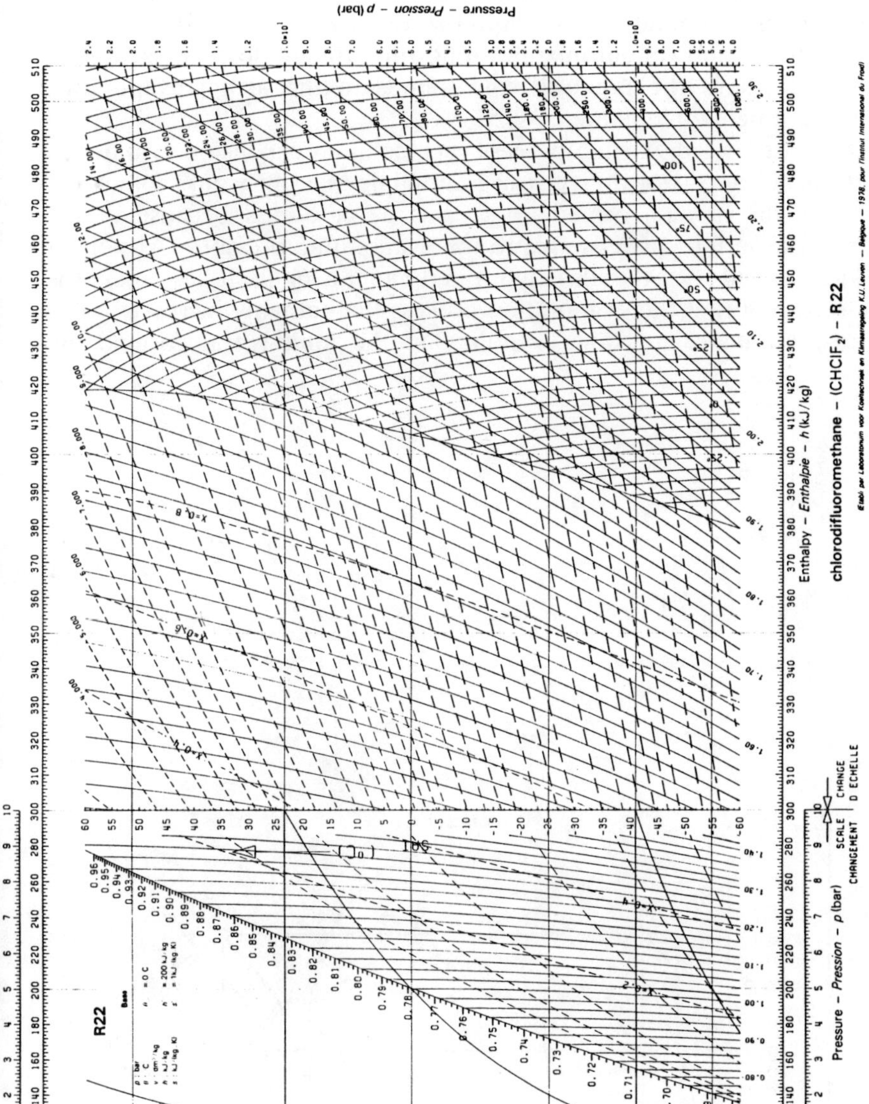

Figure 1.9 *Mollier diagram for R22*

Now refer to the refrigeration cycle shown on the Mollier diagram of figure 1.10. On the saturation line at point 1 the refrigerant is saturated liquid. As soon as the pressure falls below the saturation pressure at the corresponding temperature, the liquid starts evaporating and enters the region of liquid and vapour mixture, following straight down an isenthalp, to point 2 where the pressure corresponds to the actual evaporating temperature being maintained.

In the case of a reversible process such as in an expansion cylinder, it would not follow down the vertical isenthalpic line but would follow the isentrop. The vapour formed up to this point occurs because of the expansion from the condensating to the evaporating pressure. At point 2 the liquid mixed with the already-formed vapour, or flash gas enters the evaporator. During the evaporation process the enthalpy increases while the presure remains constant. At point 3 all the liquid is transformed into

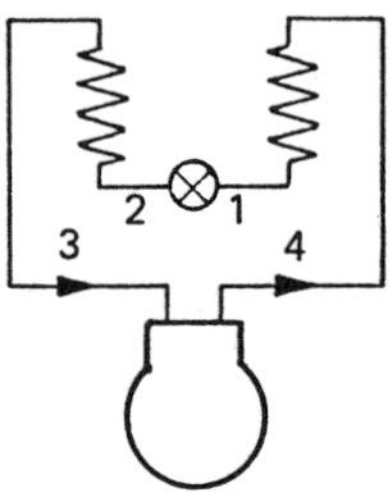

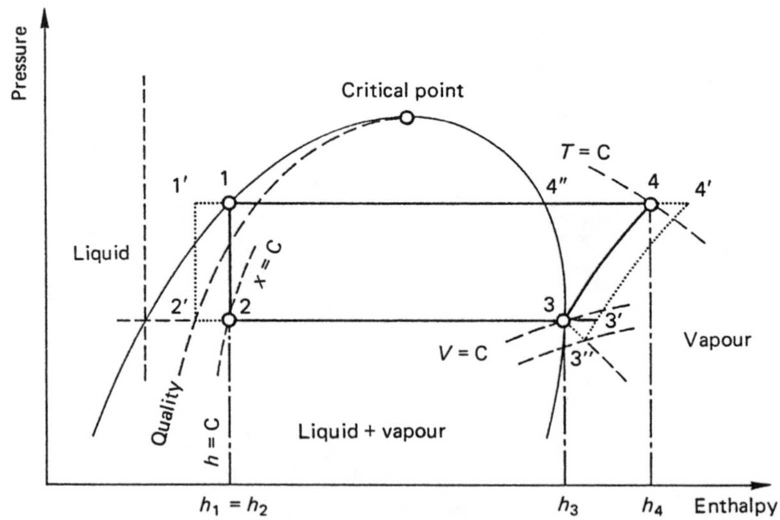

Figure 1.10 *Refrigeration cycle on the Mollier diagram showing effects of subcooling, superheating and pressure drop*

vapour which then enters the compressor. The pressure is now raised and follows the compression line s = constant to point 4, at the condensation pressure. Next the vapour enters the condenser where it undergoes first desuperheating to point 4″ then condensation until, fully transformed into liquid, it reaches point 1.

Much additional information can be obtained from a refrigeration cycle when plotted on the Mollier diagram:

(a) The mass flowrate of refrigerant in circulation: when we evaluate the enthalpy at point 2 and subtract this from the value at point 3, we can determine the quantity of enthalpy dh that 1 kg mass of refrigerant can absorb in the evaporator, given by $dh_{23} = h_3 - h_2$.

If we know the total capacity of the plant Q and divide this by dh_{23} we arrive at the mass of refrigerant per second that must circulate to absorb the necessary thermal energy.

(b) In practice we continue to remove thermal energy after point 3. We superheat the vapour to point 3′ so that dh has to be calculated between points 2 and 3′, but only when the superheat is obtained usefully by the evaporator or the pipework within the refrigerated space.

(c) By using the indicated specific volume, we can find the volume flowrate of the refrigerant at any time during the process.

(d) dh between points 3 and 4 or 3′ and 4′ identifies the amount of enthalpy which the refrigerant absorbs during compression.

(e) dh between points 4 or 4′ and 4″ determines the amount of enthalpy to be removed from the vapour before it starts condensing, the so-called 'discharge superheat'. If we need to determine the temperature of the superheated vapour leaving the compressor we can find this on the diagram.

We can also calculate this temperature using the equation:

$$T_2 = T_1(p_2/p_1)^{(k-1/k)}$$

which is true when s = constant.

It can be seen that when a large difference exists between the evaporating and condensing temperatures, there will be a corresponding large value for p_2/p_1. The resulting discharge temperature can be very high, so high in fact that it can cause mechanical problems. To avoid this we divide the compression process into two steps or stages, so creating what is called a two-stage compression system. Between the two stages the vapour is cooled down, resulting in a lower final temperature.

(f) Subcooling and superheating: the value that we determine for dh between points 1 and 1′ gives the enthalpy removed from the liquid when we continue to cool it without changing the pressure. The condition of the liquid as it enters the evaporator will in practice be 1′ and not 1. This is termed *subcooling* and is very important, especially in plants using CFC refrigerants. One reason for this is that the liquid which enters the evaporator subcooled has a greater d*h* available for the evaporation process. In addition, subcooling results in less creation of the useless flash gas during expansion and reduces the probability of it occurring in the liquid line. The fraction of flash gas formed by expansion can be read easily from the Mollier diagram. This is correctly termed the 'quality' of the mixture; it is given the symbol x and can be calculated as follows. At point 1 we have the same total enthalpy as at point 2. However, at point 2 there is a mixture of liquid and vapour. Liquid at the saturated pressure corresponding to the evaporation temperature has a smaller enthalpy; the difference is the part of the total enthalpy of the mixture which is not used to do any external work, as in an expansion cylinder – in this case it is used to produce vapour. With x fraction of vapour, xL_V kJ/kg is required to produce this change.

From $h = xL_V$ kJ/kg, we can now calculate the quality x of the mixture. Remember that L_V = latent energy, the difference between the enthalpy of the liquid and the enthalpy of the vapour at saturation temperature.

When we do this calculation under the same conditions for different refrigerants, we find that CFC refrigerants produce larger quantities of flash gas than does ammonia. When we compare the value x for the same refrigerant, but at different condensing and evaporating temperatures, we see that the greater the difference between those two temperatures, the greater the value of x.

Table 1.1 Quality x of flash gas for a liquid subcooled at 15°C

Temperature	*Ammonia*	*CFC*	
(°C)	*R717*	*R12*	*R22*
−10	8.9%	14.7%	14.4%
−20	12.1%	19.8%	18.9%
−30	15.2%	24.6%	23.4%

Subcooling, however, means that point 1 and therefore point 2 move to the left to 1′ and 2′, as shown on the Mollier diagram in figure 1.10, with the result that the value of x is smaller, an additional reason to subcool the refrigerant.

We can in fact subcool either by utilizing the final part of the condenser, by adding more tubes than are necessary for condensation or by installing a liquid suction heat interchanger into the system, through which the cold suction vapour passes on one side and the warm liquid line coming from the condenser passes on the other. The result is to cool the liquid by warming up the vapour, in other words by adding superheat to the vapour.

This, as we will see later on, requires extra energy on the compression side. The reason for this is that the superheated vapour becomes higher in specific volume, so we have to compress a larger volume. However, the gain on one side can be greater than the loss on the other side, since we can guarantee to dry the vapour and so protect the compressor against liquid slugging.

Note that, owing to pressure drop in the suction line and through compressor valves, the compression will never start at point 3 or 3′ but at a point situated for instance at 3″. More details about the practical compression process will be given in chapter 4.

2

An Introduction to Psychrometrics

2.1 Introduction

So far we have mainly described the processes which take place within the system; however, processes external to the system – processes that in the main involve the removal of thermal energy from the surrounding air, and the circulation of the air by the installation – also need to be understood.

Refrigeration of coldstores involves extracting the thermal energy from the stored goods and their surroundings into the refrigerant and, after 'pumping it up' to a higher temperature level, releasing this energy into the external atmosphere.

Psychrometrics is the study of the properties of air when it contains various quantities of water vapour. Dry air, that is air containing no water vapour, is composed of a mixture of 78.09% by volume of nitrogen, and 20.95% by volume of oxygen, the remainder being made up of 0.93% argon and minute quantities of carbon dioxide and other rarer gases. For practical purposes, atmospheric air at sea level can be considered as a mixture of dry air and a relatively small amount of water vapour varying between a maximum of 0.1% at −15°C and a maximum of 2% at 25°C, calculated on a mass basis.

In this context air is analogous to a sponge – a molecular sponge. When the temperature increases, the sponge expands and extra space is available between the molecules to absorb more water molecules (see figure 2.1). When the temperature decreases there is less space so water molecules drop out of the air; when this occurs the result is condensation.

There are different methods of expressing the quantity of water vapour in the air. *Moisture content* is defined as the mass of water vapour m_s which is contained within the mass of dry air m_a of a water vapour/air mixture. It is sometimes referred to as the *specific humidity* or *humidity ratio*, and has

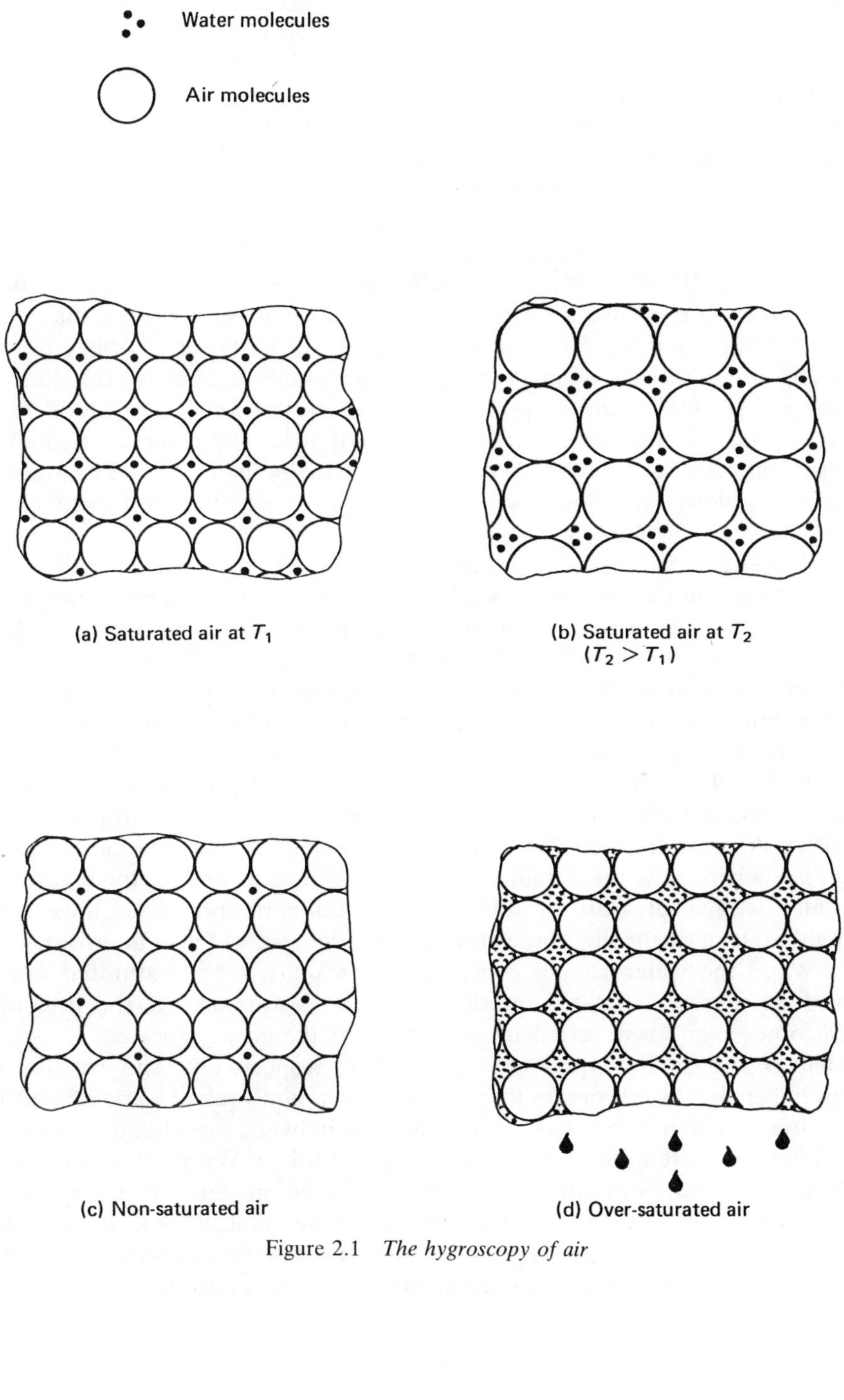

Figure 2.1 *The hygroscopy of air*

the symbol x or g. Mathematically this can be expressed as $x = m_s/m_a$. The degree of saturation μ is defined as the ratio of the moisture content of the moist air x to the moisture content of saturated air x_{ss}, both at the same temperature and pressure. Expressed mathematically, $\mu = x/x_{ss}$.

Relative humidity (RH or ϕ) is defined as the ratio of the partial pressure p_s of the water vapour in the water vapour/air mixture to the partial pressure p_{ss} of the water vapour in the saturated water vapour/air mixture at the same temperature. Expressed mathematically, RH or $\phi = p_s/p_{ss}$.

In practice the values of RH or ϕ and μ are equal to 0 and 1 respectively, or to 0 and 100 respectively if they are measured as a percentage. When measured in percentage terms the values never differ by more than one percentage point for air temperatures up to 28°C. Often people confuse them, but with such a small difference, as mentioned above, this can be ignored for refrigeration applications. For the conservation of food products, the two most important properties of atmospheric air are temperature and relative humidity. Note that it is actually the degree of saturation which is plotted on most psychrometric charts, even though they may be labelled 'relative humidity'.

Remember what was said previously about the hygroscopic effect of air – that water vapour is not formed by the addition of thermal energy. It is not what could be termed a heat-activated evaporation process but rather a pressure-activated evaporation process. To change water into vapour without adding latent energy, the pressure must be lowered. So in non-saturated air, the water vapour pressure must be lower than the saturated water pressure at the given air temperature. If the air is not saturated there will be more space for the water vapour molecules, since the pressure is low, so we see that the value of the water vapour pressure is also a measure for the relative humidity. Indeed, as was stated above, ϕ is p_s/p_{ss}, where p_s is the actual water vapour pressure and p_{ss} the saturated water vapour pressure at this given air temperature. If we lower the temperature of the air, the water vapour pressure will also be lower.

When the water vapour pressure corresponds to this saturated water vapour pressure, we are on the so-called 'dew point'. If the pressure becomes even lower, condensation starts. If the air is saturated, it means that ϕ and μ are equal to 1 and 100% respectively, and we are in equilibrium. As we saw before, temperature and pressure are related in such a way that there must be equilibrium between liquid and vapour. If the temperature falls a little, equilibrium is broken. We are now no longer at the boiling or saturation point and condensation starts. If the temperature rises a little there is more space for water molecules in the air and 'boiling' can start. Water vapour pressure is at that moment lower than the saturated water vapour pressure at the given temperature.

2.2 The Gas Laws

Before taking this subject further it is useful to have an understanding of some of the physical laws relating to ideal gases, which are also useful in calculations carried out on atmospheric air.

Firstly Boyle's law, which can be stated as follows. During a change in the conditions of any ideal gas in which the mass and the temperature remain constant, the volume varies inversely as the absolute pressure, or

$$pV = \text{constant}$$

It is worth noting that a Frenchman, Edme Mariotte, discovered this law independently at about the same time.

Secondly Charles's law, which can be stated as follows. During a change in the conditions of any ideal gas in which the mass and the pressure remain constant, the volume varies indirectly with the absolute temperature, or

$$V/T = \text{constant}$$

Another Frenchman, Joseph Gay-Lussac, made the same discovery at about the same time.

The existence of these two laws encourages investigation into the possibility of a general law applicable under all conditions when none of the parameters is fixed.

Consider a gas at pressure p_1, volume V_1 and temperature T_1. Let this gas change to the final condition, p_2, V_2, T_2. Examination of figure 2.2 will show that there are an infinite number of paths connecting point 1 to point 2. One particular path can be taken that consists of following Boyle's law to point A and then Charles's law to point 2.

Taking the change 1 to A first, the following is true:

$$p_1V_1 = p_AV_A \quad \text{where } T_1 = T_A$$

Taking the change A to 2 next, the following is true:

$$V_A/T_A = V_2/T_2 \quad \text{where } p_A = p_2$$

Rewriting the first equation and substituting $p_A = p_2$:

$$V_A = p_1V_1/p_2$$

Rewriting the second equation and substituting $T_1 = T_A$:

$$V_A = T_1V_2/T_2$$

Equating these two values for V_A we derive the following equation:

$$p_1V_1/p_2 = T_1V_2/T_2$$

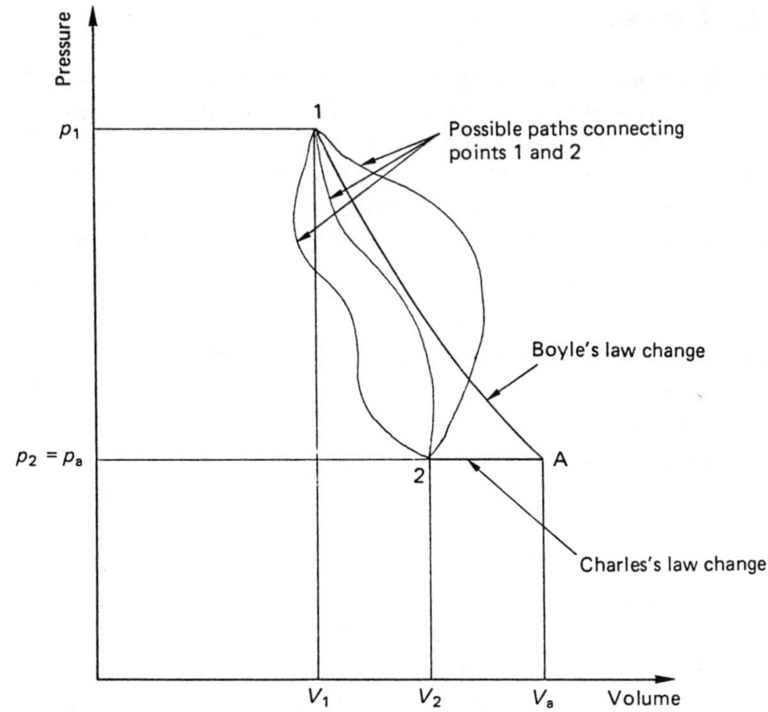

Figure 2.2 *Representation of gas law changes on a p/V diagram*

This becomes:

$$p_1V_1/T_1 = p_2V_2/T_2$$

Similarly, we can prove

$$p_2V_2/T_2 = p_nV_n/T_n$$

and so arrive at the general equation pV/T = constant. When we use this equation for 1 kg mass of a gas, then it becomes $pV/T = R$, where the constant R is termed the *characteristic gas constant*. When we consider the case where m kg of the gas are involved, then the equation becomes $pV/T = mR$ or as it is more generally written:

$$pV = mRT$$

This is known as the *characteristic equation of state* of a perfect gas.

What is the significance of the constant R? R is another form of specific heat or energy capacity, a point we have already mentioned in section 1.3 of chapter 1 when we spoke of the energy of a substance. Remember that there is internal energy and external work. R is the amount of internal energy converted into external work. As the volume increases, the distance between the molecules increases and at the same time the external work will increase. R is the increase of energy, in kJ, of 1 kg during an increase in the temperature of 1K.

In Chapter 1 we gave a value for c, the specific heat capacity, for water and another for air. For gases we need two values for c, one for a change which takes place at constant volume, c_v, and one for a change which takes place at constant pressure, c_p, and so by implication with a changing volume. When we raise the temperature of 1 kg of a substance at constant volume, the amount of heat required Q is given by $Q = c_v dT$, where c_v is the specific heat capacity at constant volume. When we raise the temperature of the same mass of the substance by the same amount at a constant pressure, the specific heat is not the same – it is called c_p and the value of c_p is higher than that of c_v. The difference $(c_p - c_v)$ is the amount of energy R. One method of proving this is to consider 1 kg of gas at p_1, V_1, T_1 and let it be heated by 1K at constant pressure. The new state will be p_1, V_2, $(T_1 + 1)$. Now from the characteristic gas equation:

$$p_1V_1 = RT_1 \quad \text{and}$$

$$p_1V_2 = R(T_1 + 1)$$

Subtracting the first equation from the second gives

$$p_1(V_2 - V_1) = R$$

Thus we can see that R is equal to the area under the graph of the process plotted on a p/V diagram, which has been shown previously to be equal to the work done.

The small test illustrated in figure 2.3 should prove helpful in further understanding this part of the subject.

Expansion by Adding Thermal Energy to a Gas

Leave the piston free to move (figure 2.3a) so that

$$p_1 = p_2 \qquad H_2 > H_1$$
$$V_1 < V_2 \qquad T_2 > T_1$$

Specific heat c_p is the quantity of thermal energy necessary to raise the temperature of 1 kg of gas by 1K at constant pressure.
Keep the piston in place (figure 2.3b) so that

$$p_1 < p_2 \qquad H_2 > H_1$$
$$V_1 = V_2 \qquad T_2 > T_1$$

Specific heat c_v is the quantity of thermal energy necessary to raise the temperature of 1 kg of gas by 1K at constant volume.

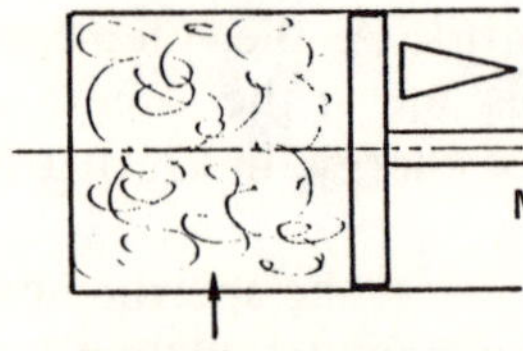

(a) Input of thermal energy: $p_1 = p_2$, $V_2 > V_1$

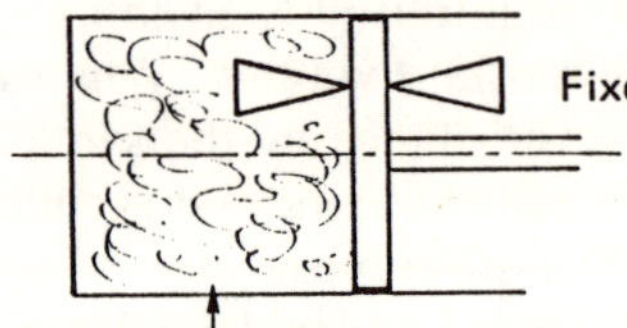

(b) Input of thermal energy, $V_1 = V_2$, $p_2 > p_1$

Figure 2.3 *Input of thermal energy to gas in a cylinder*

We have discussed above why R equals $(c_p - c_v)$ and that c_p is greater than c_v. Let us see if study of figure 2.3 can make these relationships clearer.

In figure 2.3a the addition of thermal energy results in part of the energy becoming internal energy and the remainder external work. In figure 2.3b the addition of thermal energy results in conversion of all the thermal energy into internal energy. So in figure 2.3a the process uses more specific heat c, or $c_p > c_v$.

We can also prove this more rigorously as follows. We know that the total energy Q = internal energy ($c_v dT$) + external work (pdv) in a normal process with the volume remaining constant, whereas at constant pressure $Q = c_p dT$ only. The expression for external work pdV under these conditions is $p(V_2 - V_1)$. Since $pV_2 = RT_2$ and $pV_1 = RT_1$, then external work can be expressed as

$$RT_2(RT_2/p - RT_1/p)V_2$$

Substituting V_2/T_2 for R/p the expression becomes

$$R(T_2 - T_1) = RdT$$

If Q is the same in the two cases, then

$$Q = c_p dT = c_v dT + RdT \quad \text{or}$$

$$c_p = c_v + R \quad \text{or} \quad c_p - c_v = R$$

In an isothermal process where $T_1 = T_2$, c_p must equal c_v, so

$$c_p/c_v = 1 \quad \text{or} \quad pV^n = pV^1 = \text{constant} \quad \text{or} \quad pV = \text{constant}$$

Therefore, during such a process the volume varies inversely as the pressure. In adiabatic compression the influence of the external work is shown during compression by the compression line rising more steeply on the p/V diagram. In other words, the volume decreases less when the pressure increases during adiabatic compression compared with isothermal compression; the fact that no thermal energy escapes means that all the energy put into the gas remains there. This process requires more work to be input as shown by the fact that the area under the graph is greater. The degree to which this is affected depends on the relationship between c_p and c_v, which in turn depends on the characteristics of the gas. This explains how the slope of the compression line is determined by the ratio c_p/c_v.

As volume is the inverse of mass, so specific volume is the inverse of specific density ρ. We can rewrite the simple form of the characteristic equation of state $pV = RT$ as $R = pV/T$ and then substituting $1/\rho$ for V, the equation becomes $R = p/\rho T$. This shows how R depends on ρ.

As every gas has a different value of ρ and a different relationship between p and T it is obvious that every gas has a different value of R.

Consider Avogadro's law, which states that all gases contain the same number of molecules in a mole when held at the same pressure and temperature; a mole is taken to be that quantity of matter of a substance with the same number of molecules as the molecular mass Avogadro number. From this we can see that the specific density ρ is the number of molecules, n, multiplied by the molecular mass, M.

Therefore the equation can now be rewritten as $R = p/MnT$. Multiplying both sides by M we get $MR = p/nT$; since p, n and T are constants then MR is a constant called the absolute gas-factor or *universal gas constant* with the symbol R_0. R_0 has been determined as 8.314 kJ/kmol K.

Whenever we know the molecular mass M of a gas we can always find the value of R for that particular gas.

It is easy to understand that R_0 depends on ρ, because ρ depends on the molecular mass and the energy depends on the movement of and the attraction between the molecules.

Note that the characteristic gas equation is only valid for an ideal gas in which there is no attraction between the molecules.

There are other laws, empirically derived by such people as van der Waals, Linde and others, in which allowance is made for the intermolecular forces. However, the attraction between molecules of gases is so small that we can normally use the characteristic gas equation for this type of work without any problems.

This introduction to the gas laws will help in the study of the rules governing the behaviour of the mixture of dry air and water vapour that we know as atmospheric air.

Dalton's law states that the total pressure exerted by a mixture of gases or vapours or both is equal to the sum of the individual pressures of each gas or vapour considered separately if each gas or vapour occupied the whole volume alone. The pressure of each constituent is known as the 'partial pressure'.

So the atmospheric pressure p_{at} is given by $p_a + p_s$ where p_a is the partial pressure of dry air and p_s is the partial pressure of water vapour.

We say that the air is saturated with water vapour when the partial pressure p_s of the water vapour equals the pressure p_{ss} at which water boils at that temperature. For air which is not saturated there is a temperature T, called the dew-point temperature. This is the temperature at which condensation starts when non-saturated air is cooled.

Applying the above-mentioned laws, the following will be true for a mass m_a of dry air and m_s of water vapour, occupying a volume V at a common temperature T:

$$p_a V = m_a R_a T \quad \text{and}$$
$$p_s V = m_s R_w T$$

2.3 Relative Humidity, RH or ϕ

Relative humidity and temperature are the two most important properties of air both in the fields of air conditioning and food storage, as mentioned above.

We defined ϕ as p_s/p_{ss} where p_s is the partial pressure of the water vapour and p_{ss} is the partial pressure of the water vapour at saturation, both pressures being expressed at the same temperature.

2.4 Humidity Ratio, Specific Humidity or Moisture Content, x or g

We defined x above as m_s/m_a where m_s and m_a are the masses of the water vapour and dry air respectively.

Using Dalton's law and the characteristic gas equation for both the water vapour and the dry air:

$$p_s V_s = m_s R_s T_s \quad \text{and}$$

$$p_a V_a = m_a R_a T_a$$

Then moisture content $x = p_s V_s R_a T_a / p_a V_a R_s T_s$
$= R_a p_s / R_s p_a$

which given that $R_0 = R_a M_a = R_s M_s$
and $R_a/R_s = M_s/M_a = 18/28.97 = 0.622$, shows

$$\text{moisture content } x = 0.622 p_s/p_a$$

Since we may need to calculate values at different atmospheric pressures it is more useful to express x in the following terms:

$$x = 0.622 p_s/(p_{at} - p_s) \text{ kg moisture/kg of dry air}$$

2.5 Enthalpy of Moist Air

The enthalpy of moist air is made up of the enthalpy of the dry air plus the enthalpy of the water vapour contained by the dry air. The empirical formula $h = 1.005t + x(2491 + 1.84t)$ kJ/kg enables values of specific enthalpy to be determined for different values of t and x.

2.6 The Factor dh/dx

This factor is interesting because the way in which the change of humidity is related to the change of enthalpy is an indication of the process taking place in the air. For instance, if we cool the air by passing it over a dry, non-hygroscopic surface, the amount of water vapour does not change so long as the surface temperature is higher than the dew point of the air. In this case $dx = 0$ and dh is negative, therefore $dh/dx = -\infty$. Such a cooling process will in fact be represented by a vertical line on the Mollier diagram for water vapour/air mixtures. However where water vapour condenses as the air is cooled then x obviously changes; in fact dx is negative and the dh/dx line is no longer vertical. This will be made clearer when we deal later with the cooling line.

In a sensible heating process the relationship is positive. Where $t = 0$°C, then $h = 2491x$ so $dh/dx = 2491$. We shall see the significance of this particular number later when we deal with the Mollier diagram shown on figure 2.4, on which curves for all the characteristics of water vapour/air mixtures have been drawn.

2.7 The Wet-bulb Temperature

The normal temperature of air is called the dry-bulb temperature. We measure this temperature with a thermometer with the sensing element kept dry and shaded from thermal radiation. On the other hand when we put water on the sensing element we measure a lower temperature because

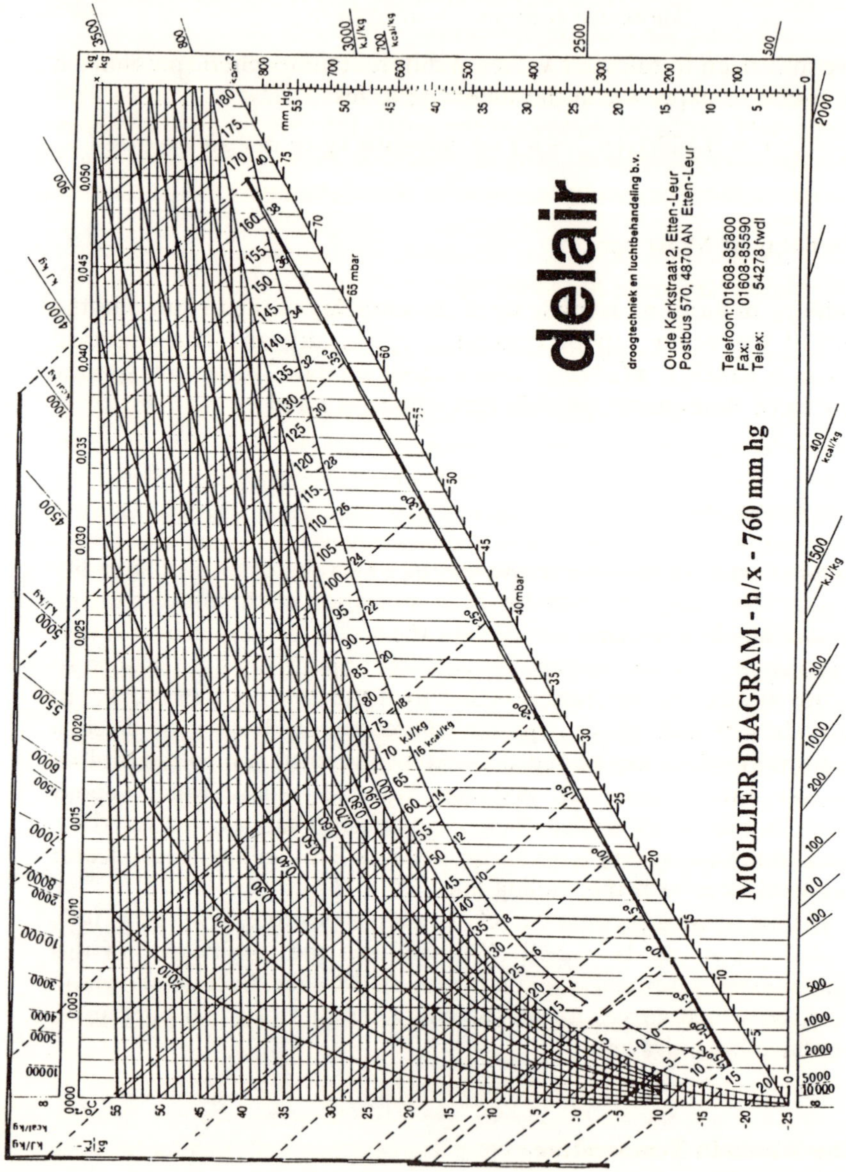

Figure 2.4 Mollier diagram for humid air (Courtesy of Delair)

some of the water on the bulb evaporates and thus cools it down. The quantity that is able to evaporate depends on the RH of the air; so the temperature we measure is an indication of the RH of the air. This lower reading is called the wet-bulb temperature.

Tables, charts and slide rules are available to determine the RH or ϕ for a given combination of wet-bulb and dry-bulb temperature. We can also measure the RH with a psychrometer and control the RH of an installation with a hygrostat or humidistat. Both are based on the degree of absorption of humidity by the hygroscopic element of the apparatus.

2.8 The Mollier Diagram for Moist Air

Mollier diagrams are normally constructed with enthalpy and pressure as coordinates and with the coordinate axes at right angles. When the thermodynamic properties of moist air are drawn on the Mollier diagram the moisture content x and the specific enthalpy are the coordinates, but in this form of the diagram the coordinate axes are oblique rather than rectangular; x on the horizontal axis and specific enthalpy on an oblique axis running from the bottom left to the top right of the chart.

On the diagram we find the following lines and curves:

(a) straight lines h = constant
(b) straight lines x = constant
(c) straight lines t = constant
(d) a saturation curve ($\phi = 1$) – note that this line varies with pressure, so different charts are available for different values of atmospheric pressure; mostly, charts use standard atmospheric pressure
(e) curves of RH or ϕ = constant
(f) lines of different $\mathrm{d}h/\mathrm{d}x$ slopes
(g) straight lines with specific volume or v = constant
(h) s = constant, straight lines.

We can use this diagram to represent the conditioning of the air, whether it be a cooling, heating, humidifying or dehumidifying process, or a combination of any two of these. We can also determine other conditions such as wet-bulb temperature, dew point, specific enthalpy and specific volume.

When we wish to represent the processing of a water vapour/air mixture on a Mollier diagram it is usual to start from the point $x = 0$ and $t = 0$°C and draw a straight line with the slope $\mathrm{d}h/\mathrm{d}x$ that we have already calculated. Next we find and mark the point representing the condition of the air, usually at the start of the process. Finally, we draw a line through this point parallel to the slope $\mathrm{d}h/\mathrm{d}x$ of the process.

When we cool air over a surface with a temperature below the dew point, water is extracted owing to condensation on the surface of the heat exchanger; obviously t, RH and x all change.

When we know that air has a certain dry-bulb temperature and a certain RH condition, we can find the wet-bulb temperature by following the line of constant enthalpy to the saturation line. Similarly, under the same conditions we can find the dew point by going vertically straight down from this point to the saturation line. Both of these constructions are shown on figure 2.5. Note that in the case of the wet-bulb temperature, by lowering the temperature far enough, the water vapour will start to condense when the temperature corresponds to the saturation pressure.

2.9 The Shape of the Cooling Line

So long as the cooling surface has a temperature above the dew point of the air to be cooled, only sensible heat is extracted. The absolute moisture content x remains unchanged. In that case the cooling line is a straight line – a vertical line from the point indicating the air inlet condition to the point indicating the cooling surface temperature marked on the saturation line. The condition of the outlet air is a point somewhere on this line above

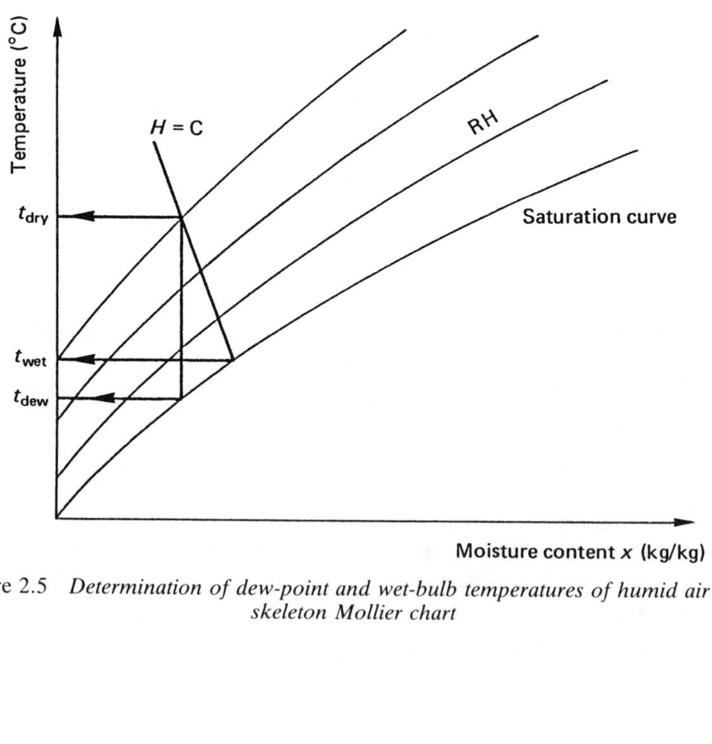

Figure 2.5 *Determination of dew-point and wet-bulb temperatures of humid air on a skeleton Mollier chart*

the cooling surface temperature; if it were to reach the cooling surface temperature on this line it would mean that the cooling surface was infinitely large and all the air had been in contact with the surface. In fact only the boundary layer of the air touches the surface and reaches a temperature so low that it approaches the surface temperature.

The actual outlet condition will depend on the construction of the cooler and the temperature of the cooling surface. When the surface temperature lies lower than the dew point, condensation of the moisture in the air will start, and when the surface temperature lies below 0°C this condensed water will freeze.

Figure 2.6a shows how the relative humidity increases during the cooling process. Point 1 indicates the inlet condition of the air and point 2 the cooling surface temperature. The boundary layers of the air in contact with the surface will be cooled down almost to point 2, possibly point 3, but they only constitute a small part of the air stream. This small part mixes with the rest of the air stream and the resulting mixture reaches condition 4. When this mixture is cooled, the boundary layers reach condition 5, and the mixture 7, and so on until the final condition of 2′ is reached.

The real situation is more complicated as the evaporation temperature and hence the surface temperature are not constant. At surface temperatures below 0°C the situation is even more complicated. In industrial refrigeration, as distinct from air conditioning, we do not use such a rigorous treatment as this; in practice a simple straight line is drawn between 1 and 2′, even between 1 and 2.

This cooling line is used in order to calculate the air outlet temperature when air inlet condition, evaporation temperature, surface temperature and the mass flowrate of the air stream are known. The calculation is as follows. Draw a straight line on the Mollier chart between the condition of air at the inlet of the cooler and the evaporation temperature on the saturation line. Determine dh over the cooler, from $\mathrm{d}h = Q/\dot{V}\rho$ in which $\dot{V}$ = air volume flowrate and ρ = density of the air at the average given temperatures. Mark off dh along the cooling line, beginning at the inlet condition, so that we can read the condition of the outlet air. Sometimes this calculation is needed when comparing the given capacities from manufacturers in which some are using $\mathrm{d}t_1$ and others dt. More information is given on this in chapter 6 on evaporators and condensers.

Sometimes the relative humidity at the inlet condition is high and the surface temperature is very low. In that case the cooling line should pass through the wet region of the chart, under the saturation line $\phi = 1$. It can, however, be proven that the cooling line will always be a tangent to the saturation line. After any cooling process the RH never reaches 1; it is lower the greater the temperature difference between cooler surface and the air entering.

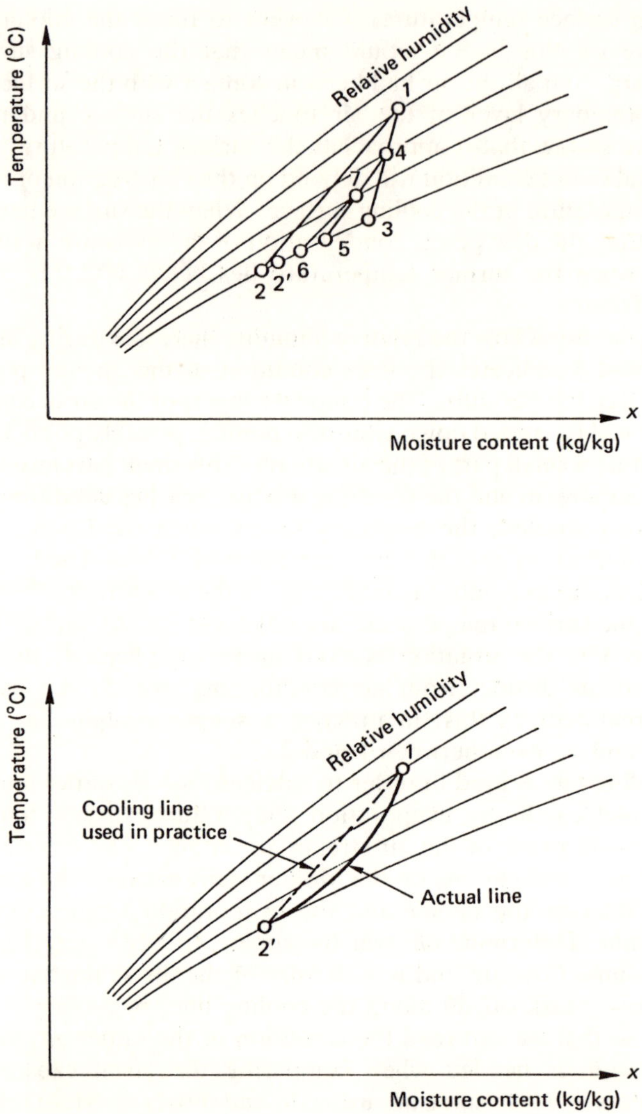

Figure 2.6 *Theoretical and practical cooling curves*

2.10 The Carrier Chart

Mollier charts are not the only charts available for use in psychrometric calculations. Carrier, in the USA, designed a different chart using similar equations, with the moisture content plotted on the vertical axis. It can be seen from figure 2.7 that the Carrier chart is a mirror image of the Mollier

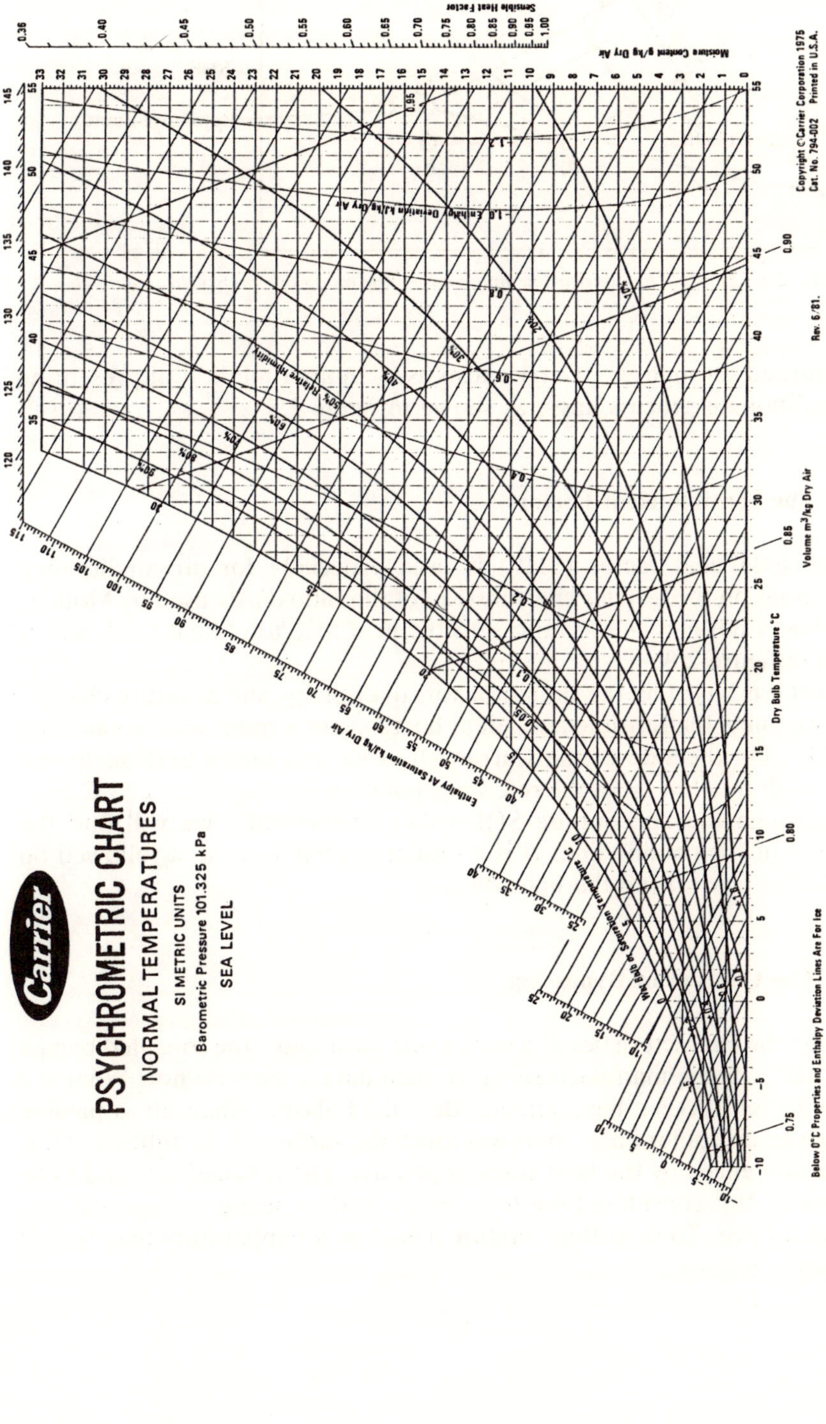

Figure 2.7 Psychrometric chart for humid air

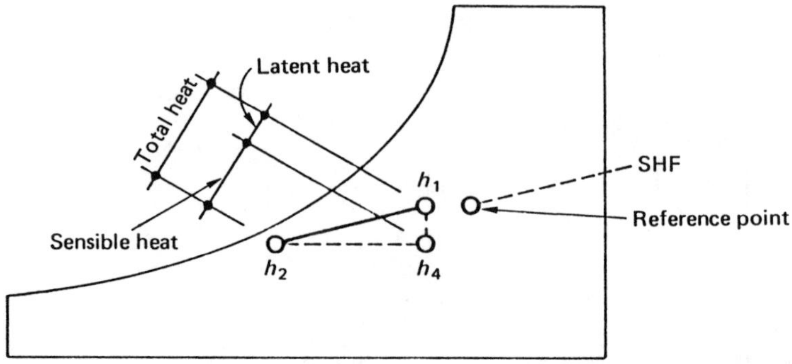

Figure 2.8 *Sensible heat factor line and cooling line on a skeleton psychrometric chart*

chart turned through 90°. Carrier also used another method of presenting air-conditioning process lines, as shown in figure 2.8.

2.11 The Sensible Heat Factor

The sensible heat factor (SHF) is used especially for air-conditioning calculations on the Carrier chart instead of the ratio dh/dx used by Mollier. The SHF is defined as the ratio sensible heat load/total heat load, and is shown in figure 2.8.

In fact it is based on the similar factors of enthalpy and moisture change. Once we know the sensible and latent heat load of a room we can calculate the SHF. On the side of the Carrier chart we can find a SHF scale and when we draw a line from the reference point on the diagram, at 24°C/50% RH, through the appropriate SHF value on the scale, we will find the direction of the cooling line. It is a similar procedure to that followed on the Mollier chart using the lines dh/dx.

2.12 The Concept of Bypassing

To judge the characteristics of a cooler one must know the so-called bypass factor (BF). Some manufacturers give such data, others do not. What is a bypass factor? As we have already described above, when air is passing through an air cooler only a part will touch the surface of the tubes and fins and be influenced by the heat-transfer process. The remainder is said to be 'bypassed'. It is considered not to be influenced by heat exchange with the cooling surface. The resulting mixture has a lower temperature than that of the inlet condition.

The bypass factor can best be expressed in mathematical terms in conjunction with figure 2.9. We will take the inlet condition of the air to be A, the outlet condition as B and the surface temperature as C.

Then BF can be expressed as $(h_B - h_C)/(h_A - h_C)$, $(t_B - t_C)/(t_A - t_C)$ or $(x_B - x_C)/(x_A - x_C)$ since the position of point B on the cooling line A–C depends on the composition of the outlet air mixture, or in other words on BF.

In the ideal case when no air is being bypassed, $h_B - h_C$ or $t_B - t_C = 0$. In practice, $h_B > h_C$ in proportion to the amount of air being bypassed.

When cooling coils with large BFs are used, with the resulting large amount of bypassed air, the outlet temperature is high and the RH low. This type of coil is usually characterized by a shallow coil, with few tubes and wide fin spacing. Alternatively if a low outlet temperature and a high RH are required then the cooling coil must be chosen with a deep section and closer fin spacing, resulting in a low BF. When a large temperature difference is required between inlet and outlet air, in other words with point B close to point C, then again we call for a high BF or a larger cooler with a lower flowrate of air through it.

2.13 The Concept of Apparatus Dew Point

At the beginning of the chapter we mentioned the variation of the surface temperature of the cooling coil. In fact not only does the fin temperature vary from base to top but it is also different from the tube temperature. When we refer to the surface temperature we mean the average surface temperature, usually called the apparatus dew point (ADP). In air-conditioning calculations the ADP is the terminal point on the saturation line of the SHF line, which in those cases has a steep slope.

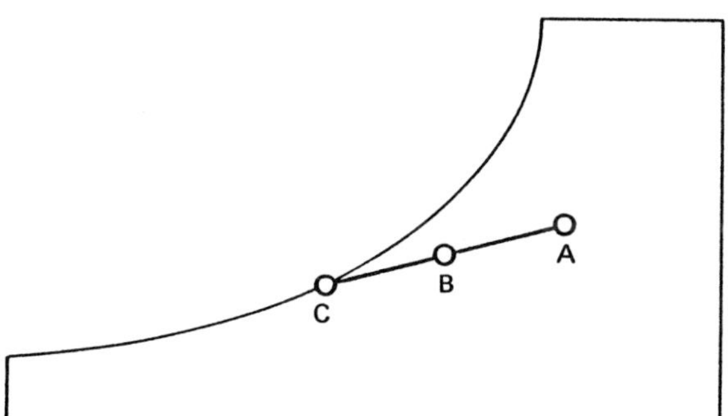

Figure 2.9 *Bypass factor shown on a psychrometric chart*

In refrigeration the ADP is taken to be the same as the evaporating temperature for two reasons:

1. in the temperature range involved the slope of the SHF line is quite shallow
2. the ratio of latent heat load to total heat load at temperatures below 0°C is usually smaller than in air-conditioning applications.

How can we apply these values? First we need to determine the conditions of temperature and RH of the cold room. Next we need to calculate the total heat and so derive x.

We now know SHF, and can draw the cooling line, h_A, h_C, x_A, x_C and x_B. From these we can calculate the BF from the equation previously used, $BF = (x_B - x_C)/(x_A - x_C)$, and hence we can determine the outlet condition of the air.

The cooler coil must be selected so that all the conditions demanded by BF and ADP are fulfilled. The manufacturer must supply the information to enable the evaporating temperature to be related to the desired ADP.

3

Refrigerants

3.1 Primary Refrigerants

Refrigerants can be divided into two groups: primary and secondary. Primary refrigerants produce coldness by changing from liquid to vapour. Secondary refrigerants transfer the thermal energy from the object to be cooled to the primary refrigerant.

The earliest primary refrigerants were water and diethylether; later came ammonia, sulphur dioxide, carbon dioxide, and certain of the hydrocarbons. Carbon dioxide was replaced because of the high pressures needed and sulphur dioxide because of its odour and toxicity; the others fell into disuse because they were too flammable. Anhydrous ammonia continued to be used and this is still the case today, though mainly for large industrial installations.

In small installations methyl chloride was used for a short time, but during the Second World War it was replaced by other halocarbon compounds known under such commercial names as Freon, Frigen, Arcton etc. This group of refrigerants contains fluorine and chlorine atoms. Nowadays these substances are generally referred to as chlorofluorocarbons compounds and abbreviated to CFCs. it is intended to use the term CFC in this book as a general term covering one or more of these categories unless the specific category requires designating.

The first one developed for commercial use was R12, or Freon 12, Dupont's registered trademark – known alternatively as Arcton 12 (ICI), Daiflan 12 (Daikin Kogyo), FCC 12 (Akzo), Forane 12 (Atochem), Frigene 12 (Hoechst), Geneton 12 (Allied Chemicals), Kaltron 12 (Kali Chemie), and many others.

R12 should really be designated R012. The three figures of this code follow the formula: first, the number of carbon atoms minus 1; second, the number of hydrogen atoms plus 1; and finally, the number of fluorine atoms – in that order. Any unfilled positions are normally occupied by chlorine atoms, but occasionally by a bromine atom, which is denoted by B1. This gives the following:

R12 (R012) denotes CCl_2F_2
R22 (R022) denotes $CHClF_2$
R11 (R011) denotes CCl_3F
R13 (R013) denotes $CClF_3$
R115 denotes $CClF_2 - CF_3$

There is also a group whose code starts with the digit 5 – R502 for instance. Those refrigerants are the azeotropic mixtures of different refrigerants. An azeotropic mixture is a mixture in which the vapour and the liquid have the same composition under all conditions. R502 is an azeotropic mixture of R22 and R115, while R503 is an azeotropic mixture of R13 and R23.

Non-azeotropic mixtures also exist; however their utilization in industrial refrigeration is not yet general. Probably the first use of these refrigerants will be in heat pumps; as the liquid and the vapour of these mixtures do not have the same composition under all conditions, these refrigerants will have a variable boiling point.

Finally, there is a group of primary refrigerants, in which the chlorine atoms are replaced by bromine atoms; an example is R13B1 ($CBrF_3$). Substances in this group are commonly known as halons.

Although many industrial installations continue to be built that use ammonia as refrigerant, there are many that now use R502, R22 and R11, the last in conjunction with centrifugal or turbo compressors. Air-conditioning installations mainly use CFCs as do most of the marine installations today. This situation will significantly change in the future as the conditions laid down by the Montreal Protocol take effect and as more stringent conditions are specified.

3.1.1 The Refrigerant of the Future

Even while this book is being written, the world is in the middle of a most significant scientific and ecological discussion regarding the environment.

It is now accepted that, when CFCs are emitted into the atmosphere, those containing chlorine and bromine atoms have the ability to attack the protecting ozone layer in the stratosphere and deplete it considerably above the Polar regions. The ozone layer protects against ultra-violet rays. It is these rays that are responsible for much of human skin-cancer. To protect the ozone layer, the USA, the European Community and 23 other nations signed, on 16 September 1987, the *Montreal Protocol on Substances that Deplete the Ozone Layer*. The agreement was designed to control the production and consumption of certain CFCs and halon compounds. The protocol calls for the freezing of R11, R12, R113, R114 and R115 and their mixtures at 1986 consumption levels, beginning in approximately July 1989, or 90 days after the agreement came into force. Reduction of 20% from 1986 levels is required by 1 July 1993, with reductions of 50% by 1 July 1998.

Halon 1211, 1301 and 2402 were fixed at 1986 consumption levels in 1992, or 3 years after the agreement came into force. In some countries the measures taken on this account are even more severe, as partially halogenated chlorofluorocarbons may also influence the ozone layer.

The members of the European Community were advised to reduce the use of fully halogenated chlorofluorocarbons by at least 25% by the end of 1991, and by at least 50% by the end of 1993, referred to the 1986 production. Complete phasing out is to be achieved by the year 2000. The 1990 conference held in London took the earlier recommendations a stage further, specifying that CFC production should cease by the year 2000 instead of being reduced to 50%; and that R22, R23, R123, R124, R141b and R142b be added to the list of proscribed substances although their use as replacements for the CFCs will be permitted until 2040 (with a possible restriction to 2020). A group of countries decided unilaterally to apply even more rigorous conditions than those above. In addition financial help was made available to Third World countries to help meet their costs in implementing the recommendations of the Montreal Protocol.

Bromofluorocarbons (BFCs) have the same effect as CFCs on the ozone layer. These compounds include the above-mentioned halons 1211 and 1301. They are known as R12B1 and R13B1. Chlorofluorocarbons and bromofluorocarbons also act as ‘greenhouse gases’ contributing to the reduction of heat radiation from the earth. This could result in far-reaching climatic changes, including melting of the Polar ice-caps and thus a considerable rise in sea level. These climatic effects could be just as serious as those resulting from the increase of ultra-violet radiation.

Note that, while the term CFC has been used to cover all the common refrigerants derived from methane and ethane, in fact the term CFC refers correctly only to hydrocarbons fully halogenated with a combination of chlorine, bromine and fluorine atoms. R22 and others with unsubstituted hydrogen atoms are termed more correctly hydrochlorofluorocarbons (HCFCs); when refrigerants contain no chlorine atoms they are referred to as hydrofluorocarbons (HFCs). The term hydrofluoroalkene (HFA) covers both HCFC and HFC refrigerants. The term CFC will be used throughout this book to include all the common refrigerants. When it is felt necessary to indicate the make-up of a particular refrigerant or group of refrigerants then the correct term will be used.

3.1.2 The Choice of Refrigerant

Whatever refrigerant is being used, care must be taken to adapt the design of the installation to the characteristics of the refrigerant used. The use of chlorofluorocarbon compounds in a big refrigeration installation demands a very different approach to that when using ammonia. Unfortunately, too many chlorofluorocarbon compound installations are being built as extended commercial installations, which results in cheap installations with low efficiency.

As table 3.1 shows, there are numerous chlorofluorocarbon compounds used as refrigerants. However, the above-mentioned R11, R22 and R502 are of most interest in industrial refrigeration.

Table 3.1 CFC refrigerants and their fields of application

R11 Trichlorofluoromethane CCl_3F Boiling point +23.8°C	Used as a refrigerant in air conditioning, generally in centrifugal compressors in industrial and other large buildings and for cooling water or brine, especially to maintain industrial processing temperatures. Used as brine at temperatures as low as −100°C. It was also used up to 1989 as a solvent for system degreasing or cleaning, for example, after burn-out of semi-hermetic systems
R12 Dichlorodifluoromethane CCl_2F_2 Boiling point −29.8°C	Up to 1989 the most widely used fluid in domestic appliances: it is used also in commercial and industrial cooling installations, for example, frozen food display cabinets, ice-making appliances, cooling fountains, coldstorage at temperatures above −15°C, refrigerated trucks, railway wagons or containers. Used for air conditioning, especially in automobiles and buses. Used with all types of compressors: hermetic and open; piston, rotary, centrifugal and screw
R13 Chlorotrifluoromethane $CClF_3$ Boiling point −81.4°C	For temperatures down to −80°C in cascade freezing installation using R22 or R502 in the first stage
R13B1 Bromotrifluoromethane $CBrF_3$ Boiling point −57.8°C	Refrigerant fluid mainly for temperatures between −45°C and −60°C in multi-stage piston compressors and in special centrifugal compressors. Eliminates the need for cascading at temperatures above −60°C

R14 Tetrafluoromethane CF_4 Boiling point −128°C	Has the lowest boiling point of all CFC refrigerants. Used for temperatures as low as −125°C with R22 or R502 in the first stage and R13 or R503 in the second
R22 Chlorodifluoromethane $CHClF_2$ Boiling point −40.8°C	Used as refrigerant in domestic, commercial and industrial air conditioning; commercial and industrial refrigeration including coldstorage and food processing; refrigeration and air conditioning aboard ships; and for heat pumps. Permits use of smaller equipment than does R12. Used in piston, screw and centrifugal compressors
R113 Trichlorotrifluoroethane CCl_2FCClF_2 Boiling point +47.6°C	Used as refrigerant for cooling water or brine for commercial or industrial applications in centrifugal compressors
R114 Dichlorotetrafluoroethane $CClF_2CClF_2$ Boiling point +3.6°C	Used as refrigerant fluid for large industrial air conditioning or processing temperature control; in multi-stage centrifugal compressors; for air conditioning at high temperatures and in aircraft
R500 Azeotropic mixture of R12 and R152a CCl_2F_2/CH_3CHF_2 Boiling point −33.5°C	Intended for refrigeration appliances and air-conditioning installations designed for 60 Hz but used at 50 Hz. Increases compressor capacity by between 10% and 15% more than R12
R502 Azeotropic mixture of 48.8% by mass R22 and 51.2% by mass of R115 $CHClF_2/CClF_2CF_3$ Boiling point −45.6°C	The most widely used refrigerant for frozen food display cabinets, also in freezing chambers and coldstores at temperatures of the order of −35°C where it is used as an alternative to ammonia. Has greater cooling capacity and a lower discharge temperature than R22; often avoids the need for water cooling. Used in air-to-air heat pumps and for industrial applications, generally in piston compressors.
R503 Azeotropic mixture of R13 and R23 $CClF_3/CHF_3$ Boiling point −88.7°C	Uses are similar to those of R13, but has an even lower boiling point and 50% greater freezing capacity at −85°C. Used in the second stage of a cascade system with two or three stages in scientific research, test chambers, metal hardening, pharmaceutical and other processes

3.1.3 New Refrigerants

New refrigerants are now under development, with the intention of replacing the harmful CFC refrigerants in the near future. Two examples of new refrigerants just coming into commercial use are R134a and R123, substituting for R12 and R11 respectively. Full details of these refrigerants are given in Table 3.2. Others are at a later stage of development and some of these are listed in table 3.3 with the refrigerant they are likely to substitute.

Table 3.2 Comparison of physical properties of two new and two traditional refrigerants

	R123	*R11*	*R134a*	*R12*
Chemical formula	CF_3—$CHCl_2$	CCl_3F	CF_3—CH_2F	CCl_2F_2
Molecular mass	152.9	137.4	102.0	120.9
Boiling point (°C)	27.9	23.8	−26.5	−29.8
Vapour pressure at 20°C (bar)	0.75	0.89	5.72	5.67
Heat of evaporation at 0°C (kJ/kg)	181.8	190.6	197.3	152.4
Critical temperature (°C)	184	198	101	112
Critical pressure (bar)	37.3	44.1	40.7	41.1
Critical specific volume (dm^3/kg)	1.804	1.804	1.952	1.792
Flammability	nil	nil	nil	nil

Table 3.3 Traditional CFC refrigerants and their probable substitutes

Original refrigerant	*Substitute*
R11	R123, R141b
R114	R124a, R142b
R501, R502	R125
R501, R502	R143a
R12	R134a, R152a

Present indications are that R152a has a doubtful future, owing to its flammability. R123, one substitute for R11, has good thermodynamic properties but is both corrosive and narcotic. R125, another substitute for R11, has a high global warming potential (GWP) – as has R134a. R123 suffers from the fact that the toxic R123a is difficult to separate from it.

The CFCs and HCFCs are all ultimately to be phased out more for their greenhouse effect than their ozone depletion potential (ODP) (see table 3.4). The refrigerants and refrigerating systems of the future will be chosen

Table 3.4 Ozone depletion potential (ODP) and Global warming potential (GWP) values

Refrigerant		*ODP*	*GWP*
CFCs			
	R11	1	1
	R12	1	2.8–3.4
	R12B1	3	
	R13	0.45	6.0
	R13B1	10–13	0.8
	R113	0.8–0.9	1.2–2.0
	R114	0.6–0.8	3.5–4.5
	R115	0.3–0.5	5.0–9.0
	R500	0.74–0.87	3.38–4.87
	R502	0.17–0.29	2.66–4.78
HCFCs			
	R21	0.04	
	R22	0.04–0.06	0.2–0.35
	R123	0.013–0.022	0.17–0.025
	R124	0.016–0.024	0.08–0.12
	R141b	0.07–0.11	0.08–0.11
	R142b	0.055	0.3–0.5
	R125	0	0.42–0.84
	R134a	0	0.23–0.29
	R143a	0	0.6–0.95
	R152a	0	0.024–0.033

Note: Bearing in mind the lack of consistency in the values available at the time of writing, the reader is advised to compare the information given in this table with the latest data available from refrigerant manufacturers.

predominantly for their energy efficiency. The improvement in overall coefficient of performance (COP) may well be achieved by a plant with an indirect ammonia system incorporating new heat-exchanger design with a 'cold' accumulation secondary refrigerant system. Evidence suggests that the timescale involved in developing and testing, and finding suitable lubricating oils, will be lengthy. Meanwhile ammonia, long held by many to be the best refrigerant for industrial installations, seems the obvious alternative for CFCs in this field.

What is the 'best' refrigerant? It does not yet exist. Every refrigerant has its advantages and disadvantages. The choice of refrigerant depends on the circumstances of the project and the preferences of the engineer. A refrigerant should be chosen that will satisfy most of the requirements of the project. Unfortunately, this choice is not always made in a 'cool', objective way. Commercial and psychological reasons, the effects of insurance conditions and local byelaws can result in a less efficient refrigerant being chosen than would be from a purely technical point of view. This is frequently the case when the choice is being made between ammonia and chlorofluorocarbon compounds. As there are no general rules about making such a decision, we can only draw your attention to a comparison of the characteristics of different refrigerants. This, we hope, will enable you to make up your mind.

The first question to ask is: what are the characteristics of a good refrigerant? They are as follows:

1. harmless to the environment – zero ODP and low GWP
2. high latent thermal energy (L_V) per unit mass, so that only a small quantity of refrigerant need be circulated. For comparison of different refrigerants, see table 3.5.

Table 3.5 Latent heat capacity (L_v) of some common refrigerants at −10°C

Refrigerant	L_v*(kJ/kg)*
R717	1296.4
R22	213.12
R502	153.45
R12	157.28
R11	193.77

3. A large enthalpy difference per unit volume, between the liquid and the vapour state: $dh/v = (h_3 - h_2)/v_3$ kJ/m^3 (see table 3.6).

Table 3.6 Volumetric refrigeration effect q_{ov} at various conditions

Refrigerant	q_{ov} (kJ/m^3)			
	−10/+35°C	*−10/+45°C*	*−25/+35°C*	*−45/−10°C*
R717	2591.9	2474.0	13979.3	619.7
R22	2370.0	2067.8	1313.7	866.4
R502	2587.8	2411.0	1501.8	817.8
R12	1486.2	1354.1	816.6	464.1
R11	251.95	237.63	121.98	

4. A small isentropic exponent k (formerly referred to as γ) where $k = c_p/c_v$. The power consumption depends on this exponent. Most of the power used in an installation is consumed by the compressor. We saw earlier that the compression process can follow different curves on

the indicator diagram. We can go from low pressure to high pressure by:

(a) Isothermal compression: the equation of this process is pV = constant or pV^n = constant, where $n = 1$. This is impossible to realize in practice in refrigeration applications because at $n = 1$, c_p would be equal to c_v, and so $(c_p - c_v)$ and R would then be 0. The area under the p/V diagram is a minimum, and the lowest possible power consumption results.

(b) Adiabatic compression: the equation of this curve is pV^k = constant, where k is greater than 1. This results in a larger area of the indicator diagram, hence high power consumption and low efficiency – even worse than in the practical process.

(c) Polytropic compression: these processes are processes where pV^n = constant, where n is greater than 1 and smaller than k. These are the practical processes. On the p/V diagram we see that the practical area lies between (a) and (b). In calculations we use the exponent k. This exponent k is different for each refrigerant and varies under different conditions as the tables of physical properties below show. The lowest values come nearest to the ideal isothermal value of 1. From this point of view, the best refrigerant is that with the lowest k value.

5. A good efficiency ratio η_c (or COP_{ref}) Carnot showed that the thermodynamic process with the highest efficiency has the smallest difference between the highest and lowest temperatures of the process. He compared the processes by their Carnot Coefficient of Performance $\phi_c = T_2/(T_1 - T_2)$. This is a comparison between the energy that is used and the resulting energy that is transferred. The T/s diagram can explain this ratio as being the efficiency ratio of a thermodynamic process.

   Another factor is the cold factor ϕ_0 that judges the practical process: $\phi_0 = (h_2 - h_3)/(h_3 - h_4)$. It is interesting to test this practical ratio ϕ_0 against the theoretical ratio of Carnot ϕ_c for a process with fixed values of T_1 and T_2, but with different refrigerants. You will see that the results are different for the different refrigerants. But the results also show differences that depend on the values of T_1 and T_2 used for the comparison. ϕ_0 can be considered as the efficiency ratio of a practical process, compared with the ideal Carnot process.

   In figure 3.1 we can appreciate that the best result should be obtained if we could move from T_2 to T_1 by the shortest route between the two points, which is a straight vertical line. This is of course impossible.

   Now let us consider the isothermal compression process. This is not realizable as we know, and would not be very effective. Carnot

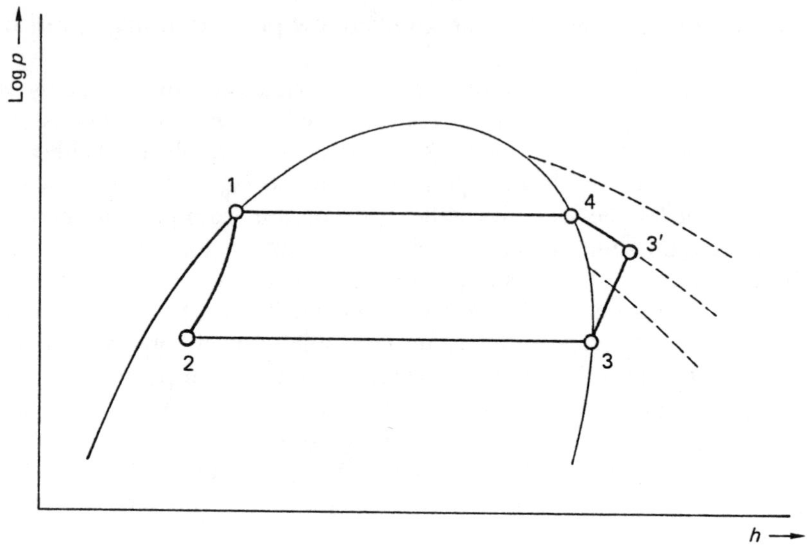

Figure 3.1 *The Carnot cycle*

demonstrated the practical processes: first the adiabatic, then the isothermal process.

For an effective process, we must come as close as possible to the Carnot process. Therefore, a refrigerant with a straight saturation line, and an almost straight adiabatic, gives the highest efficiency from this point of view. A small difference between suction pressure and discharge pressure or a low pressure ratio p_c/p_e (which means a small difference between evaporation temperature and condensation temperature) is desirable for good efficiency. Also on this point the refrigerants differ greatly, as seen in table 3.7.

6. Good heat-transfer characteristics: the refrigerant must have a good conduction coefficient in the liquid as well as in the vapour state. The

Table 3.7 Compression ratio p_c/p_e for different refrigerants

Refrigerant	*−10/+35°C*	*−10/+45°C*	*−25/+35°C*	*−45/−10°C*
R717	4.64	6.1	8.91	5.65
R22	3.57	4.5	6.1	3.78
R502	3.71	4.9	6.85	4.06
R12	3.71	4.95	6.84	4.12
R11	5.8	8	12.2	—

Note: These factors must not be confused with the coefficient of performance of an installation, or overall COP:

$$\text{Overall COP} = \text{Refrigerating capacity/Power consumption}$$

This is calculated on practical values measured on a working installation.

dynamic viscosity must be as low as possible, and the specific heat of the liquid must be high.

7. A high molecular mass M results in a small value of the exponent k, k being approximately

$$c_p/c_v = (c_v + R)/c_v = 1 + R_0/(Mc_v)$$

This has however disadvantages: for instance, low thermal conductivity, high density, high viscosity and low latent thermal energy. CFCs have high molecular masses compared with R717.

8. Moderate pressure at the condensing temperature – see table 3.8.

Table 3.8 Refrigerant saturation pressures at $t_c = 35°C$

Refrigerant	*Pressure (bar)*
R717	13.50
R502	14.50
R22	13.68
R12	8.48
R11	14.72

9. At low evaporating temperatures the pressure should still remain above the atmospheric pressure. There will thus be no introduction of air into the system should there be leakages – see table 3.9.

Table 3.9 Refrigerant saturation temperatures at $p_e = 1$ bar

Refrigerant	*Temperature (°C)*
R717	−34
R502	−46
R22	−42
R12	−30
R11	+23

10. The refrigerant should not be expensive. It should also be readily and widely available.
11. It should be odourless and non-toxic to human beings. However, having no odour, leakages are more difficult to detect and locate.
12. The refrigerant should not be corrosive or attack metals, other materials of construction or oil.

Examine tables 3.10, 3.11, 3.12 and 3.13 before we take a closer look at the chemical characteristics of the commonly used refrigerants in the following sections.

Table 3.10 Physical properties of R717

	t	ρ	c_p	c_v	k	η	ν	λ	P
	(°C)	(kg/m^3)	($kJ/(kg\ K)$)			($\mu Pa\ s$)	(mm^2/s)	($W/(m\ K)$)	(bar)
Saturated liquid	−60	713.90	4.37			380	0.532	0.550	0.219
	−40	690.14	4.44			280	0.406	0.550	0.718
	−20	665.13	4.51			230	0.346	0.547	1.902
	0	638.58	4.60			190	0.298	0.539	4.294
	+20	610.12	4.72			154	0.252	0.520	8.572
	+40	579.16	4.90			126	0.218	0.490	15.54
	+60	544.76	5.15			103	0.189	0.453	26.13
Saturated vapour	−60	0.21205	2.079	1.565	1.316	7.112	33.54	0.0176	0.219
	−40	0.64306	2.035	1.582	1.320	7.892	12.27	0.0191	0.718
	−20	1.6074	2.271	1.644	1.320	8.860	5.400	0.0201	1.902
	0	3.4805	2.504	1.751	1.319	9.482	2.724	0.0232	4.294
	+20	6.7695	2.862	1.903	1.317	10.31	1.523	0.0259	8.572
	+40	12.156	3.391	2.103	1.314	11.19	0.921	0.0291	15.54
	+60	20.611	4.171	2.352	1.306	12.14	0.589	0.0329	26.13

Table 3.11 Physical properties of R12

	t (°C)	ρ (kg/m^3)	c_p (kJ/(kg K))	c_v	k	η ($\mu Pa\ s$)	ν (mm^2/s)	λ (W/(m K))	P (bar)
Saturated liquid	−60	1573.55	0.868			482.6	0.307	0.1030	0.226
	−40	1507.94	0.883			384.1	0.253	0.0970	0.642
	−20	1459.06	0.901			316.8	0.217	0.0905	1.509
	0	1396.3	0.925			269.1	0.193	0.0840	3.086
	+20	1327.59	0.960			231.7	0.175	0.0775	5.673
	+40	1251.65	1.014			198.4	0.158	0.0707	9.606
	+60	1164.40	1.109			164.8	0.142	0.0637	15.26
Saturated vapour	−60	1.5646	0.512	0.439	1.150				0.226
	−40	4.1200	0.546	0.439	1.150				0.642
	−20	9.1456	0.583	0.496	1.112	10.65	1.165	0.00790	1.509
	0	17.961	0.626	0.525	1.084	11.62	0.647	0.00898	3.086
	+20	32.318	0.677	0.554	1.049	12.56	0.391	0.01017	5.673
	+40	54.762	0.744	0.584	1.003	13.81	0.252	0.01159	9.606
	+60	89.548	0.844	0.614	0.945	15.26	0.170	0.01343	15.26

Table 3.12 Physical properties of R22

	t	ρ	c_p	c_v	k	η	ν	λ	P
	(°C)	(kg/m³)	(kJ/(kg K))			(μPa s)	(mm²/s)	(W/(m K))	(bar)
Saturated liquid	−60	1464.98	1.004			471.2	0.322	0.1230	0.378
	−40	1409.08	1.048			369.5	0.262	0.1159	1.052
	−20	1349.23	1.100			302.1	0.224	0.1085	2.444
	0	1284.27	1.160			252.5	0.197	0.1007	4.963
	+20	1212.34	1.233			212.2	0.174	0.0925	9.087
	+40	1130.16	1.328			173.6	0.154	0.0832	15.36
	+60	1030.82	1.478			136.5	0.132	0.0728	24.41
Saturated vapour	−60	1.8636	0.551	0.445	1.214				0.378
	−40	4.8833	0.602	0.482	1.194	10.05	2.058	0.00756	1.052
	−20	10.868	0.664	0.523	1.171	11.06	1.018	0.00872	2.444
	0	21.483	0.744	0.567	1.144	12.10	0.563	0.00999	4.963
	+20	39.032	0.853	0.616	1.111	13.18	0.338	0.01147	9.087
	+40	67.119	1.013	0.669	1.069	14.58	0.217	0.01337	15.36
	+60	112.78	1.291	0.728	1.017	16.43	0.146	0.01603	24.41

Table 3.13 Physical properties of R502

	t	ρ	c_p	c_v	k	η	ν	λ	P
	(°C)	(kg/m³)	(kJ/(kg K))			(μPa s)	(mm²/s)	(W/(m K))	(bar)
Saturated liquid	−80	1587.3	1.022			600	0.378	0.150	0.144
	−60	1530.6	1.038			450	0.294	0.143	0.488
	−40	1466.9	1.076			355	0.242	0.133	1.309
	−20	1398.1	1.125			288	0.206	0.121	2.937
	0	1323.7	1.178			229	0.173	0.105	5.764
	+20	1241.4	1.235			180	0.145	0.090	10.23
	+40	1141.7	1.304			137	0.120	0.075	16.81
Saturated vapour	−80	1.0320	0.536	0.458	1.159	8.412	8.151	0.00501	0.144
	−60	3.1743	0.577	0.493	1.140	9.418	2.967	0.00604	0.488
	−40	7.8729	0.625	0.530	1.118	10.471	1.330	0.00714	1.309
	−20	16.901	0.682	0.567	1.089	11.560	0.684	0.00822	2.937
	0	32.490	0.754	0.608	1.051	12.736	0.392	0.00970	5.764
	+20	57.762	0.853	0.650	1.002	14.094	0.244	0.0114	10.23
	+40	100.25	1.009	0.696	0.938	15.839	0.158	0.0137	16.81

3.2 Toxicity and Other Hazards

Are refrigerants toxic or in other ways a threat to human life? Many people think that CFCs are not dangerous but that ammonia poses a threat. However this is too simple a statement. First of all the majority of refrigerants can burn the skin, or more importantly, eye tissue on contact at low temperatures. At high temperatures, for instance in contact with a naked flame or surface at high temperature, CFCs and HCFCs produce phosgene, a lethal gas which has been used in warfare. It kills at a concentration of 0.02–0.05% by volume. A concentration of 0.1–1.0 ppm is dangerous. The maximum safe concentration for CFCs is 1000 ppm, bearing in mind that 4500 ppm will attack the central nervous system. It should not be forgotten, that all CFCs are dangerous after 2 hours of inhalation at concentrations of 10–20% by volume. Higher concentrations in the air kill because of lack of oxygen. Long before this concentration is reached, however, a human would already be unconscious, with the possibility of brain damage. As CFCs and HCFCs have no smell, they are difficult to detect, and in the event of a leak, concentrations can increase up to danger levels before anyone would be aware of the fact. Leakages are also difficult to detect, so servicing is costly both because of the high price of CFCs, and because the installation may work for a long period with poor performance before the leak is detected. The use of copper tubes in CFC installations in which vibration often occurs makes leakages more likely, since such tubes are relatively fragile. As leakages are so difficult to detect, an additive (about 10%) can be put in the refrigerant, which colours the leakage points red and is very obvious except where the pipe is covered with insulation. This dye actually goes into solution in the oil. When the refrigerant passes through the leak the red oil carried with it indicates the location of the leakage. The dye can be used with R12 and R22 and is stable up to 370°C. However most manufacturers of hermetic compressors forbid its use in their compressors. Another method is to introduce a solvent which responds to ultra-violet radiation in the event of a leak.

In order to reduce the emission of CFCs into the atmosphere to an acceptable amount it is necessary that only competent qualified staff install, maintain and repair CFC and HCFC plants, and that regular inspection and preventative maintenance are carried out.

Every gas under pressure can cause the rupture of a vessel or tube under extreme circumstances, with disastrous consequences. The regulations covering the manufacture and modification of liquid storage vessels have been made much more stringent recently.

There is no doubt that ammonia is very toxic and that it smells unpleasant. Contact with food products is disastrous. An unpleasant smell, however, provides a good warning system. A concentration as low as 5 ppm is detectable by nose. Ammonia may kill at a concentration of

1700 ppm after half an hour, at 3500 ppm it is considered very dangerous and it is lethal at 5000 ppm. The threshold limit value (TLV – maximum concentration allowable for an eight-hour working shift) is 25 ppm. Concentrations around 200 ppm are tolerable, any irritation of the mucous membrane or irritation of the eyes then being only short term. In the range 400–700 ppm it produces additional eye, nose and throat problems which are not serious so long as exposure is less than one hour.

One cubic metre of water can absorb 120 kg of ammonia. Without cooling, water can absorb 18% by mass; by higher concentrations, too much heat of solution is generated.

Pure ammonia does not burn, but a mixture of ammonia and air at a concentration of 16–20% is somewhat explosive when ignited by a spark. The ignition temperature is 630°C. The ignition energy required is 0.1 J. For hydrogen the energy required is only 10^{-4} J.

Research shows that:

- Ammonia is very difficult to ignite
- In the free atmosphere it only burns while it is boiling and exposed to a strong spark
- When liquid ammonia at atmospheric pressure is stabilized it cannot be ignited
- A mixture of dry air and ammonia at ignition temperature can be ignited in a closed vessel or room in the presence of a powerful electric spark or an open flame
- Thermal decomposition (breakdown into N_2 and highly flammable H_2) occurs at temperatures above 458°C
- High humidity can prevent ignition
- The flame velocity for ammonia is one-fiftieth of that of natural gas
- The 'Brisanz' factor, that is the energy generated during burning, is only one-seventh of natural gas.

Summarizing, the high ignition temperature, small air/ammonia ignition concentration range, low 'Brisanz' factor, and inhibiting effects of humidity make the probability of an explosion very low.

Ammonia vapour is lighter than air. When liquid is released to the atmosphere the natural low temperature causes the water vapour in the air to condense into a mist of water droplets.

It is generally stated that all refrigerants are chemically stable; this is only correct under certain circumstances. It is often said that ammonia has the disadvantage of a relatively low temperature limit at the end of compression. The highest permissible operating temperature is limited to about 130°C, because of stability problems with ammonia and the oil.

Let us now compare the situation with that for CFCs and HCFCs. At certain temperatures these refrigerants lose their stability as well, namely under the influence of certain unavoidable 'catalysts'. If those catalytic effects are to be avoided, the limits would be:

R12	480°C (753K)
R22	250°C (523K)
R502	309°C (582K)

Such 'catalysts' are

- degenerated oil
- metal swarf, oxides in general, all kind of impurities
- oxides formed by the presence of water.

Water can enter into the circuit during repair. It accelerates the catalytic action of all metal parts and makes the refrigerant and oil acidic. Acidic refrigerant and oil form salts when in contact with metal; the resulting salts then oxidize and break down the oil. By further catalytic action of oil, R12 will break down at 175°C (488K) into sludge and chlorine derivatives, forming R22. R22 reacts between 205 and 230°C (478 and 503K) with metal parts and dirty oil and generates, for example, R32. The creation of other fluorocarbon compounds can lead to problems with thermostatic expansion valves, changes in pressure and capacity.

The speed of the reactions depends on the refrigerant used. If we specify the reaction speed of R11 as unity, then it can be shown that the reaction of

R12	is 10	times faster
R502	is 1000	times faster
R22	is 5000	times faster.

The maximum allowable temperatures at the end of compression are as follows:

R717	130°C (403K)
R11	110°C (383K)
R12	120°C (393K)
R22	150°C (423K)
R502	150°C (423K)

When discussing oil in the compressor, not only ammonia but also CFC systems, especially those using semi-hermetic and hermetic units, it is important to stress the need for constant control of the oil quality and quantity.

In all installations, purging of air is necessary for other reasons too, which we will discuss later. In CFC systems, water and all other impurities must be scrupulously avoided at all costs.

A note of warning: high temperatures can occur by friction at some spots in the compressor. These temperatures are even higher than those mentioned as limits above, sometimes up to 200°C (473K). These so-called hot-spots can occur, for example, on crankcase heaters.

3.3 The Advantages of Ammonia

There is no doubt that ammonia will remain the 'best' industrial refrigerant for the foreseeable future, particularly because it has a zero ODP. It is worthwhile to provide more information on ammonia at this point.

Ammonia or R717 is produced by combining free nitrogen and hydrogen under high pressure and temperature in the presence of a catalyst. The process most commonly used is the Haber–Bosch method. The molecular mass of ammonia is 17.03, while that for R22 is 86.47.

The nitrogen component is inert in the combustion reaction and accounts for the limited flammability of ammonia. Ammonia's high lower limit of flammability and low heat of combustion substantially reduce its combustion/explosion hazards. Ammonia/air mixtures are flammable by spark ignition at concentrations of 16–27% by volume in air. However oil carried by ammonia lowers this level considerably, so that a figure of 4% by volume in air is considered the practical safe limit to prevent explosion.

Industrial-grade anhydrous ammonia is the most economically abundant and efficient heat-transfer medium available for industrial refrigeration. Its pungent odour provides an extremely effective self-alarming and leak-detecting characteristic.

All designers, operating and maintenance personnel involved with ammonia plant must carefully follow the recommendations of the appropriate standards and guides published by the various institutions and associations; a list of relevant publications and where they can be obtained is given at the end of Chapter 13.

One of the advantages of R12 compared with R717 is the relatively low condensing pressure. However, since the only CFC refrigerant with unrestricted use is R22, this advantage is lost.

In virtually all refrigeration systems, the use of pressure vessels involves the risk of the vessel rupturing, regardless of the fluid it contains, whether it be ammonia, a CFC or even air. In fact the nature of the risk presented by the fluid can only be secondary to the mechanical risk. However, it is obvious that the more rigorous standards applied to vessels, pipework and plant for ammonia, compared with CFC, systems make an ammonia system inherently more safe. Better reliability and lower maintenance costs for ammonia systems result from the use of: steel rather than copper pipework; welding rather than brazing for jointing; bolted flanges rather than flared connectors; and measures to prevent the ingress of impurities such as dirt, moisture and metallic particles, which adversely affect the refrigerant and oil. Furthermore, the management of oil in ammonia systems is simpler than that for CFC systems.

However, any malfunctioning system will always result in damage to the product or failure of the process, if prompt intervention is not made. Preventative maintenance, regular testing and continuous inspections, either by operating or maintenance staff or by means of tele-monitoring, are essential to prevent breakdown, whatever the refrigerant being used.

Finally, here are some advantages and disadvantages of ammonia as a refrigerant:

1. zero ODP and low GWP values
2. high latent thermal energy (L_v) per unit mass
3. large enthalpy difference per unit volume
4. high COP
5. good heat-transfer characteristics caused by the high thermal conductivity, high latent thermal energy, low viscosity and low liquid density compared with the CFCs and HCFCs
6. low molecular weight, $M = 17.03$, provides another advantage of low pressure-drop through compressor valves and ports – the disadvantage is the high value of k compared with those for CFCs and HCFCs
7. high compression ratio compared with that for most CFCs
8. low purchase price and low maintenance costs
9. toxicity and unpleasant smell
10. leakages easy to detect.

It is useful to draw some comparisons between R717 and R22 at this juncture when considering point 5. above:

Characteristic	*Ratio R717/R22*
Specific heat (liquid and vapour)	4:1
Latent thermal energy (L_v)	6:1
Thermal conductivity	5.5:1
Viscosity	0.8:1
Liquid density	0.5:1

Source: W. F. Stoecker, University of Illinois, USA.

3.4 Oil and Refrigerant Relationships

The refrigerants that we use in the system are never pure, they are always mixtures of refrigerant and oil. In reality the pressure/enthalpy charts for refrigerants do not represent the real situation in the installation. The charts used for the refrigerant/oil mixture should be similar to those available for water vapour/air mixtures. Scientists at the Polytechnic High School in Mons, Belgium, are working on this subject.

Always use the oil recommended by the manufacturer of the compressor; never mix two oils with different characteristics. Furthermore, make sure that the oil stays where it is needed, namely in the compressor. Oil causes problems in heat exchangers, where the oil film decreases the value of the heat-transfer coefficient. In refrigerant liquid pumps it creates cavitation, and in float-operated systems it leads to malfunction; however in automatic controls it acts as a lubricant.

In installations with oil-free compressors, trouble with control equipment can be expected if proper greasing of moving parts is not maintained; however the use of nylon parts can avoid this problem. Oil-free compressors are not popular today, mainly because of their high cost, the low efficiency rate caused by internal leakage, bypass problems and fragile construction.

Where installations need an almost oil-free system, as for instance in the case of skating rinks, ice-cream freezers or other production plants, high-efficiency oil separators can be installed. It has been discovered that modern piston and screw compressors allow more oil to pass than the earlier ones. Using standard oil separators based on lowering the gas velocity and passing of the gas stream over filters, the oil carry-over can still be between 20 and 50 ppm or even more. Special oil separators can bring down the carry-over to between 5 and 10 ppm. Today more and more synthetic oils, which exhibit a much smaller carry-over, are being used.

In fact there are various kinds of oil carry-over:

(a) Drops of oil leaving the compressor outlet and easily separated in standard oil separators
(b) Aerosols, very fine droplets, which have sizes in the range 0.01–0.8 micron and are created by the high shearing action of the compressor's moving parts. An aerosol could be described as an oil mist
(c) Oil vapour: this is oil in gaseous form not as a mist.

The amount of oil vapour depends on many factors: the type of oil, the pressure of the oil/refrigerant mixture, the age of the oil, but also on the temperature. By applying Dalton's law of partial pressures, it is possible to calculate the theoretical amount of oil vapour that can be present in a system, under given conditions, as follows:

$$M_0/M_g \times p_0/p_g \times 10^6 \text{ ppm/wt}$$

where M_0 = molecular mass of the oil
M_g = molecular mass of the refrigerant
p_0 = oil vapour pressure in operating conditions
p_g = refrigerant pressure of the system.

Figure 3.2 shows the importance of the gas temperature during separation. In the special oil separator, the discharge vapours are passed through a bath of cool liquid refrigerant. Vapours are cooled from 80 to 100°C down to 40°C and bubble through the refrigerant to the surface of the bath. Vapour leaves the separator from the top of the vessel and oil from the bottom.

There are also other more complicated oil separators of this type, which are claimed to separate the oil out completely.

As oil is heavier than ammonia, it is easy to purge it from the lower parts of vessels. This, however, is not the case with CFCs. Oil remains in solution in liquid refrigerant at all times. It must be removed by 'bleeding'.

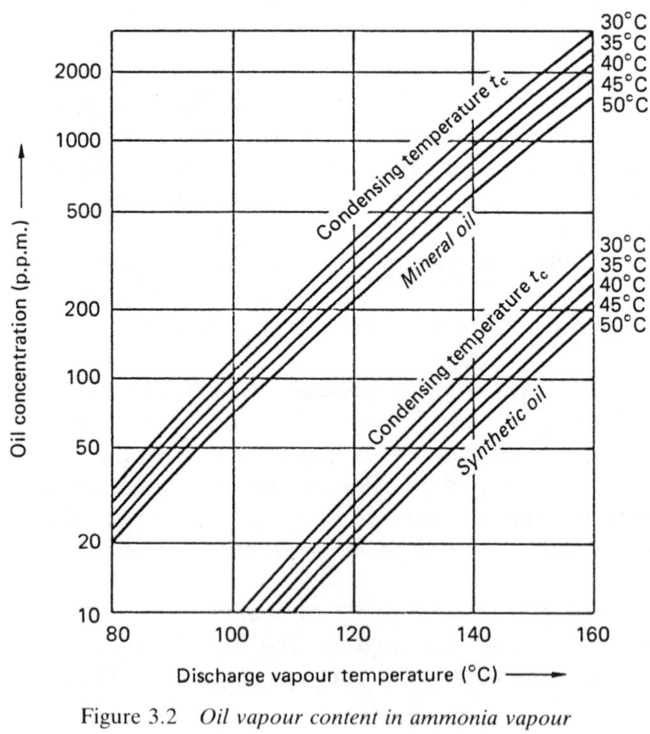

Figure 3.2 *Oil vapour content in ammonia vapour*

This involves the purging of a small flow of oil-rich refrigerant from the evaporator. This mixture is then passed to a special heat exchanger in which the refrigerant is evaporated out of the oil by heating, and then sucked away by the compressor. The bleeding of an R22/oil mixture or R502/oil mixture is not complicated, because in the evaporator the richest refrigerant/oil mixture is floating on the refrigerant surface; furthermore, the level of this surface is not constant. Connections to enable bleeding to be carried out must be provided on different levels.

Most oils also contain wax. This impurity can solidify on to valve seats and other devices under certain circumstances, and so cause problems.

With modern machines, we can work with very high temperatures in the system. These reach 200°C in some places in the compressor, for instance on crankcase-heaters. This can create oil problems: for instance, carbon formation because of reaction with R12, or too high temperatures in ammonia systems together with the presence of oxygen. This, however, only occurs when the oil is too old. Always note the maintenance instructions for oil changes or oil tests, as periodic tests for acid levels are necessary.

3.4.1 Lubricating Oils

The choice of lubricating oil for a refrigeration compressor is dependent on the refrigerant and the operating conditions. Refrigeration oils are characterized by being highly refined lubricants containing only small quantities of additives.

The main standard covering refrigeration oils in the United Kingdom is BS2626 with a comprehensive revised edition due in the near future. In Europe the most widely used standard is DIN 51503, Part 1 for new oils and Part 2 for used oils.

The oil types are only slightly miscible with ammonia and they can be divided into the main groups shown in table 3.14.

M – Mineral oil is extracted from crude oil. Naphthenic oil or mixtures with this oil base are generally used. Mineral oil is characterized by comparatively low miscibility with CFCs at low temperatures.

A – Synthetic oil based on alkyl benzene is mainly extracted from natural gases and is characterized by fairly high miscibility with CFCs at low temperatures. Alkyl benzene based oil is generally more stable under the influence of moisture and the oxygen of the air than is mineral oil. This fact can be utilized in R717 installations where problems of oxidation might occur.

MA oil is a mixture of mineral oils with resulting greater stability and less tendency to foam than the individual mineral oils. This stability is an advantage in CFC installations too.

Table 3.14 Different oil groups and their miscibility with CFCs

Main group	*Oil type*	*Viscosity index*			*Miscibility with CFC*		
		low	*medium*	*high*	*low*	*medium*	*high*
M	Mineral oil	x	x		x	x	
A	Synthetic oil based on alkyl benzene	x					x
MA	Mixture of M and A	x	x			x	x
P	Synthetic oil based on polyalphaolefin			x	x		
AP	Mixture of A and P			x		x	
G	Synthetic oil based on polyglycol			x	x		

P – Synthetic oil based on polyalphaolefin is a newly developed oil type, distinguished by its chemical and thermal stability. Consequently, it is used with advantage in compressors operating at high discharge gas temperatures, as, for example, heat pumps

P oil is remarkably stable in the presence of oxygen in the air. The oil has a very low solidifying point, thus it is suitable for use in R717 installations at low temperatures. On the other hand, miscibility with CFCs is low at low temperatures. Therefore, this oil should never be used in such installations. The oil is miscible with other commonly used refrigeration oils, but a change to P oil should only take place after efficient draining of the compressor and oil separator. P oils have a high viscosity index. Consequently, they are less sensitive to temperature changes than other oil types.

However, note that the P oils have a high aniline point, which may cause leaks at gaskets and O-rings, if changing from another oil type to P oil. They have a lower vapour pressure than M and A oils. The amount of carry-over from the compressor to the plant with R717 gas is therefore reduced significantly, thus resulting in a marked reduction in oil consumption.

AP – Mixtures of alkyl benzene and polyalphaolefin synthetic oils have a higher miscibility with CFCs than does P oil, and thus the oil is suitable for installations with low evaporating temperatures. Furthermore, the aniline point is lowered by mixing, so the risk of leaks at gaskets and O-rings is less than with pure P oil.

G – Synthetic oil based on polyglycol is extracted mainly from the natural gases ethane and propane and is distinguished by low miscibility with LPG. The polyglycol based oil is used in LPG installations.

3.4.2 Specific Gravity

The specific gravity contributes to the determination of the oil type. The alkyl benzene oils, for example, are lighter, and the glycol based oils heavier, than mineral oils.

A mineral oil with increased paraffin content will have a lower specific gravity than a naphthene based oil.

3.4.3 Viscosity

The choice of the viscosity depends strongly on the design of the compressor and on the refrigerant used. An oil with the lowest possible viscosity must be chosen, provided that it enables a satisfactory tightness to be maintained when used in combination with the refrigerant, over the entire temperature range that could occur. It is therefore necessary to know the viscosity of the oil/refrigerant mixture at different working temperatures.

3.4.4 Viscosity Index

This refers to the dependence of the viscosity on temperature. The larger the value, the less the dependence. The viscosity index is used in nomograms for comparison of oils.

3.4.5 Flash Point

This is the temperature at which the oil vapour from an open, heated vessel is ignited by a flame. The methods of measurement are specified in ASTM Standard D92–57 (Cleveland Open Cup), BS2000 Part 34 and DIN Standard 51376 (Flame point).

The flash point is used to determine the applicability of the oil at high temperatures. Oil with a high flash point has a low vapour pressure. This gives a better possibility for separation of the oil from the discharge gas in the oil separator, thereby reducing the oil transfer from the compressor to the refrigeration installation.

3.4.6 Pour Point

This is the temperature at which the oil has become sufficiently viscous so as not to move for 5 seconds when the reservoir in which the oil has been cooled is tilted. The standards define the pour point temperature as 3°C lower than the temperature measured. The method of measurement is specified in ASTM D97–66, DIN 51597 and BS2000, Part 15.

Pour point is of special interest for oils used in R717 installations, since oil with a low pour point is easier to drain from the low-pressure side of the installation. It appears that, for R717, it is possible to use some oils at evaporating temperatures lower than the pour point, without problems. In R717 installations with an evaporating temperature lower than −40°C, it is recommended that an additional, high-efficiency oil separator be installed in the discharge line from the compressor to the condenser. In this way the oil transfer to the low-pressure side of the installation is reduced.

3.4.7 Floc Point

This is the temperature at which R12 mixed with 10% oil, when chilled, becomes foggy with wax particles which have separated from the oil. The method of measurement is specified in DIN 51351. The floc point is significant where oil and refrigerant are miscible, as, for example, in refrigerating installations working with CFCs.

A low floc point indicates that the oil has a low wax content, and is thus suitable for CFC refrigeration installations at low temperatures. Separation of wax may cause plugging of the expansion valves of the installation.

3.4.8 Colour Number

The colour number expresses the translucency with a standard light source. A visual comparison is used with coloured glasses, graduated from 0.5 as the lightest colour to 8.0 as the darkest one. 'L' used in connection with the number means that the oil is more lightly coloured than the number indicates. The method of measurement is specified in ASTM D1500 and DIN 51578. Normally, refrigerating machine oils are characterized as being rather light in colour.

3.4.9 Aniline Point

This indicates the temperature in °C at which the oil in question combines with pure aniline in a homogeneous mixture. Aniline point is an indication of the quantity of non-saturated hydrocarbons which is present in the oil, and this value is important in assessing the likelihood of the oil attacking the types of rubber that it may come into contact with. The method of measurement is specified in ASTM D611 and DIN 51775. Since most of the refrigeration machine oils have a comparatively low aniline point, they cause neoprene rubber to swell without dissolving the rubber or otherwise affecting the sealing ability. This might require replacement of rubber gaskets, such as O-rings, during dismantling. P oils do not have this ability, because they have a high aniline point. Refer to section 3.4.1 on P oils.

3.4.10 Neutralization Number

This is a measure of the degree of acidity of the oil, determined by titrating potassium hydroxide against the oil sample.

The value is stated in mg potassium hydroxide per 1 g of oil sample. The method of measurement is specified in ASTM D974, BS2000, Part 139 and DIN 51558.

Refrigeration oils are highly refined without any additives. Generally, they have a low neutralization number.

For used oils, the neutralization number is called TAN (Total Acid Number) in accordance with the above-mentioned standards. The number indicates the content of acid particles in the oil, such as oxidizing products, which cause decomposition of the oil.

3.5 Water in Refrigerants

Finally, water is another impurity which will separate from the oil, causing problems. It is not significantly soluble in CFCs. Figure 3.3 shows the degree of solubility of water in R12 and R22, and points A and B indicate that 100 kg of R12 at 27°C (300K) can hold 10.2 g of water, whereas at −8°C (265K) it can hold only 0.17 g. Commercial-quality refrigerants contain less than 1 g of water per 100 kg.

Water can freeze on spindles and block the seats of thermostatic expansion valves, block up filters, degenerate oil and create acids. Another problem that could occur is the dissolving of copper into the oil. This dissolved copper may afterwards be deposited on hot-spots, such as the moving parts of the compressor, so making the running tolerances smaller and finally resulting in seizure. Sulphur, air, acid and water in the oil under high operating pressures can also encourage this action. Until recently it was assumed that water was the main cause of this copper deposition, but today it is believed that the problem lies also in the instability of the CFCs at very high local temperatures. The presence of water in ammonia is less harmful but at concentrations of more than 400 ppm problems may occur with controls; it is permissible to use ammonia containing not more than 200 ppm water. Water can also create problems with the oil. European

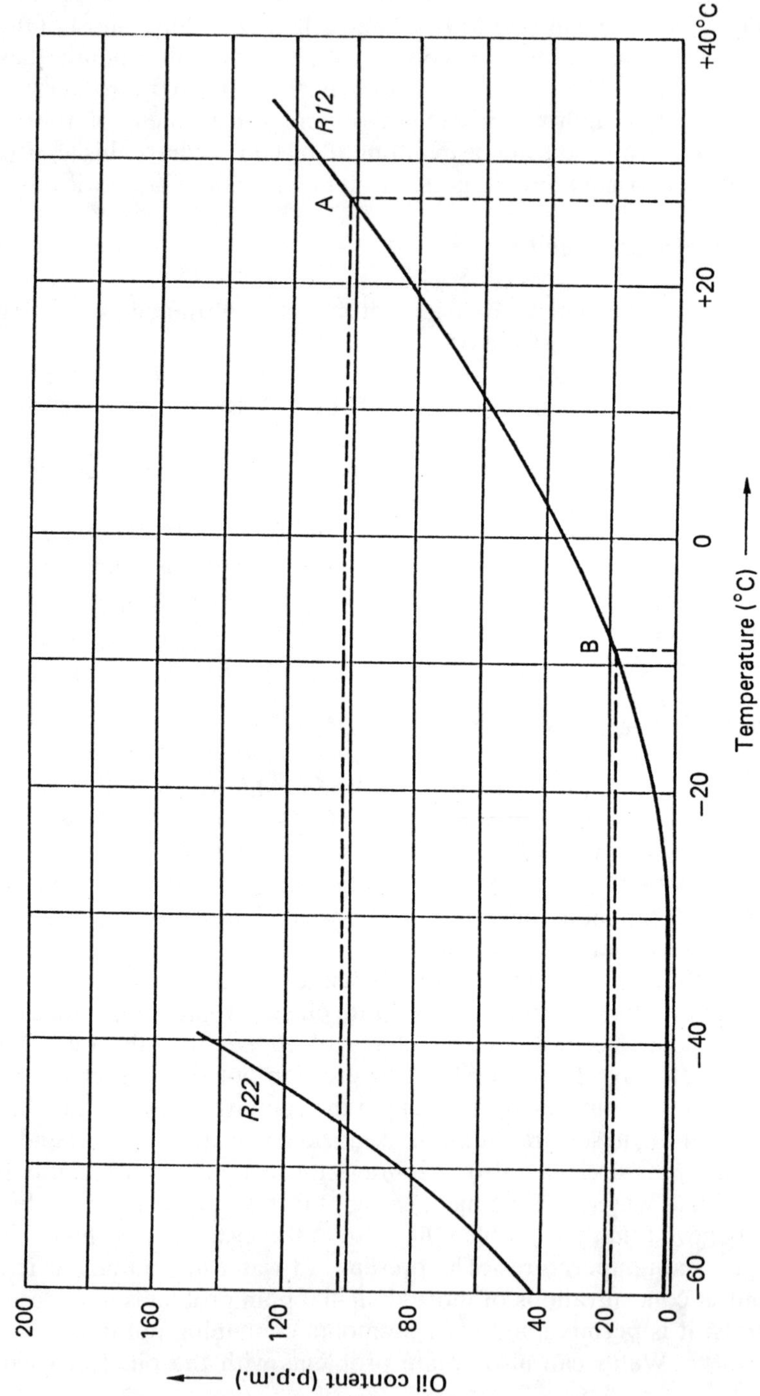

Figure 3.3 *Solubility of water in R12 and R22 liquid*

standard EN 378 allows 2000 ppm. Consult DIN 8960 and ARI 700–88 for characteristics of good refrigerants.

3.6 Secondary Refrigerants

In the past, only water/salt solutions were used as secondary refrigerants. Salts such as sodium chloride and calcium chloride were the best known ones. They are still used today, but as they are very corrosive, engineers looked for other methods of reducing maintenance costs.

Nowadays, glycol solutions are generally used. For special applications there are also dichloromethane (CH_2Cl_2), trichloroethylene (C_2HCl_3), alcohol solutions and acetone.

A good secondary refrigerant is expected to:

(a) be non-corrosive
(b) be inexpensive
(c) be stable
(d) have good heat transfer characteristics
(e) have a sufficiently low freezing point at a not-too-high concentration
(f) have a low density
(g) have a high specific heat.

Point (g) is a disadvantage to heat transfer but an advantage for the quantity of refrigerant that must circulate, and that is generally more important.

Nowadays, calcium chloride ($CaCl_2$), ethylene glycol ($C_2H_6O_2$) and propylene glycol ($C_3H_8O_2$) are mostly used as secondary refrigerants (see tables 3.15, 3.16 and 3.17). The latter are non-corrosive and non-toxic but more expensive.

Table 3.15 Comparison of brine concentrations at −50°C

Product	*Freezing point (°C)*	*Product in 100 l water (kg)*	*Concentration in solution*
Calcium chloride	−51.6	62	29.9% by weight
Ethylene glycol	−51.2	134	55% by volume
Propylene glycol	−49.4	150	59% by volume

Note: The eutectic point of propylene glycol is −73°C at a volumetric concentration of 60%

Table 3.16 Comparison of brine characteristics at 20°C

Solution	*Specific heat c*	*Density*	*Brine quantity needed to accumulate 1000 kJ/K*	
			Weight	*Volume*
	(kJ/kg K)	*(kg/dm³)*	*(kg)*	*(l)*
Calcium chloride	2.77	1.289	360	278
Ethylene glycol	3.28	1.070	305	286
Propylene glycol	3.38	1.042	295	283

Table 3.17 Comparison of brine viscosities

Product	*Weight needed in 100 l of water (kg)*	*Freezing point (°C)*	*Dynamic viscosity at 20°C (mPa s)*
Calcium chloride	62	−51.6	3.5
Ethylene glycol I	62	−20.5	2.6
Ethylene glycol II	134	−51.2	4.4
Propylene glycol I	62	−18	3.9
Propylene glycol II	150	−50	8.6

Calcium chloride is very corrosive but can be neutralized with additives. It is cheap, non-toxic and has a high specific density and a low viscosity. Ethylene glycol is slightly corrosive and toxic, and its cost lies between the other two.

Speaking of costs, one must be careful in making comparisons because, although the price of calcium chloride is the lowest of the three, it should be borne in mind that in relation to the accumulation of thermal energy, calcium chloride requires the highest concentration. However, as table 3.16 shows, the smallest concentration of calcium chloride is needed to reach a certain low freezing point, compared with ethylene glycol and propylene glycol. However, the final choice depends again on the circumstances of the project.

3.7 Heat-transfer characteristics

As we will see in chapter 5, these characteristics influence the calculation of the heat-transfer surfaces needed. Most important from this point of view are viscosity, specific heat and coefficient of thermal conductivity.

Changing from one refrigerant to another produces changes in the heat-transfer surface needed, a fact which is explained in chapter 5.

This change depends on a ratio denoted by p':

$$p' = (1/\alpha_1)/(1/\alpha_2)$$

$$= \left(\frac{\eta_1 c_{p1}}{\eta_2 c_{p2}}\right)^{0.47} \times \left(\frac{\lambda_2}{\lambda_1}\right)^{0.66} = x^{0.47}y^{0.66}$$

where λ = coefficient of thermal conductivity, η = viscosity and c_p = specific heat capacity. Here calcium chloride is indicated by the suffix 1 and glycol by the suffix 2. If p' is < 1, changing from glycol to calcium chloride results in a smaller surface.

The smaller the ratio p', the bigger the gain. Tables 3.18 and 3.19 show the results at different temperature levels.

Table 3.18 Replacement of ethylene glycol by calcium chloride solution

Temperature of refrigerant (°C)	x	y	p'
−10	0.570	0.875	0.585
−20	0.538	0.825	0.536
−30	0.274	0.819	0.389
−45	0.157	0.795	0.289

Table 3.19 Replacement of propylene glycol by calcium chloride solution

Temperature of refrigerant (°C)	x	y	p'
−10	0.307	0.847	0.424
−20	0.134	0.796	0.278
−30	0.055	0.765	0.171
−45	0.025	0.733	0.113

Source: Solvay & Cie SA, Belgium. G. della Faille d'Huysse.

Calcium chloride always provides the possibility of reducing heat-transfer surfaces, especially at low temperatures.

To conclude this chapter, simple equations are provided that can be used to convert the concentration ratio *n* in degree Baumé, or *Tw* in degree Twaddle, into the density ρ of a solution:

$$\rho = 145.88/(145.88 - n) \text{ kg/dm}^3$$

or

$$\rho = 1 + Tw/200 \text{ kg/dm}^3$$

Figure 3.4 illustrates the relationship in graphic form.

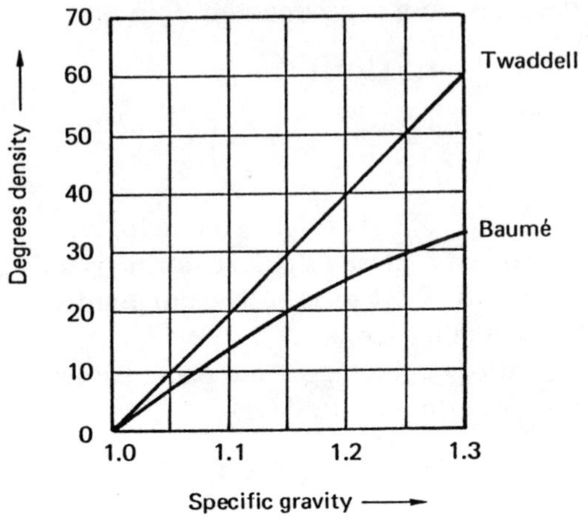

Figure 3.4 *Comparison between Baumé and Twaddell hydrometer readings*

References

G. Lorentzen, *Ammonia, an Excellent Alternative*, Butterworth, 1988; *IIR Int. Refr.*, Vol. 11, July 1988.

W. F. Stoecker, *Growing Opportunities for Ammonia Refrigeration*, University of Illinois and Urbana Champaign, IIAR Meeting, Austin, Texas, March 12–15, 1989.

4

Compressors

This book does not deal with the mechanical construction of compressors but attempts to describe the applications of the different compressor types in the field of industrial refrigeration and how they perform when using the refrigerants available to the refrigeration system designer.

Two principal groups can be identified:

(a) piston or reciprocating compressors, subdivided into open, semi-hermetic and hermetic types
(b) rotary compressors, subdivided in the same way.

4.1 Open and Semi-hermetic Compressors

The terms ‘open’, ‘semi-hermetic’ or ‘hermetic’ compressor are misleading; every compressor is sealed, none is really ‘open’. When referring to open, semi-hermetic and hermetic compressor/motor combinations it is preferable to use the term ‘compressor unit’. Semi-hermetic and hermetic compressor units in which motor and compressor are enclosed in one casing were developed to obviate the shaft-seal leakage problems, which are particularly relevant when using CFCs and HCFCs. Disadvantages of this form of construction are as follows:

(a) poor cooling of the motor-windings
(b) contamination of the system pipework by the residues after a burn-out of the motor-windings
(c) difficult maintenance and repair
(d) electric motor repair is complicated – in open-type units, replacing the standard motor is simple
(e) much supplementary superheating of the suction gases
(f) not possible to match the rotational speed to the demand

(g) not possible to use with ammonia as refrigerant, because of copper contact.

In the semi-hermetic form it is possible to construct the motor casing and the compressor casing separately and bolt the two together. The alternative form is a monoblock casing with end covers and removable cylinder heads. In the hermetic form the whole mechanical and electrical system is contained within a single metal casing. This form of construction does not allow for inspection or replacement of valves, which is yet another disadvantage. Hermetic units are not often used in industrial refrigeration, however they are found in medium and even some large air-conditioning plants. Their main field of application is household and commercial refrigeration. The development of semi-hermetic and hermetic units with piston compressors has stopped at about 30 kW motor-power. Rotational speed of piston compressors today varies between 600 and 1700 rev/min.

A recent development in the field of semi-hermetic compressor design seeks to eliminate some of the problems mentioned above. In this construction the compressor and rotor are assembled on a common crankshaft and placed inside the sealed compressor housing. The electric motor stator is assembled to the outside of the housing around the rotor. In this design the motor-windings make no contact with the refrigerant or any contaminants in the system.

The results obtained from this are as follows:

- no contamination of system in the event of a burn-out
- stator windings easy to replace
- no suction gas cooling available for the stator windings
- less possibility for copper plating.

4.2 Compression Action in the Refrigeration Process

As we know, the theoretical refrigeration cycle takes place between two isobars and two isentrops. In the practical cycle this is reasonably true for the processes in the evaporator and condenser where evaporation and condensation will differ a little from those theoretical processes, mainly owing to pressure losses in the equipment. However, for the processes in the compressor and the expansion valve the difference is more significant. The practical compression process is not reversible and follows the law pV^n = constant, a polytropic process instead of pV^k = constant, the isentropic process. An additional complication is the variation in the value of n during compression.

It might be asked why the compression is not isothermal. The answer is that, during compression a part of the external energy is converted into internal energy. As the volume is decreasing the molecules are more closely packed, so the possibility of collisions and friction as well as the vibration of the molecules increase. In other words the internal energy increases, which results in an increase of temperature. In order to realize an isothermal process, every increase of temperature should be avoided by an infinite heat exchange with the environment. This requires an infinite intercooling between each compression step or an infinitely slow process in which the gas flow is laminar without friction between the molecules; this defines a reversible process, which is actually impossible to realize.

The same explanation can be given for the expansion process. In that case all loss of internal energy should be compensated by infinite external heat supply and by introduction of the gas flow very slowly in a laminar stream, without friction between the molecules – once more a reversible process which of course is impossible to realize. In practice, only a small supply of heat will take place. The aim should be to try to make the process reversible or to let it take place at an infinitely slow rate, so that the internal energy will not decrease. Internal energy decreases during expansion because as the volume is increased and the distance between the molecules increases, so do they contact and influence each other. In a very slow process this should be avoided by complete conversion into additional potential energy.

For the sake of clarity we must remind the reader that we are referring to the expansion of the gas in the clearance volume of the compressor or to expansion in the refrigerating cycle between condenser and evaporator in an expansion cylinder, and not in an expansion valve. In an expansion valve the expansion is isenthalpic or adiabatic non-reversible, which means with unchanged enthalpy. The real decrease of enthalpy is used in this case to evaporate part of the liquid refrigerant. Now it might be asked why the compression process is not adiabatic reversible (or isentropic), following the law pV^k = constant. Such a process is not possible in practice because it needs complete avoidance of an energy change. In practice, heat exchange by cylinder walls, oil etc. cannot be avoided. Of course it is also impossible in this situation to create an infinitely slow continuous process. Note that in speaking of internal energy and the index k, it must be remembered that $k = c_p/c_v$, where c_p = specific heat at constant pressure and c_v = specific heat at constant volume.

In a process with a supply of thermal energy at constant pressure, the volume will increase. The distance between the molecules is increased, for which internal energy is used, which is not the case in a process at constant volume with increasing pressure. The difference $(c_p - c_v)$ is the quantity of work known as R. The value of R differs for each gas.

As we will see, in daily practice pV^k = constant is simply applied and corrected afterwards by introducing an efficiency ratio η_{is}, where the suffix 'is' stands for isentropic.

Before studying the practical vapour compression cycle it is helpful to refer to the theoretical indicator diagram shown in figure 4.1, where p_a = initial pressure, p_e = final pressure, T_a = initial temperature, V_c = maximum cylinder volume, V_0 = clearance volume, $V_s = (V_c - V_0)$ = swept volume, V_a = volume of gas induced, 1–2 = isentropic compression, 3–4 = isentropic expansion.

Let us now compare the practical indicator diagrams shown in figures 4.2a and 4.2b. What can we see? Firstly point 4′ lies below 4 and point 2′ lies above 2. This is because, in order to open the valves, there are resistances to overcome, such as:

- the valve tends to cling to the seat
- inertia of the valve ring
- pressure of the valve spring.

Furthermore, we can see on figure 4.2b that there is some throttling in the valves, which causes points 1 and 3 to be moved in a vertical direction to 1′ and 3′. Because of this, compression and expansion also no longer start at the same point. Point 2′ is displaced to the left and point 4′ to the right, giving 2″ and 4″ respectively. The throttling in the valves also results in an extra volumetric loss.

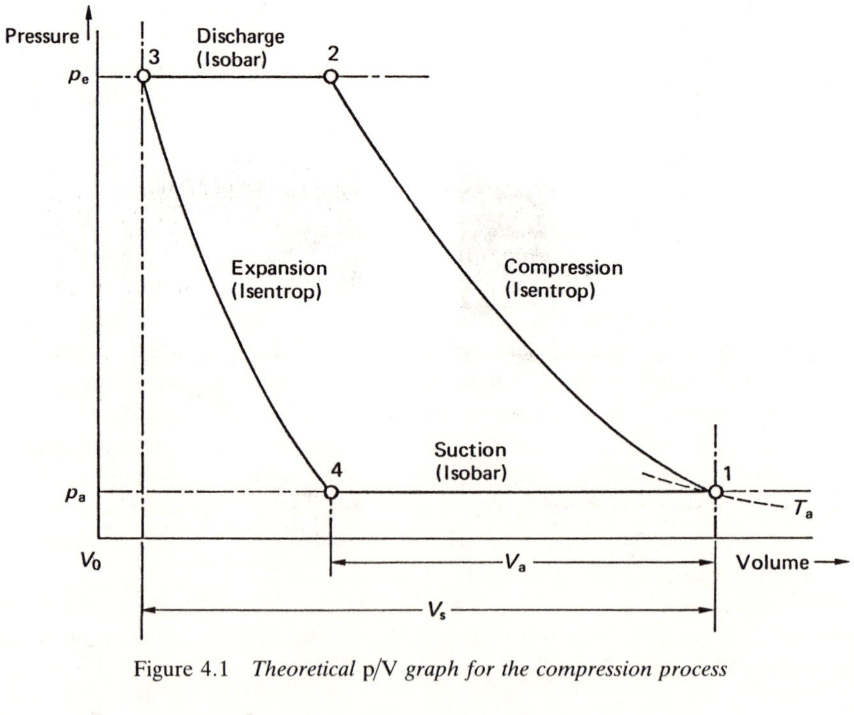

Figure 4.1 *Theoretical* p/V *graph for the compression process*

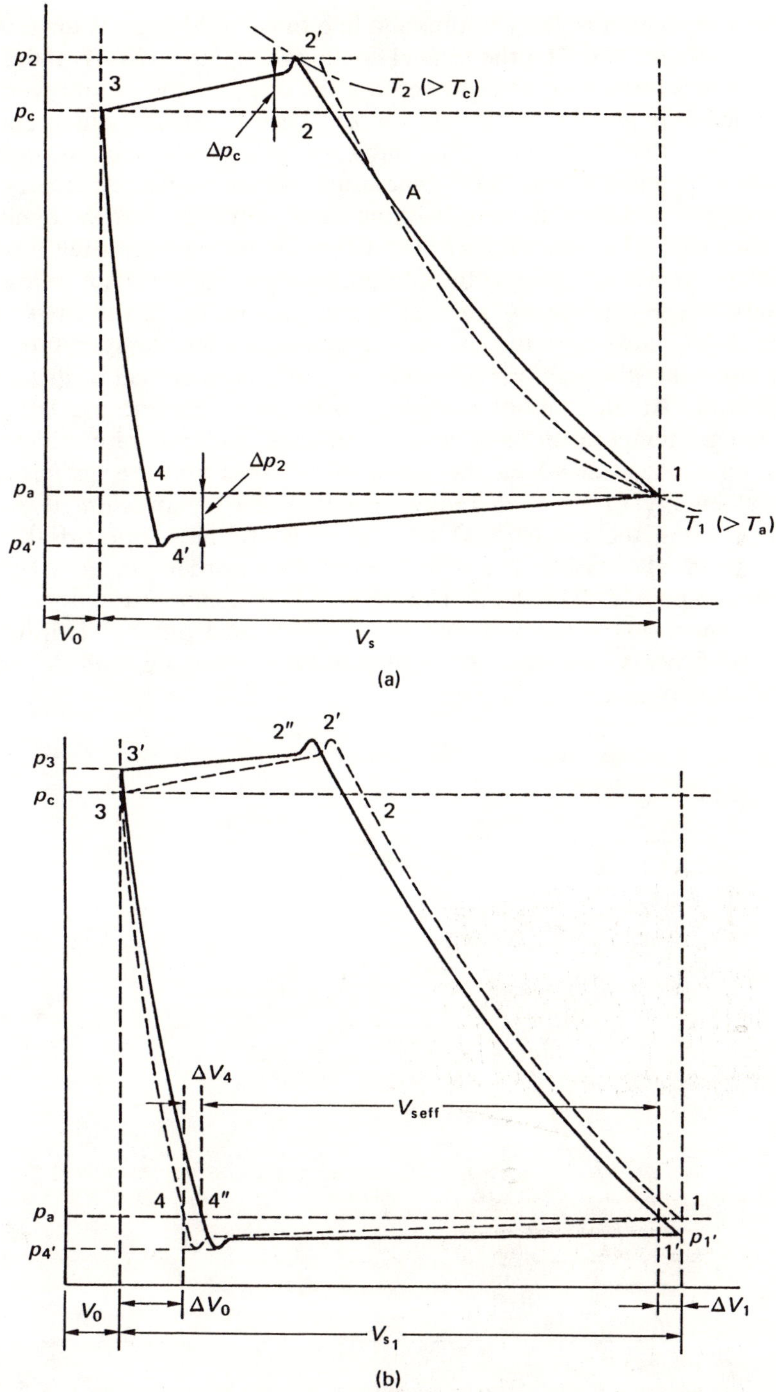

Figure 4.2 *(a) A practical indicator diagram. (b) A practical indicator diagram showing* $p_1 < p_a$ *and* $p_3 > p_e$

Let us see now how the compression line runs on the log p/h diagram as shown on figure 4.3. The theoretical compression curve is I–II. Point III, however, lies further to the right because the compression is not reversible adiabatic but polytropic, as we saw before. The final delivery temperature T_e is greater than $T_{e_{is}}$. During the induction process the suction pressure decreases to point 4′ and the temperature increases, while the pressure increases in the direction of $p_{4'}$, without reaching that point; the temperature rises to T_1, because of friction suffered by the gases passing through the valve. As we saw earlier, the compression process does not follow the theoretical isentrop because of heat exchange from the gas to the walls.

Figure 4.4 shows both the changes in the gas and wall temperatures and the gas pressure through one complete revolution of a typical reciprocating compressor. At the start of compression the wall temperature is higher than the gas temperature; as a result of this the gas temperature increases up to point A, after which the action is reversed and the gas gives up thermal energy to the wall. Note also that the compression does not proceed along the isentrop. After the point A the supply of thermal energy from the gas to the wall is now greater than the loss before, resulting in point 2′ lying further to the left. The temperature of the gas is now so high that there is a big drain of thermal energy. During the discharge process, pressure and temperature decrease through the valve; finally we arrive at point 3.

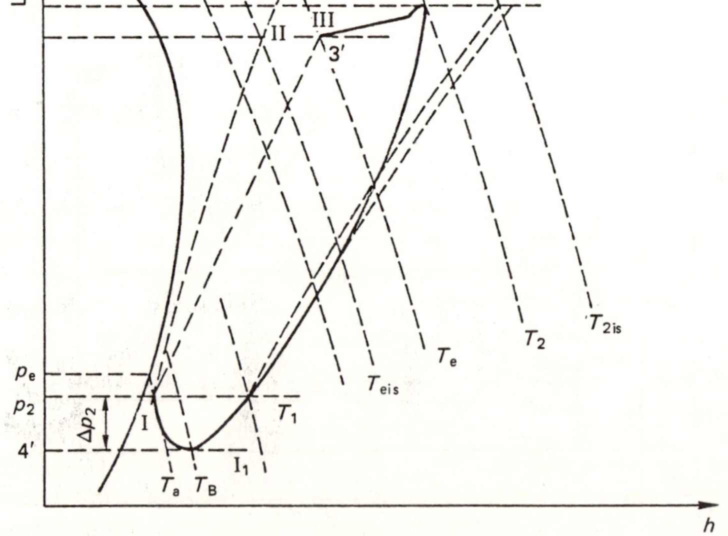

Figure 4.3 *The shape of a typical compression line on the* p/h *diagram*

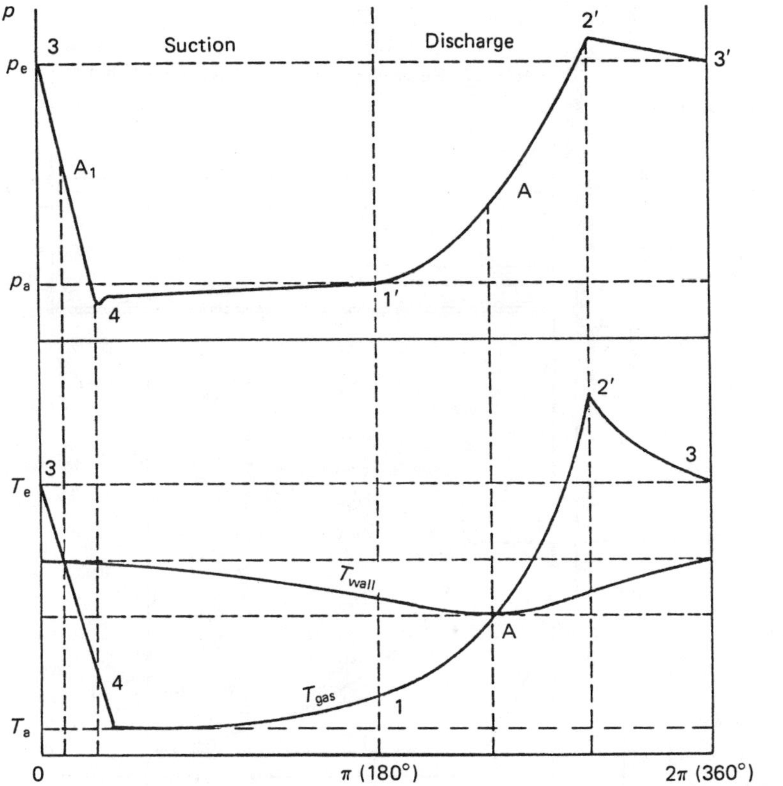

Figure 4.4 *The pressure/time and the temperature/time graphs for a typical vapour compression cycle*

4.3 Charge Coefficient λ – Overall Volumetric Efficiency

The charge coefficient λ is determined by the ratio of the actual or useful suction volume of gas or quantity of refrigerant induced, divided by the swept volume of the compressor. The overall loss of efficiency is composed of different sub-losses. These losses are made up of:

(a) expansion of gases in the clearance volume, expressed by $\eta_v = S_1/S$
(b) various leakage losses described later in detail and expressed by η_d.

Figure 4.5 should be referred to when studying this section. Losses due to expansion from the clearance volume η_v depend on the actual space occupied by the clearance volume S_0. It is usual to express S_0 as a fraction or percentage of the compressor's cylinder volume S and use the symbol ϕ_0. Thus $\phi_0 = S_0/S$. The value of ϕ_0 for large and medium-sized industrial

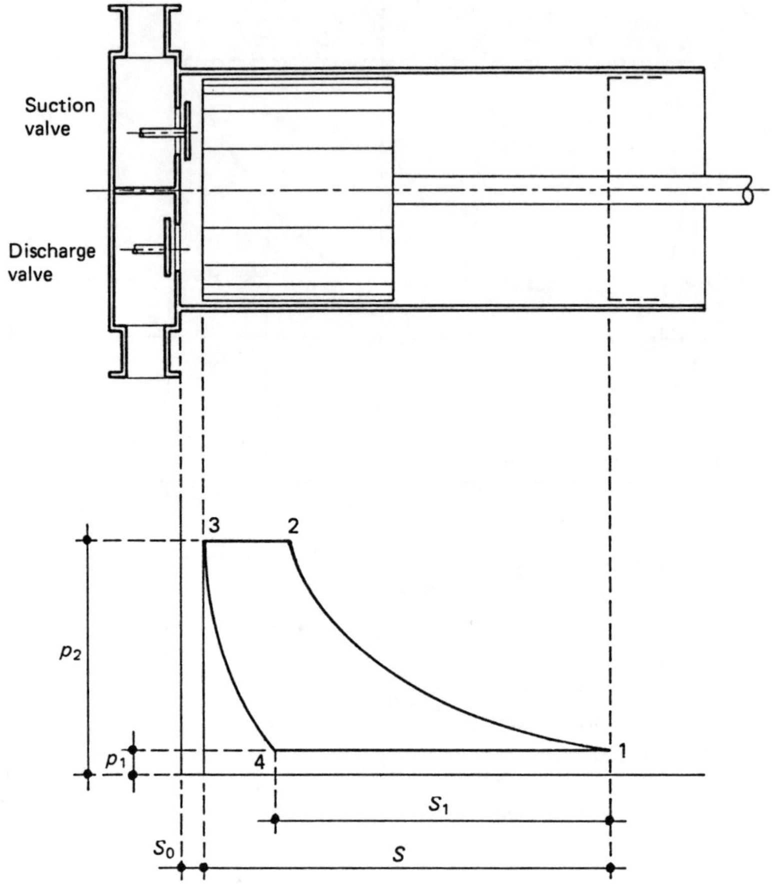

Figure 4.5 *The* p/V *diagram for a reciprocating compressor*

compressors varies from 3 to 8% and from 5 to 15% for small compressors. The actual size of this space determines how much gas is left in the cylinder at the start of the suction stroke, gas that has to be allowed to re-expand before the really useful gas volume can enter the cylinder. The amount of this expansion also depends on the ratio of discharge to suction pressure, usually expressed as $\Pi = p_c/p_e$. Figure 4.5 illustrates the effect of the clearance volume on the clearance volumetric efficiency η_v. It is calculated in the following way.

The quantity of gas entering the cylinder is represented by the length 1–4, in other words S_1 the volume of gas induced. The larger S_1 is compared with S, the smaller the compressor can be built. We defined S_1/S above as the clearance volumetric efficiency η_v.

The gas volume at 3 is V_3 and at 4 is V_4:

$$V_4/V_3 = \{S_0 + (S - S_1)\}/S_0$$

$$S_1 = \eta_v S$$

$$V_4/V_3 = \{S_0 + (S - \eta_v)\}/S_0$$

Dividing the right-hand side top and bottom by S gives

$$V_4/V_3 = \{S_0/S + (1 - \eta_v)\}/(S_0/S)$$

Now $(S_0/S) = \phi_0$, so

$$V_4/V_3 + (\phi_0 = 1 - \eta_v)/\phi_0$$

$$\eta_v = \phi_0 V_4/V_3 + \phi_0 + 1$$

and rearranging:

$$\eta_v = 1 - \phi_0\{(V_4/V_3) - 1\}$$

$$p_2 V_3^{\,k} = p_1 V_4^{\,k}$$

$$V_4/V_3 = (p_2/p_1)^{1/k} \quad \text{or}$$

$$\eta_v = 1 - \phi_0\{(p_2/p_1)^{1/k} - 1\}$$

We may calculate the value of k from the ratio c_p/c_v as is used for compression calculations; whereas we know that in reality we should use n. As the index for polytropic expansion is even lower than for polytropic compression, the difference from k is even greater. The charge coefficient λ is composed of certain other losses in addition to this loss, owing to expansion of the gas trapped in the clearance volume, for which the term η_d has been introduced above. These losses are due to leakages and other causes. We have already introduced some of these when explaining the deviation of the practical diagram from the theoretical diagram.

These losses can be summarised as follows:

(a) leakage losses past piston and valves
(b) absorption of gas in oil
(c) losses through valves or ports
(d) heat-exchange losses
(e) non-tightness losses of the construction.

Finally, the charge coefficient λ can be defined as $\lambda = \eta_v \times \eta_d$ where η_d represents all the losses listed above. In practice η_d is determined empirically, η_v is calculated and the other losses are derived from λ/η_v. The manufacturer determines a value at only one pressure ratio and assumes that λ is proportional to T_c/T_e. Theoretically $\lambda = 1$ when $T_c/T_e = 1$. For ranges of practical values of λ, refer to figure 4.6.

η_d will tend to be lower for CFCs than for ammonia because of the ease of penetration of CFCs into oil, their high solubility in oil, high molecular weight and poor heat-transfer characteristics. This is made clear from figure 4.6.

It is claimed that λ can be improved by using a higher degree of suction superheat, on the basis that suction gases should be dry and contain no oil/refrigerant droplets that will expand in the compressor cylinder.

It is understandable that η_v is higher for a longer stroke arrangement with the clearance volume unchanged, because of the decrease in the value of S_0/S with the longer stroke. High gas velocities through the valves lower the value of η_d. Comparing the losses of R22 against R717 due to heat exchange between gases and cylinder walls we see that, because of the lower discharge temperature, R22 is preferable to R717. For the same reason, screw compressors compare more favourably than piston compressors on this point. However, η_d, because of valve losses, is worse for R22 with its higher density than for R717. It is interesting to calculate η_v for the different makes of compressor, provided that the manufacturer supplies the clearance volume values. Normally we know n, D and S, so with the known capacities we can calculate λ and then derive η_d, in order to compare them.

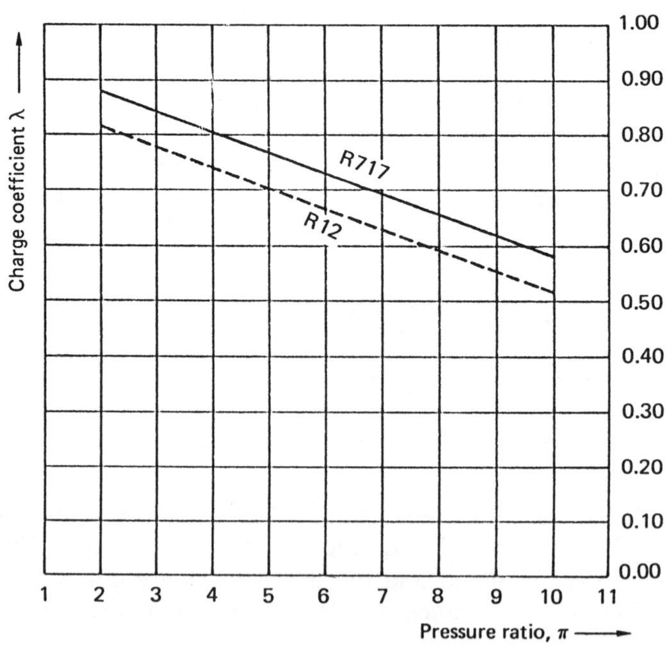

Figure 4.6 *Variation of charge coefficient with compression ratio*

4.4 Influence of Compressor Design on λ

G. Lorentzen carried out some experiments with two types of piston compressor in order to determine the importance of the different losses making up λ. These results can be seen on figure 4.7, where:

B_1 = losses due to clearance volume, i.e. η_v
B_2 = losses due to heat exchange between gas and cylinder wall during the suction stroke

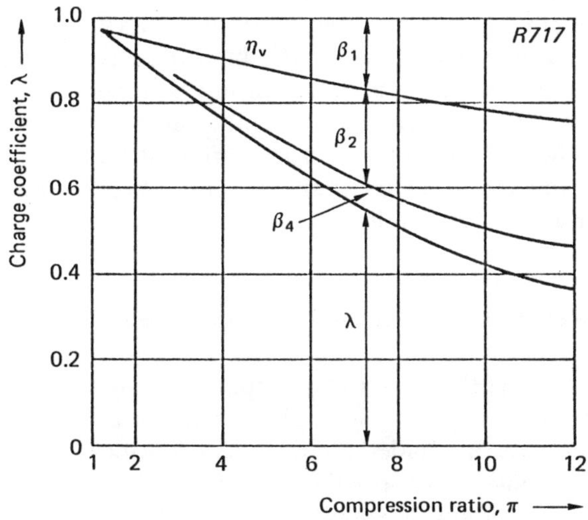

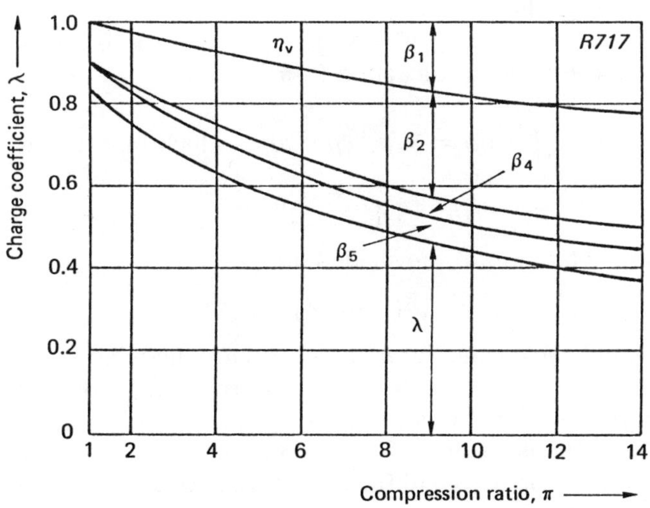

Figure 4.7 *Composition of the volumetric losses for two compressors with varying compression ratio*

B_3 = losses due to the lower pressure in the cylinder at the end of the suction stroke
B_4 = tightness losses – negligible
B_5 = losses due to pressure and volume variations during the suction stroke.

It can be seen that B_4 is less important than B_1. B_1 depends on the relative value of the clearance volume to the stroke and cylinder diameter or 'bore' (D). A large ratio S/D results in an increase in the value of η_v which is shown on figures 4.8a and 4.8b for R22 and R717 respectively. The losses by absorption of refrigerant in the oil are not shown on figure 4.7 since they are losses related to the kind of refrigerant used – not the design of the compressor. This relationship is important for other losses because of the differences in heat-exchange value and the specific heats c_p and c_v, and other characteristics for the different refrigerants. Remember when comparing compressors that we must always consider the same refrigerant and working conditions.

4.5 Power Requirement of a Piston Compressor

As discussed earlier, the power consumption during the cycle is expressed by the area of the pressure/volume diagram as shown on figure 4.9; that is the total area minus the shaded area, which takes into account the effect of the clearance volume. If we split the area of this diagram into small slices of volume V and an infinitely small pressure difference dp, each slice will approximate to a rectangle with an area Vdp. If we integrate all these slices we can derive the total area.

Since $pV^n = p_1V_s{}^n$, so $V = V_s(p_1/p)^{1/n}$ where V_s is the cylinder volume. The area can now be calculated from

$$\int V\mathrm{d}p = V_s p_1{}^{1/n} p^{-1/n} \mathrm{d}p \quad \text{which brings us to}$$

$$p_1 V_s n/(n - 1)[\{p_2/p_1\}^{(n-1)/n} - 1]$$

or after subtraction of the shaded area

$$\eta_v V_s n(n - 1)[\{p_2/p_1\}^{(n-1)/n} - 1]$$

We do not have a value for n but we can use the relationship $c_p/c_v = k$ and for this reason we have to apply a correction to the calculated power consumption. To do this we need to divide the theoretical power consumption P_{th} by the isentropic efficiency η_{is} with values in the range 0.95–0.99 depending on the design of the compressor concerned. So η_{is} is the ratio of the theoretical power consumption for isentropic compression and the

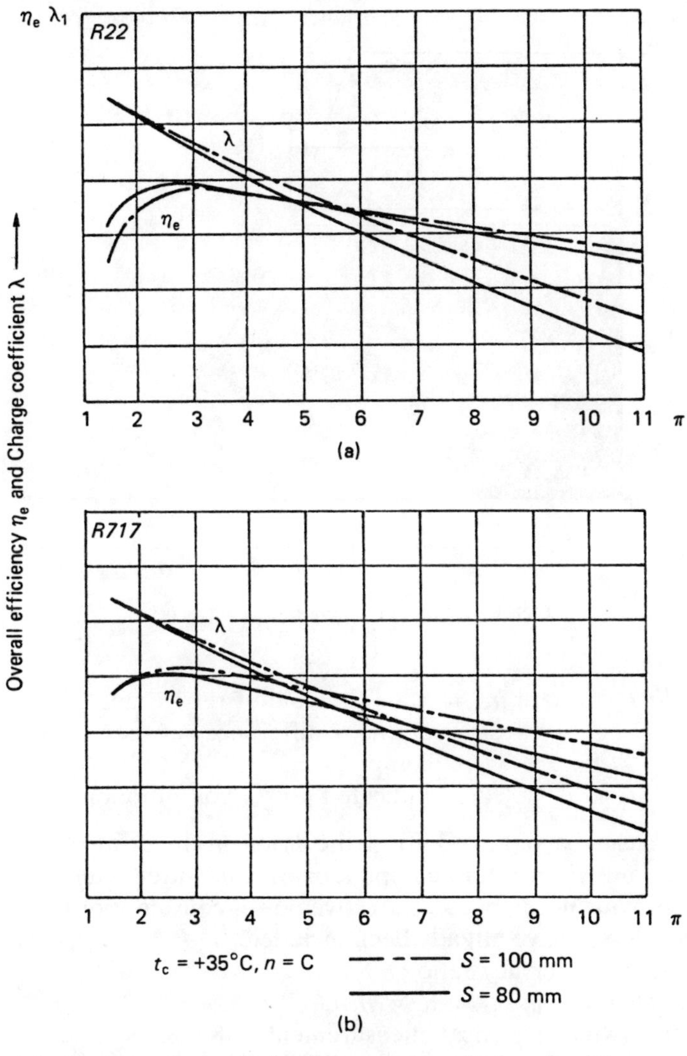

Figure 4.8 *Variation of η_e and λ for different values of* S *for a fixed* D *over a range of compression ratios*

actual absorbed power. The value of this ratio gives a good indication of energy conversion of a compressor. This figure is preferred to the COP when evaluating compressor design. The COP is the ratio refrigeration capacity/energy input, a ratio that tells us more about the efficiency of the total installation (COP = $\eta_{is}Q_0/P_{th}$). η_{is}, on the other hand, gives the relationship between the actual and ideal power consumption for a given compressor:

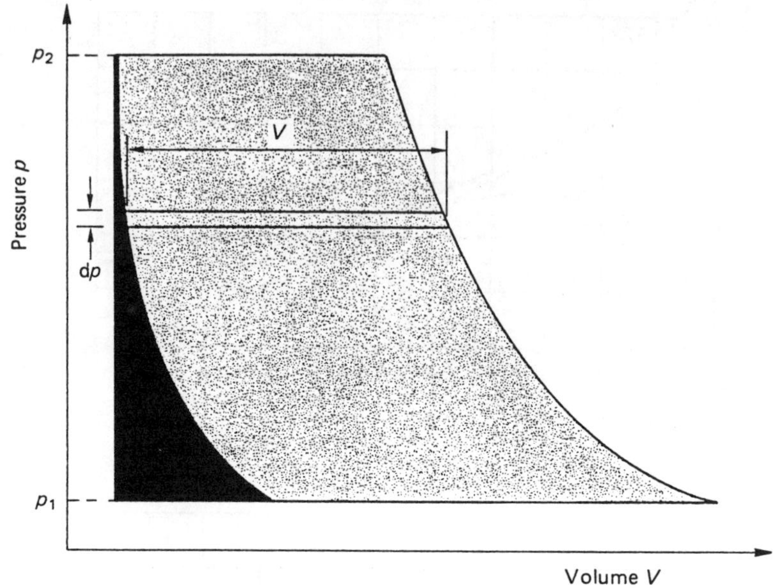

Figure 4.9 *A* p/V *diagram for analytical use*

$\eta_{is} = m'\mathrm{d}h/P_i$ where m' = mass flowrate of refrigerant
$\mathrm{d}h$ = enthalpy difference between start and end of compression
P_i = indicated power consumption.

m' and $\mathrm{d}h$ can be obtained from the $h/\log p$ chart for the refrigerant concerned, and P_i from the compressor manufacturer. Generally, manufacturers provide not P_i but P_e (effective power consumption), in which the mechanical losses have already been included. So $P_e = P_i/\eta_m$, where η_m is the mechanical efficiency ratio.

We can also use $\eta_{is}\,\eta_m = \eta_e = m'\mathrm{d}h/P_e$.

Note that when practical measurements of voltage and current are taken, the absorbed power can be estimated from $W = AV\cos\theta\sqrt{3}$.

4.6 Influences of Compressor Design on η_{is}

The differences between the theoretical and actual power consumption are, from a design point of view, due to three main energy losses which make the compression process polytropic and not reversible adiabatic:

(a) heat exchange losses gas to cylinder wall
(b) losses due to gas flow through valves and ports
(c) tightness losses (these have more effect on λ).

Losses under (a) take place during suction, compression and discharge. Heat exchange during compression, which in fact changes from positive to negative as shown on figure 4.3, is less important. Little evidence is available on its value or influence. It is strongly affected by the design, the amount of cylinder wall cooling, the ratio S/D and the characteristics of the materials of construction. The number of these factors make analysis very difficult.

An analysis is possible of the heat exchange during the period during which the gas enters the cylinder. This heat exchange causes an increase in volume owing to the increase in the temperature of the gas during this time. Although it is difficult to determine these losses analytically, Jensen produced an interesting model from which we can draw some conclusions. His equations show that these exchange losses increase because of an increased pressure ratio, but decrease with a larger cylinder and a higher rotational speed. A longer stroke in relation to the cylinder diameter also decreases these losses, as the average wall temperature decreases.

The losses occurring during the gas flow through valves and ports are due to loss of kinetic energy caused by the friction through the valves, resulting in a pressure drop. As we know, this pressure drop increases with the square of the velocity of the refrigerant gas, so these losses are also proportional to the piston speed, assuming a fixed relationship between the area of the piston and the area of the valve arrangement. So higher piston speeds result in higher losses. At higher rotational speeds the valves are hitting their seats more frequently with a higher velocity of impact. From the point of maintaining working life, the valve lift should be reduced. But this results in a higher gas velocity through the valve. If the valve section could be increased as another means of offsetting the negative effect of a high rotational speed n, the stroke S could be made shorter, since piston speed is a function of S and n. A high rotational speed does not necessarily lead to a high piston speed as will be explained later. By all these measures the negative influences on the high rotational speed can be minimized. Unfortunately, these measures negatively influence other losses and efficiencies. A large valve section increases the clearance volume, decreasing η_v. A shorter stroke also results in a proportionally larger clearance volume. We know that a short stroke encourages more heat transfer loss. Finally, a high rotational speed causes an unfavourable η_m. So in general we can say that a high rotational speed results in a poor value of η_e, a point made clear on figure 4.10. We can conclude, from a design point of view, that η_{is} is influenced by:

(a) cylinder diameter (D)
(b) stroke, especially the relationship between stroke and diameter (S/D)
(c) rotational or piston speed
(d) valve construction.

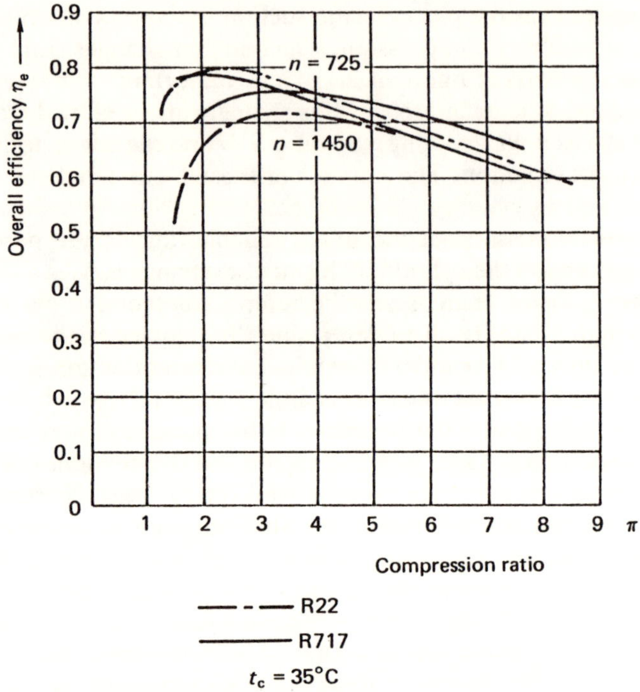

Figure 4.10 *The influence of rotational speed on η_e for R22 and R717 reciprocating compressors at differing compression ratios*

In order to reduce heat-exchange losses a large piston diameter, a high value of S/D and a high rotational speed are preferred. In order to reduce valve losses a large valve area, a small value of S/D, and a low piston speed, or better, a low rotational speed, are preferred.

When comparing the graphs of η_{is} for various compressors, the relative effects of their constructional dimensions and physical characteristics must always be borne in mind. Some manufacturers do not quote the value of η_{is} for changing speed of rotation n. Sometimes they advise in their catalogues that the power consumption P_e is proportional to the speed of rotation.

Later we will include the effects of mechanical losses calculated with values between 0.8 and 0.95 for the efficiency η_m, depending on the compressor construction; a factor of about 0.9 should be included, especially for vee belt drive, to compensate for drive losses. This ratio is designated η_{dr}.

It is interesting to compare the compressor efficiencies from different manufacturers. By dividing the power consumption given on the manufacturers' data sheets with our calculation of the theoretical power consumption, we can derive values for η_{is} and thereby compare their relative efficiencies.

4.7 Capacity Control

The capacity control of a piston-type compressor, especially in the industrial field, is usually achieved by putting one or more cylinders out of action. This is done by a valve-lifting system operating on the suction valves. Modern compressors, even the smallest, are multi-cylinder with between 2 and 18 cylinders, thereby permitting capacity control in several steps. From an energy point of view it is a relatively economical method of control, not totally loss-free, but as we will see, closest to the ideal.

Capacity control using a variable-speed motor is more efficient, as shown in figure 4.11. Its disadvantage lies in the cost of the motor and the controlling device. For most compressors, design factors limit the speed reduction to a minimum of 50%; at lower speeds there are problems with lubrication and out-of-balance forces. Every compressor is designed with a minimum rotational speed of between 500 and 600 rev/min. The maximum permissible speeds are in the range 1500–1700 rev/min. Note that there are limits to the permitted power consumption at different speeds. All compressors are also designed for limited operating conditions. There is a

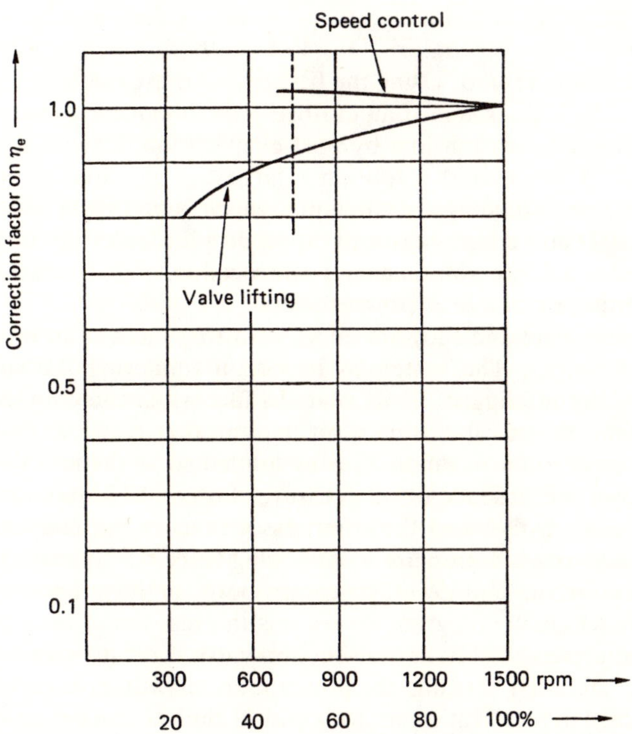

Figure 4.11 *Effects on the power/capacity ratio resulting from different methods of capacity control for piston compressors*

minimum evaporating pressure to consider, a maximum discharge pressure and a maximum pressure ratio p_d/p_e. Capacity control by suction valve lifting is not without some danger to the compressor because of the out-of-balance forces on the crankshaft. For this reason:

(a) a minimum number of cylinders must remain in operation
(b) the compressor should not operate for a long, uninterrupted period of time on reduced capacity

The losses using suction valve lifting can be explained by the fact that since the pistons continue to move, the gases still flow backwards and forwards even in the unloaded cylinders. Thus η_{is} and η_m continue to influence the system.

Before going into greater detail about the valve-lifting method of capacity control, it is necessary to mention a number of other methods that are used on smaller compressors which have no valve-lifting mechanisms. The first involves bypassing some of the discharge gas directly back into the suction inlet, thus short-cycling part of the compressor output; another method throttles the suction inlet, thereby reducing the quantity of gas reaching the cylinders; finally, there is the system in which hot gas is injected into the evaporator. This last method is really an evaporator capacity control method. Using the first two of these methods may result in excessively high discharge temperatures which can be reduced by liquid injection into the suction line by post-evaporation, a technique shown in figure 4.12. This method results in a large loss in capacity. The process utilizes the thermal energy of the liquid which is no longer available to the evaporator. It also creates a risk of liquid droplets entering the compressor cylinders. Liquid injection is also used to reduce high discharge temperatures resulting from a high pressure ratio.

Let us now examine the suction valve-lifting method in more detail to see how it works. The system is based on rendering the suction valve inoperative by holding it off its seat. In fact when the compressor is not running the suction valve is held off its seat as part of the design. This is actually achieved by pushrods, spring-loaded so as to hold the valves off their seats in the absence of oil pressure. Figure 4.13 shows this arrangement in detail. Only when the compressor reaches running speed and the oil pump delivers full pressure is the spring force on the pushrods negated and the suction valves allowed to seat properly, activated now by their own springs. A schematic diagram shows this in great detail in figure 4.14. As mentioned above, at start-up the suction valve is off its seat; this naturally results in unloaded starting on all cylinders. This fact is of considerable importance when operating motors under the low starting current conditions laid down by the electrical supply companies. By incorporating solenoid valves into the individual lines supplying oil to the unloading

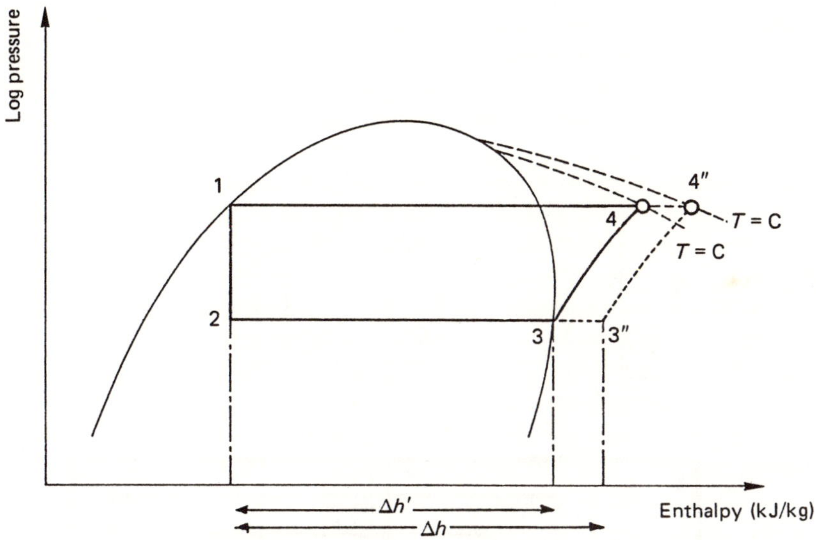

Figure 4.12 *The effects of post-injection of refrigerant shown on the* p/h *diagram*

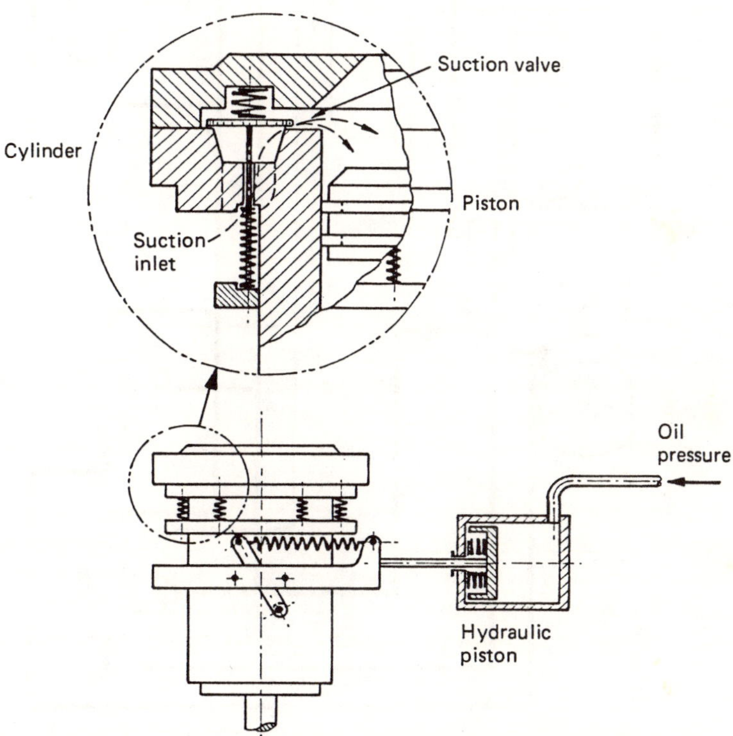

Figure 4.13 *An oil-operated suction valve-lifting system*

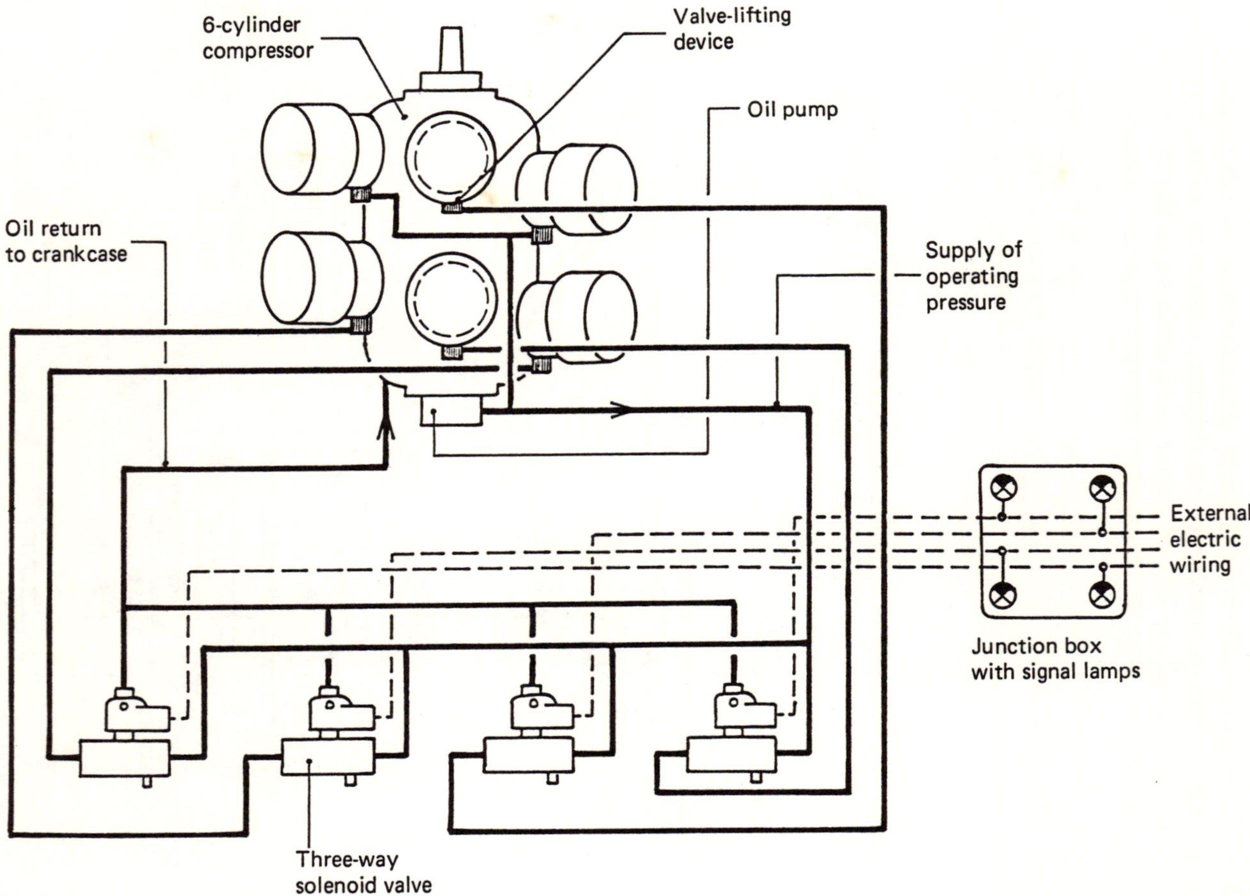

Figure 4.14 *A compressor oil system showing an oil-pressure-operated suction valve-lifting method incorporating a diverting valve control (Courtesy of Grasso, The Netherlands)*

device of each cylinder or bank of cylinders, we can put one or more cylinders out of action. The solenoid valves can be controlled by the suction pressure, or the temperature sensor in the product can activate a signal to a microprocessor or mechanical step-controller. A fall in the suction pressure, resulting from a drop in demand, causes the solenoid to close and the oil pressure to the unloading mechanism to be cut off; the cylinder unloads to reduce the compressor capacity to match the lower demand.

The variation of P_e with Q_0 resulting from capacity-controlled operation is shown in figure 4.15 for screw and piston compressors against an ideal compressor.

4.8 Main Design Features of a Piston Compressor

The main design features of a piston compressor are:

(a) cylinder bore D in m
(b) length of stroke S in m
(c) rotational speed n in rev/min.

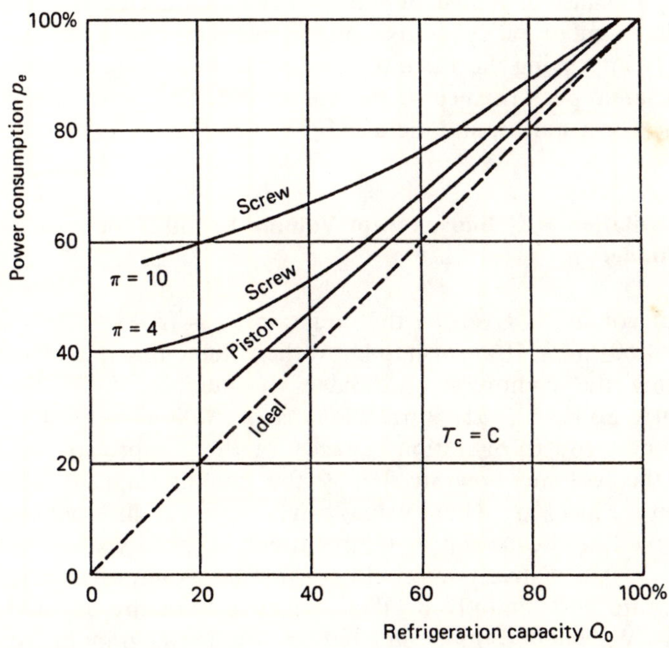

Figure 4.15 *Variation of* p_e *with reduced capacity for different compressors*

When we try to compare compressor performance and calculate the swept volume, it is important to compare piston speed rather than rotational speed. Piston speed, U m/s, is calculated from $U = 2nS/60$. We can see from this that it is possible to combine a high n with a low S, or vice versa, for a fixed piston speed. Regarding the ratio of S/D, it is interesting to note that the so-called square compressor ($S/D = 1$) yields the highest efficiency. This fact is shown in figures 4.8a and b where separate diagrams for R22 and R717 compare values of η_v and for $S = 100$ mm and 80 mm with a constant value of D.

In the past, when manufacturers increased rotational speeds they shortened the stroke. In normal practice the value of S/D lies between 0.6 and 1.0. A relatively short stroke compared with a large piston diameter results in a larger clearance volume, thus decreasing η_v and increasing the piston leakage. Modern manufacturing practice results in closer tolerances, longer strokes and thus fewer cylinders.

4.9 Effects of Valve Construction

Increasing the rotational speed makes a more solid valve construction necessary. In the past there was a tendency to reduce the gas flow by constructing the compressor with more cylinders and consequently smaller diameters. Because of the use of better valve materials it is now possible to reduce the number of cylinders and, as we saw above, to use longer strokes. By improving the valve tolerances, leakage losses are decreased. Better dynamic performance of the valves is achieved by damping their oscillation, for instance, and increasing their overall efficiency.

4.10 Calculation of Cylinder Swept Volume V_s and Number of Cylinders a

The swept volume is given by the equation $V_s = (aSn60\pi D^2/4)$ m^3/h or $(aSn\pi D^2/4.60)$ m^3/s. If we wish to know the volume flowrate of refrigerant V_n entering the compressor we have to multiply V_s by the charge coefficient λ. So $V_n = (\lambda aSn60\pi D^2/4)$ m^3/h or $(\lambda aSn\pi D^2/4.60)$ m^3/s. If we want to know the refrigeration capacity of the compressor we have to multiply this volume flowrate V_n by the cooling capacity of 1 m^3 of refrigerant q_{ov} in kJ/m^3. These values can be found at different condensing (or subcooled) and evaporating temperatures. Typical values can be seen in table 3.3. Alternatively we could convert the volume flowrate V_n into mass flowrate and multiply by the change in enthalpy dh at the given conditions. We can determine dh from the Mollier diagram or from tables for the refrigerant concerned. Now we have all the information to choose the correct compressor and its motor for a given refrigeration capacity. Before we give you some examples to calculate, we will deal first with the subject of two-stage compression.

4.11 Multi-stage Compression with Open and Closed Intercooling

Above a certain value of the pressure ratio, generally about 8, compression should be made in two steps or 'stages'; these are usually carried out in two compressors but sometimes arranged in a single compressor in which specific cylinders are allocated to different stages. Otherwise, at very high pressure ratios the value of the charge ratio becomes too low, the final discharge temperature too high and the energy consumption excessive. By including in the multi-stage system an intercooler in which gases delivered by the first or low-pressure stage are desuperheated before they go on to the second or high stage, the problem of very high final discharge temperatures is largely overcome.

In fact the two-stage process does in a way approach the ideal. It would be ideal or isothermal if we could arrange it with an infinite number of stages, but only if we could cool down the gases with an external cooling agency. In practice, the recooling between the stages is achieved directly by means of the refrigerant, but there are different ways to achieve this which we will now explain by reference to the two-stage system illustrated on the *p*/*h* diagram in figure 4.16.

The principle is to interrupt the compression process at a so-called inter-stage or intermediate pressure point 3′, and to desuperheat the refrigerant back to the saturation line at point 3″. From this point the compression in the high stage restarts and terminates at the point 4′, at a much lower temperature than at 4, where the compression would have terminated had the compression been single stage. A simple way to achieve this is to use a throttling valve to inject refrigerant between the outlet of the low stage and the inlet of the high stage. For safety reasons the gas should never be cooled down completely to the saturated line.

In all these methods the high-stage compressor has not only to compress the gases from the low stage produced in the main evaporator ($dh = h_3 - h_2$), but also the gases produced as a result of desuperheating the low-stage gases ($dh = h_{3'} - h_{3''}$) and in fact the 'flash gas' created in the expansion process through the valve down to point 2′ as shown on figure 4.17, given by $dh = h_{2'} - h_{2''}$).

A more sophisticated method of intercooling is achieved by using a specially constructed desuperheater/subcooler, usually referred to as an intercooler. The intercooler can be thought of as a supplementary evaporator placed between the low stage and the high stage, fed by liquid from the condenser and maintained at an operating level by a float-valve. It

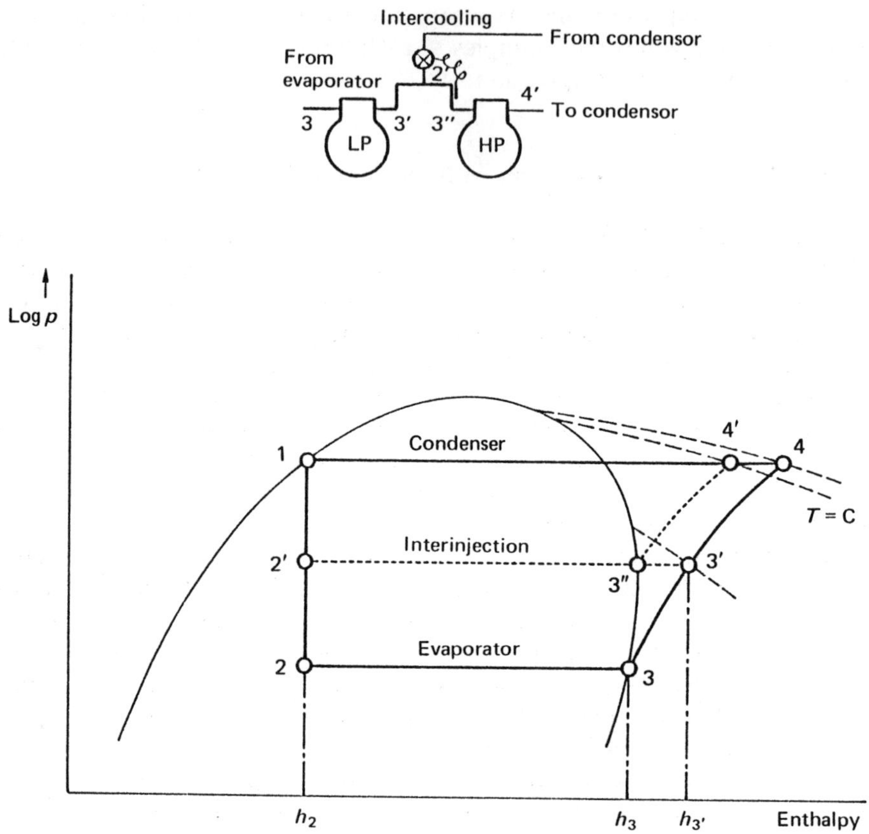

Figure 4.16 *Two-stage compression with inter-stage liquid injection*

removes the superheat from the low-stage vapour as it bubbles through the 'bath' of liquid, while subcooling the main supply of liquid to the low-stage evaporator and acting as a 'flash chamber' for the flash gas created in the first stage of the expansion process. Not only is the compression process improved thermodynamically by working in two stages, but also the expansion process. This method is called 'flash intercooling' or 'open intercooling', and is shown in detail in figure 4.17. Another method, in which the main subcooling is achieved in a coil closed to the liquid in the bath, is called 'closed intercooling' and is shown in detail in figure 4.18. In this system not all the liquid is expanded by the throttling valve of the intercooler, but only the part necessary to desuperheat the low-stage gases and the part necessary to subcool the liquid proceeding to the low stage while still at high-stage pressure. The degree of subcooling achieved in a closed intercooler is not so great as that obtained in an open intercooler, usually 10K above the 'bath' temperature. One advantage of the closed

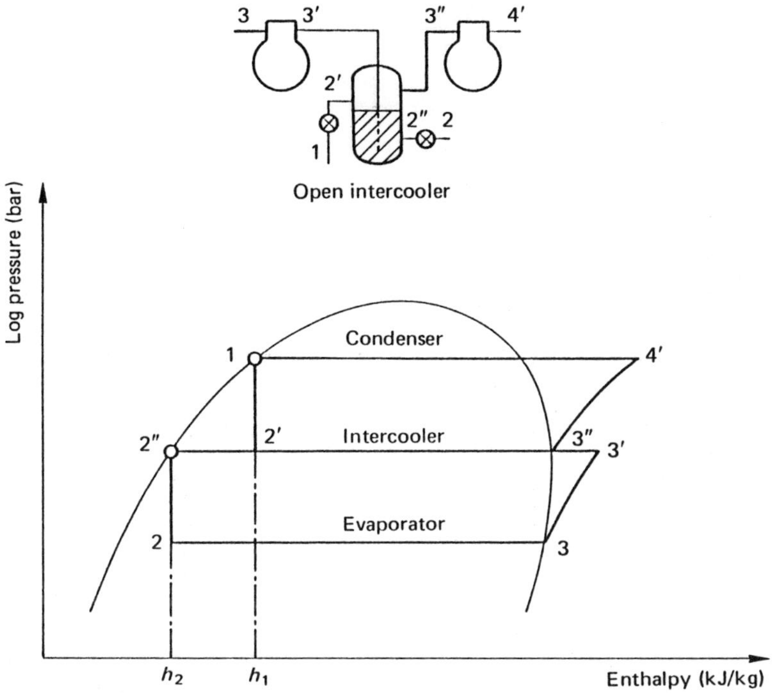

Figure 4.17 *Two-stage system with open or flash intercooler*

system lies in the fact that the liquid is at condensing pressure and provides for a more satisfactory operation of a thermostatic expansion valve control. In situations where the liquid run is abnormally long, and the use of a pumped system is not suitable, then the closed system is also to be preferred.

Some system designers use a system called the Direct Expansion Intercooler which employs a subcooling coil cooled by direct expansion from a thermostatic expansion valve; 'wet' vapour leaves the intercooler/ subcooler and is injected into the line between the low stage and the high stage in such a condition as to desuperheat the low-stage vapour to a reasonable state. This system when shown on the p/log h diagram is identical to the closed intercooler, except that point 3 does not approach the saturated line so closely. The construction of this type of heat exchanger is shown in figure 6.10. This method is more suitable for use with smaller systems using CFC refrigerants which otherwise would suffer from oil return problems. All three methods have another advantage in being able to provide additional refrigeration at the intermediate pressure.

The two stages can be achieved in a single compressor, sometimes called a compound compressor, with different numbers of cylinders devoted to

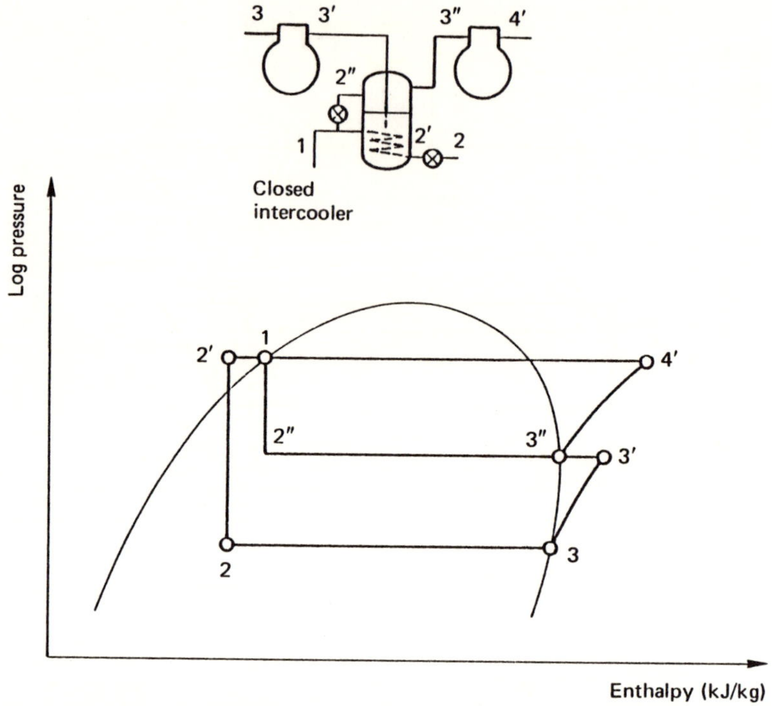

Figure 4.18 *Two-stage system with closed-type intercooler*

the stages according to the conditions; alternatively, two different compressors can be used in which the low-stage compressor is termed a booster compressor.

Finally, we should mention the cascade system which in contrast uses two separate circuits and two different refrigerants to achieve the low temperature. One advantage is the opportunity to choose different oils to suit the different refrigerants and the narrower range of their operating temperatures. For very low evaporating temperatures, three or more stages can be used. Figure 4.19 shows a simple two-stage cascade system and the interrelation between the two distinct $h/\log p$ diagrams.

4.12 Determination of the Intermediate Pressure

How can the ideal intermediate pressure of a two- or multi-stage system be determined? There are in fact several methods. The most simple one uses the formula $p_{int} = (p_c p_e)^{1/2}$.

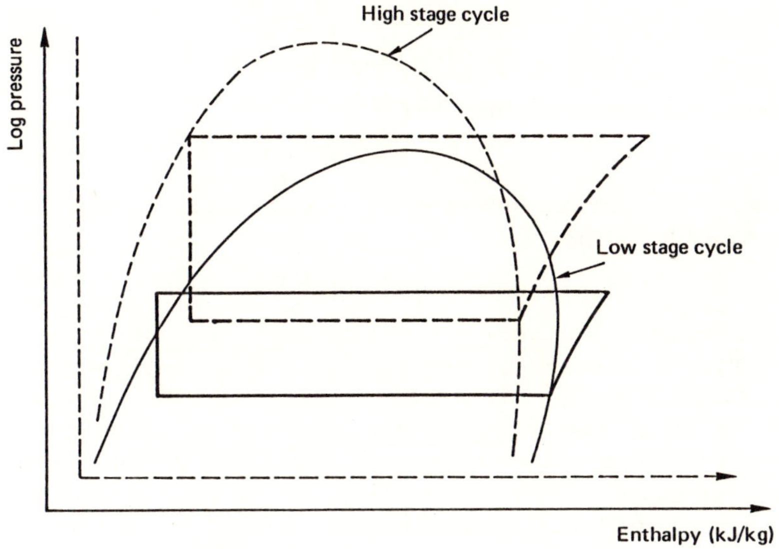

Figure 4.19 *Two-stage cascade system*

Suppose the gases are cooled between the different steps back to the original temperature, and the temperature drop is equal for every stage, then:

$$T_2/T_1 = T_3/T_2 = T_4/T_3 \text{ etc.}$$

Furthermore:

$$(p_2/p_1)^{(n-1)/n} = T_2/T_1 \quad \text{and} \quad (p_3/p_2)^{(n-1)/n} = T_3/T_2 \text{ etc.}$$

This gives

$$(p_2/p_1)^{(n-1)/n} = (p_3/p_2)^{(n-1)/n} = (p_4/p_3)^{(n-1)/n} \text{ etc.}$$

and with equal exponents

$$p_2/p_1 = p_3/p_2 = p_4/p_3 \text{ etc.}$$

We see that when the temperature rise per stage is equal, the pressure ratio per stage is equal too.

For a two-stage compressor this means that

$$p_2/p_1 = p_3/p_2$$

so

$$(p_2/p_1)^2 = (p_2/p_1)(p_3/p_2) = p_3/p_1$$

or

$$p_2/p_1 = (p_3/p_1)^{1/2} \quad \text{and} \quad p_2 = (p_3 p_1)^{1/2}$$

Formel suggests an alternative equation as follows:

$$p_{int} = (p_c p_e)^{1/2} + 0.35 \text{ (bar)}$$

where p_c = condensing pressure and p_e = evaporating pressure.

S. A. Andersen suggests

$$p_{int} = T_c (p_c p_e)^{1/2} / T_e$$

The Ramalingham Institution of Engineers, India, produced graphs illustrating the results of the equations given above and compared them with operating systems. Seven other studies have been published, each providing an alternative method for calculating p_{int}. For all practical purposes the equations above are sufficiently accurate. In both cases a higher pressure than that resulting from $p_i = (p_c p_e)^{1/2}$ is recommended since it results in a lower temperature at the end of high-stage compression, which is more critical than the temperature at the end of the low-stage compression. The amount of superheat to be extracted in the first part of the condenser will be less and its efficiency thereby increased.

The post-injection system is one method of avoiding high discharge temperatures. It is similar to inter-injection in that it carries with it a high risk of liquid slugging. The method involves injecting liquid into the suction line between the evaporator and compressor, and by post-evaporation cools down the suction gases. On the $h/\log p$ diagram we can see that point 4 moves to point 4′ which is on a lower isotherm. Figure 4.12 shows this and the decrease in the effective compressor capacity.

This method can only be used as a temporary means of correction for a critical situation which only occurs infrequently; for example, in systems in which the discharge temperature normally remains below 120°C (393K) and only rarely exceeds this critical value during abnormally low evaporating or exceptionally high condensing pressures for short periods of time in unusual circumstances.

4.13 Determination of Discharge Temperature

To determine the discharge temperature for an ammonia compressor operating with evaporating and condensing temperatures −20°C (253K) and +35°C (308K) respectively, proceed as follows.

Firstly, from the NH_3 $h/\log p$ chart, estimate the discharge temperature to be 126°C (399K). Then using Poisson's law, alternatively calculate the discharge temperature assuming that the process is isentropic:

$$T_2 = T_1(p_c/p_e)^{(k-1)/k}$$
$$= 253(13.765/1.94)^{(1.31-1)/1.31} \quad k \text{ for ammonia} = 1.31$$
$$= 253(7.1)^{0.237} = 402\text{K or } t_2 = 129°\text{C}$$

The difference is negligible because of the variation in the value of k and our inability to read the p/h chart to a greater accuracy than ±1°C. When we compare these calculated values with practical readings we find significant differences. Since the compression is not isentropic – the polytropic index n should be used, not the isentropic value k; when this is done, the resulting final temperature is higher than the theoretical by some 20°C and is in fact different for each refrigerant. One quick rule-of-thumb method uses the formula $t_2 = 2.5(T_c - T_e)$ which in this case yields $2.5(308 - 253) = 137$°C.

4.14 Determination of Compressor Speed and Power Consumption

To calculate the power of the electrical driving motor, two further efficiency ratios must be introduced: η_{dr}, which takes account of the losses in the driving mechanism – belts or coupling; and η_{el} which takes account of losses in the electric motor. Electric motor power can be expressed as

$$P_{el} = P_e/\eta_{dr}\eta_{el}$$

The overall COP of the compressor/motor combination is finally $(Q_0/P_{is})\ \eta_{is}\eta_m\eta_{dr}\eta_{el}$.

Note that when calculating the overall COP for the refrigerating plant, the consumption of the pump and fan motors to the compressor power consumption must be added.

4.14.1 Single-stage Compression

Now that we have discussed all the factors that enable us to calculate the refrigeration capacity of a compressor, we will calculate the speed and power consumed for a practical example, using the conditions illustrated in figure 4.20.

Data:

Cooling capacity of refrigeration plant: $Q_0 = 350$ kW

Evaporating temperature: −14°C (259K)

Condensing temperature: +35°C (308K)

Refrigerant: NH_3

Bore d: 160 mm

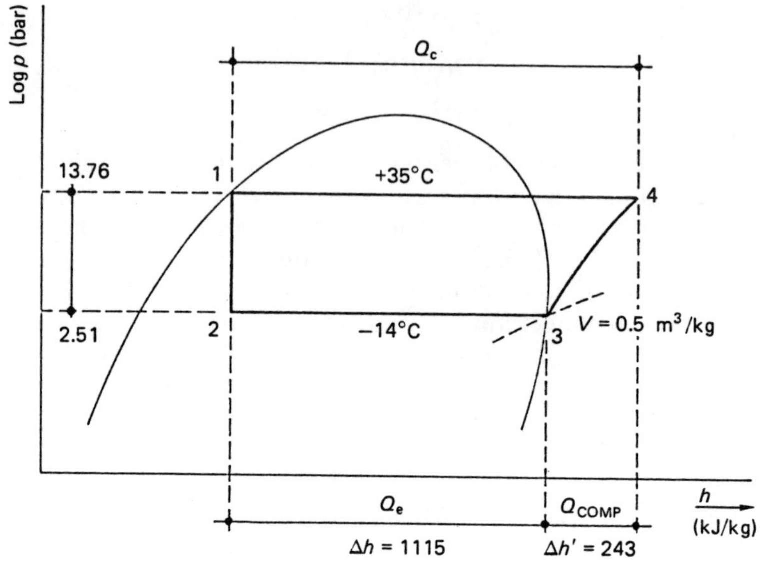

Figure 4.20 *Example of a simple refrigeration cycle depicted on a* p/h *chart*

Stroke S:110 mm

Number of cylinders a: 6

Clearance volume ϕ_0: 3%

η_m: 0.9

η_{is}: 0.89

η_d: 0.82

To be calculated:

(a) the speed of rotation n (1/min)
(b) the necessary motor power

(a) *The speed of rotation*
First draw the process on to the $h/\log p$ chart. We are already familiar with the factor q_{0v} kJ/m^3, the refrigeration effect per unit volume of refrigerant.

The difference in enthalpy dh between the liquid that enters the evaporator, point 2 on the figure, and the gases entering the compressor, point 3 on the figure, can be found from a chart or table. For this example we assume that there is no subcooling or superheating. dh is 1674 − 559 = 1115 kJ/kg, so the refrigeration effect q_{0v} is 1115/0.5 = 2230 kJ/m^3. We can find this value immediately from a table.

The rate at which the gas must be compressed for a cooling load of 350 kW is 350/2230 = 0.157 m^3/s or 565 m^3/h. We saw above that the suction volume is determined by the formula $V_s = aSn60\pi d^2/4 = 6 \times 0.11 \times n \times 60 \times \pi \times 0.16^2/4$ or $0.79n$. We saw also that the charge ratio reduces the real volume. In order to permit 565 m^3/h of gas to enter the compressor the cylinder capacity must be $565/\lambda$ m^3/h. First we must calculate λ.

The pressure ratio $p_c/p_e = 13.76/2.51 = 5.5$ so

$$\begin{aligned}\eta_v &= 1 - \phi_0[(p_c/p_e)^{1/(k-1)} - 1] \\ &= 1 - 0.03[5.5^{1/0.31} - 1] \\ &= 1 - 0.03[3.67 - 1] \\ &= 1 - 0.03 \times 2.67 \\ &= 0.92\end{aligned}$$

so

$$\lambda = \eta_v\eta_d = 0.92 \times 0.82 = 0.75$$

Finally we can see that

$$n = 565/\lambda 0.79 = 565/(0.75 \times 0.79) = 953\ 1/\text{min}$$

In practice we take 960 1/min corresponding to a direct motor drive of 1000 1/min electric motor.

(b) *The required motor power*
This can be calculated in two ways:

(i) Using the $h \log p$ diagram
At point 3 we find the enthalpy value at the beginning of compression and at point 4 the enthalpy value at the end of compression.

The specific enthalpy increase during compression dh is given by

$$dh = 1917 - 1674 = 243 \text{ kJ/kg}$$

We next determine the mass flowrate m' of the refrigerant which has to circulate in the installation. For this calculation we use the value of dh resulting from evaporation as determined above:

$$m' = Q_0/dh = 350/1115 = 0.314 \text{ kg/s}$$

The ideal power P_{is} is then given by $m'dh'$, where dh' is the isentropic work done during compression:

$$P_{is} = m'dh' = 0.314 \times 243 = 76.3 \text{ kW}$$

Now we have to introduce the efficiency rates in order to obtain the true values. The so-called indicated power P_i is given by

$$P_{is}/\eta_{is} = 76.5/0.89 = 85.73 \text{ kW}$$

Finally we find the effective power P_e from the relationship

$$P_e = P_i/\eta_m = 85.73/0.9 = 95.25 \text{ kW}$$

(ii) Using the area of the pV diagram

$$P_{th} = \lambda V_s p_e \text{ kW}$$

where

$$\begin{aligned}\lambda &= k/(k-1)[(p_c/p_e)^{(k-1)/k} - 1] \\ &= 1.31/0.31\ (5.5^{0.31/1.31} - 1) \\ &= 1.31/0.31(1.46 - 1) \\ &= 2 \\ P_{is} &= 0.157 \times 251 \times 2 = 78.14 \text{ kW} \\ P_i &= 78.14/\eta_{is} \\ P_e &= 78.14/\eta_m\eta_{is} \\ &= 78.14/0.9 \times 0.89 \\ &= 97.5 \text{ kW}\end{aligned}$$

There is a small difference between the results of the two methods b(i) and b(ii) because of the small error introduced by assuming the value of k used is valid for compression as well as for expansion, which in fact is not correct. A second error is introduced by using λ instead of η_v.

4.14.2 Two-stage Compression (see figure 4.21)

For the low stage of a two-stage system the calculation is the same as that for a single-stage compressor. First determine the intermediate pressure p_{int}, or the end of compression pressure of the low stage using the formula:

$$p_{int} = [p_c p_e]^{1/2} + 0.35$$

The circulating mass in the high stage m_h is the addition of m_1 and m_t, being the circulating mass in the low stage and the circulating mass required for the intercooling respectively:

$$m_1 = Q_0/(h_3 - h_2) \text{ kg/s} \quad \text{and} \quad m_t = Q_t/(h_{t''} - h_{t'}) \text{ kg/s}$$

During intercooling in an open system, the heat load per kg of refrigerant in the high stage, which has to be removed, is $dh_{t'}$, where $dh_{1'} = (h_{1'} -$

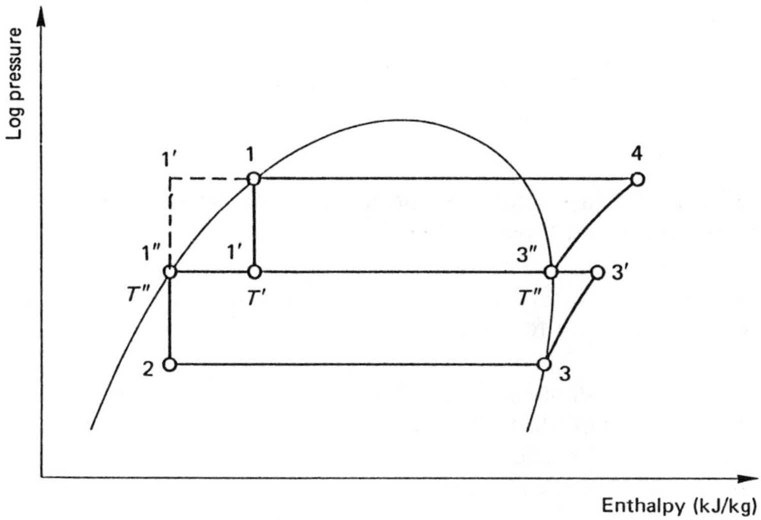

Figure 4.21 *Two-stage refrigeration cycle with open intercooler shown on a* p/h *chart*

$h_{1''} + (h_{3'} - h_{3''})$, the first part being the subcooling of the liquid refrigerant and the second part the desuperheating of the low-stage compressed gases close to saturation temperature. In the case of a closed system the subcooling part is $h_1 - h_{1'}$.

The heat load to be removed from the intercooler is Q_t where

$$Q_t = m_1 \mathrm{d}h_{t'}$$

The circulating mass of refrigerant in the high stage is given by

$$m_h = m_1 + m_t$$

The volume induced at the suction of the high stage is then

$$m_h \times \text{specific volume } v \text{ at point } T''$$

In daily practice we work in a slightly different way, which involves introducing a small error. In order to find the refrigeration capacity of the high stage we need to add the power consumption of the low stage to Q_0. When using closed intercooling we also need to include the heat load of subcooling the liquid refrigerant.

Finally, we can also calculate the heat-load rejection of the condenser Q_c. Note that this is composed of three parts: the superheat, the condensation heat and the subcooling. We have to reject the thermal energy in the condenser added during evaporation and during compression and eventually the thermal energy from subcooling in the condenser:

$$Q_c = Q_0 + P_1 + P_h$$

or using the $h/\log p$ chart:

$$Q_c = m_h \frac{(h_4 - h_1)}{\eta_{is}} + \text{subcooling}$$

4.15 Effects of Evaporating Temperature, Condensing Temperature, Subcooling and Superheating

Using figures 4.22, 4.23, 4.24 and 4.25 we can now examine the effects on the process due to the following:

(a) increase of condensing temperature
(b) decrease of evaporating temperature
(c) superheating of suction gases
(d) subcooling of liquid refrigerant.

Causes of (a) may be:

1. polluted condenser surface
2. poor water or air supply to condenser
3. air inside the condenser tubes
4. increased cooling water or outside air temperature

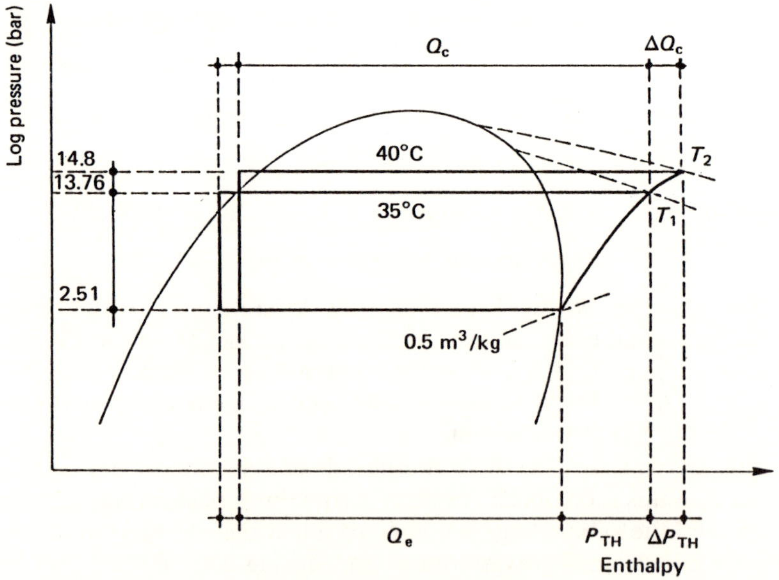

Figure 4.22 *Effects of increased condensing temperature ($T_2 > T_1$, p_{TH} increased by Δp_{TH}, Q_c increased by ΔQ_c)*

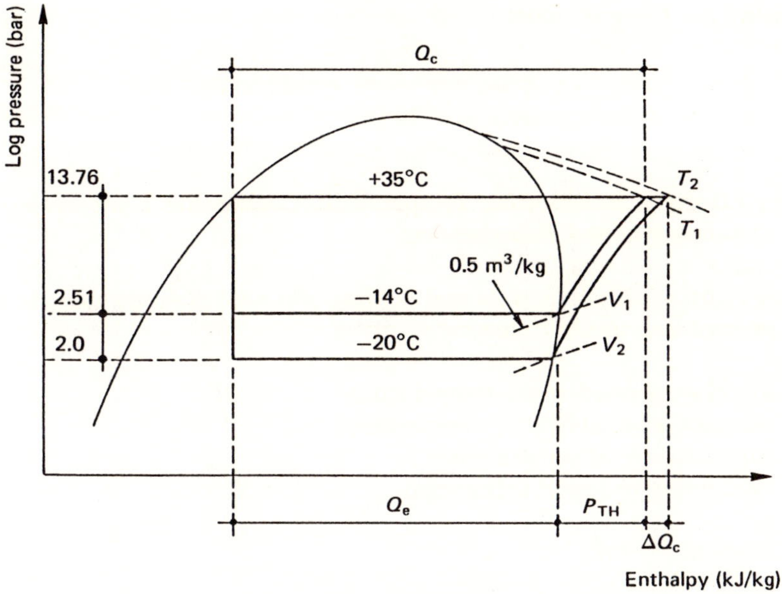

Figure 4.23 *Effects of decreased evaporating temperature* ($T_2 > T_1$, p_{TH} *increased by* Δp_{TH}, Q_c *increased by* ΔQ_c)

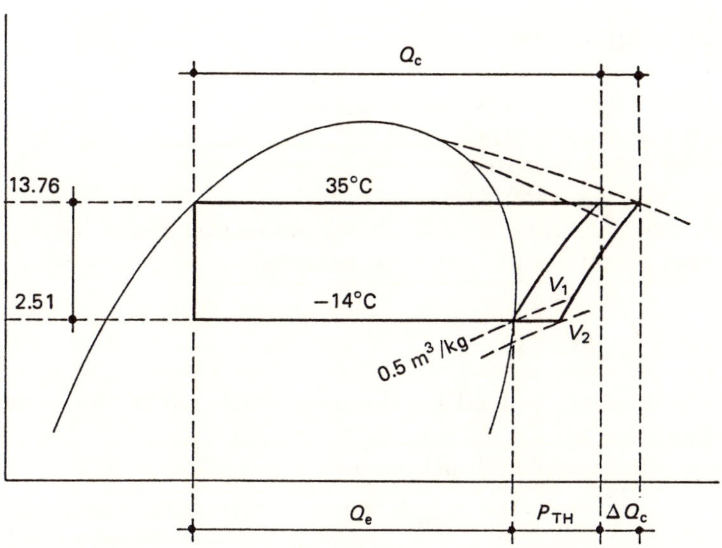

Figure 4.24 *Effects of increased suction superheat* ($T_2 > T_1$, $V_2 > V_1$, Q_c *increased by* ΔQ_c)

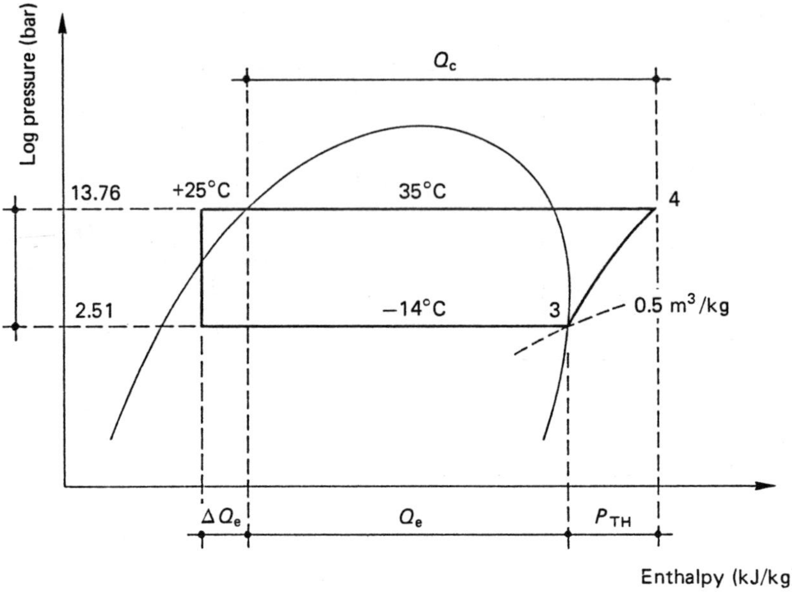

Figure 4.25 *Effects of subcooling (increase of* Q_c *by* ΔQ_c*)*

5. overcharge of refrigerant
6. high relative humidity.

Causes of (b) may be:

1. too little refrigerant because of leakages
2. frozen evaporator surface
3. polluted evaporator surface
4. poor refrigerant supply to evaporator
5. partly closed suction valve or obstructed suction line
6. changed conditions of controlled temperature.

Causes of (c) may be:

1. superheat setting needed for the proper function of the thermostatic expansion valve
2. poorly or non-insulated suction line
3. poor refrigerant supply to evaporator
4. evaporator surface too small
5. effects of heat exchanger
6. heating effects of semi-hermetic or hermetic compressor motor.

Causes of (d) may be the wish to subcool in order to reduce the formation of flash gases to increase the efficiency. Flash gases are created during expansion in the expansion device. But they can also be created at certain points before the expansion valve in the liquid line. This situation is caused by static head losses or frictional pressure losses in the lines and their conversion into thermal energy. This may lead to the risk of poor supply of refrigerant to the expansion device, resulting in a poor supply of refrigerant to the evaporator, low suction pressures, too much superheat, too high discharge temperature, and also too much superheat of the compressed gases and an increase in the part of the superheat rejection in the condenser. This means that there is a decrease in available condensation surface which results in an increase of the condensing temperature and pressure. An attempt should be made to avoid all of this by adequate subcooling, so that the losses in the liquid lines will be neutralized. Using R12 we need even more subcooling than if we were using R22. When using ammonia very little subcooling is necessary, the subcooling being normally obtained by the use of a heat exchanger, an oversize condenser or a special subcooler.

The resulting effects of the foregoing can be summarized briefly as follows:

(a) (i) increase of end-compression temperature and pressure
(ii) increase of pressure ratio
(iii) increase of specific volume at the start of compression
(iv) heat-ratio reduction
(v) reduction of the capacity of the compressor

(b) as for (a), except point (iii)

(c) as for (a), except point (ii)

(d) d*h* increases, efficiency increases, capacity increases.

4.16 Rotary Compressors

4.16.1 Vane Compressors

We will briefly describe the construction and give the formula for the calculation of capacity for this type of compressor. Compression is obtained between the cylinder wall and the vanes slotted into the rotor, the centre of which is usually located eccentrically in the cylinder casing. Only small pressure ratios can be obtained owing to poor sealing between blade tips and the cylinder wall.

The hydraulic efficiency is poor, regular maintenance is required because of the fragile nature of the blades, and their normal application until recently was only found in the low stage of two-stage systems.

The suction volume V can be calculated using the following formula:

$$V = l(D - Sz)2zmn \text{ m}^3/\text{s}$$

where l = cylinder length
D = cylinder bore
S = vane thickness
z = number of vanes
m = eccentricity
n = speed of rotation per second.

The rotary vane compressor can be considered as a piston compressor in which the piston area is $l(D - Sz)$ and the stroke $2m$. This type of compressor has one main advantage – its low investment cost.

4.16.2 Screw Compressors

The screw compressor holds a most important position in the industrial refrigeration field today. In the double-screw compressor, compression is achieved by two intermeshing rotors housed in a suitable casing whereas in the single-screw compressor, compression is achieved by the intermeshing of a rotor with two star wheels, the shafts of which are at right angles to the rotor shaft.

Construction

In both cases there is no metal contact between the moving parts, either rotors or wheels. The running clearance between the male rotor lobes and female rotor flutes, or rotor lobe and teeth of the gear wheels, is obtained by injecting a large amount of oil into the compression chamber. The oil stream absorbs a large part of the compression heat, which results in a very low discharge temperature. It is possible to work with a pressure ratio of 1:20, compared with a piston compressor which is limited to a pressure ratio of 1:8. Of course, we need extra motor power to drive a suitable oil pump, but with most system designs this is only necessary at start-up.

Figure 4.26 shows views of a double-screw compressor and figure 4.27 shows the internal moving parts. The main, or male motor-driven rotor, is usually constructed with four lobes; the matching female rotor has six flutes. Other combinations, such as 3 with 4 and 6 with 8, and recently 5 with 7, are possible. 3 with 4 gives a big suction volume with big lobes but little 'meat'; 6 with 8 gives the opposite result. So 4 with 6 or 5 with 7 seem reasonable compromises. However, the combination of 5 with 6 and others

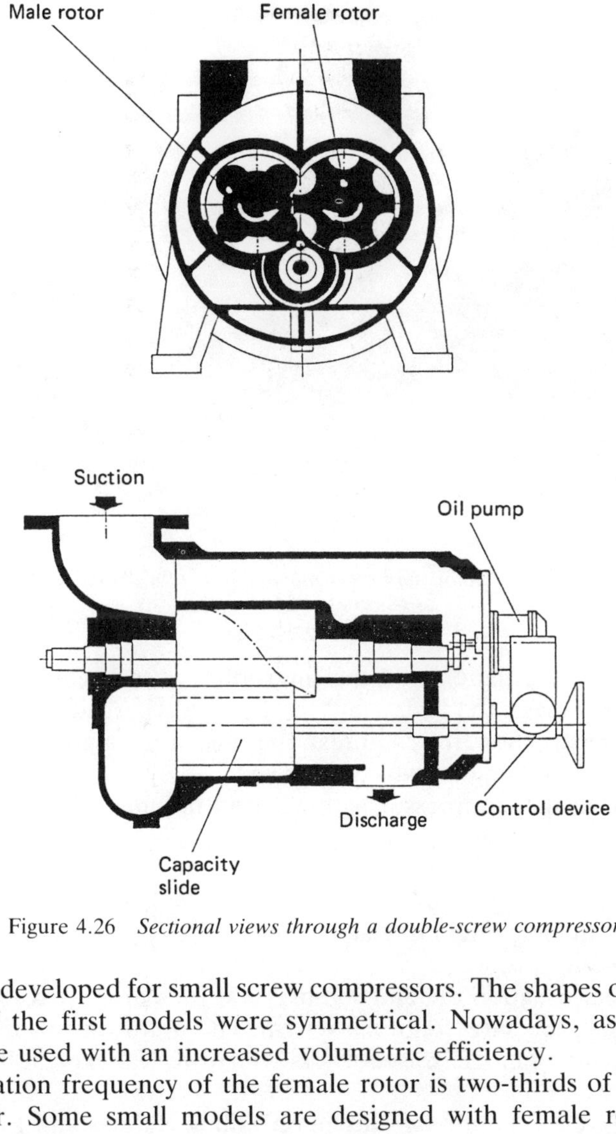

Figure 4.26 *Sectional views through a double-screw compressor*

have been developed for small screw compressors. The shapes of the screw profiles of the first models were symmetrical. Nowadays, asymmetrical profiles are used with an increased volumetric efficiency.

The rotation frequency of the female rotor is two-thirds of that of the male rotor. Some small models are designed with female rotor-driven machines, resulting in an increase in capacity.

As the rotors turn, gas is drawn through the inlet port to fill the space between adjacent lobes and flutes. When the inter-lobe space along the rotor length is filled, the rotation of the rotors moves the end of the lobes past the inlet port, sealing the inter-lobe space. As the rotors continue to rotate the intermeshing of the lobes on the discharge side of the compressor progressively reduces the space occupied by the gas, causing compression.

Figure 4.27 *Details of rotors and other moving parts of a double-screw compressor (Courtesy of Sabroe)*

Compression continues until the inter-lobe space becomes exposed to the discharge port in the casing and the gas is discharged. Because of this method of construction, the compression ratio, or the amount by which the original suction volume is reduced, is fixed.

We describe the compressor as having an in-built volume or pressure ratio:

$$\psi = V_m/V_p$$

where V_m is the maximum inter-lobe space and V_p the minimum space at the end of compression.

In the ideal situation the end compression pressure, p_1, corresponds to the condensing pressure, p_c, of the system. This unfortunately is not always the case, resulting in an efficiency loss which we will explain later.

Effects of Screw Compressor Design on Filling Ratio λ and Isentropic Efficiency η_{is}

Although there are no pistons, valves or clearance volume to affect the filling ratio, the leak losses along the rotors are important. The gas that leaks back to the suction side not only occupies space but is also at a higher temperature and so effectively increases the suction gas temperature. It is obvious that the tolerances between rotors and rotors, and rotors and the barrel, are very important. An increase in tolerance of 0.01 mm results in an increase of 1% in the volumetric losses. The dimension of the oil stream also plays an important role. The rotor diameter D and the speed of rotation n, which together determine the tip speed, have a large influence. Remember that tip speed $U = f(n \times D)$ m/s. There is an optimum tip speed to achieve maximum efficiency for each in-built volume ratio, which is shown in figure 4.28. When the rotor speed n is increased while D remains constant, or vice versa, the displaced gas volume per unit time increases, so the losses become relatively smaller. This could lead to the conclusion that the overall efficiency always increases if the rotor diameter D or the speed of rotation n increases. However, this is only true for the filling ratio. As the losses due to friction and turbulence of the gas increase, so do the mechanical losses and there is a fall-off in the value of η_{is}.

The relationship between rotor length and diameter has its influence as well. Particularly at a high pressure ratio, the discharge port section becomes very small and a short rotor will give less discharge losses. The shape of the rotor section has a considerable influence on the filling ratio also. The first rotors had symmetrical profiles. Using asymmetrical profiles, filling ratios can be increased. Piston compressors experience significant heat-exchange losses, whereas with screw compressors this is less so. This is because gas temperatures inside the screw are moderated by the artificially cooled oil stream.

For a piston compressor λ is lower than 0.5 at a compression ratio of 10, while a screw compressor still has a λ of 0.7 at a compression ratio of 20.

An important design influence comes from the so-called in-built volume ratio ψ.

In-built Volume Ratio or Compression Ratio ψ

As the in-built volume ratio is an essential difference between the screw compressor and the piston compressor, it is useful to take a closer look at this subject.

Refrigeration plants have to work under variable conditions, because evaporation and condensation temperatures can change during the operation. In a piston compressor, suction volume varies according to the varying conditions. For instance, when the condensing pressure increases, the valve opens later; when it decreases the valve opens sooner, since the pressure necessary to open the discharge valve is reached more quickly. The compressor adjusts its compression ratio to the prevailing conditions.

In a screw compressor the compression process always ends at the same point, and at the same moment; this is in fact when the inter-lobe space comes into open contact with the discharge port. Under certain conditions the pressure of the gas arriving at the discharge port may not have reached the actual condensing pressure. In this case the condition is known as undercompression. Alternatively, when the condensing pressure in the

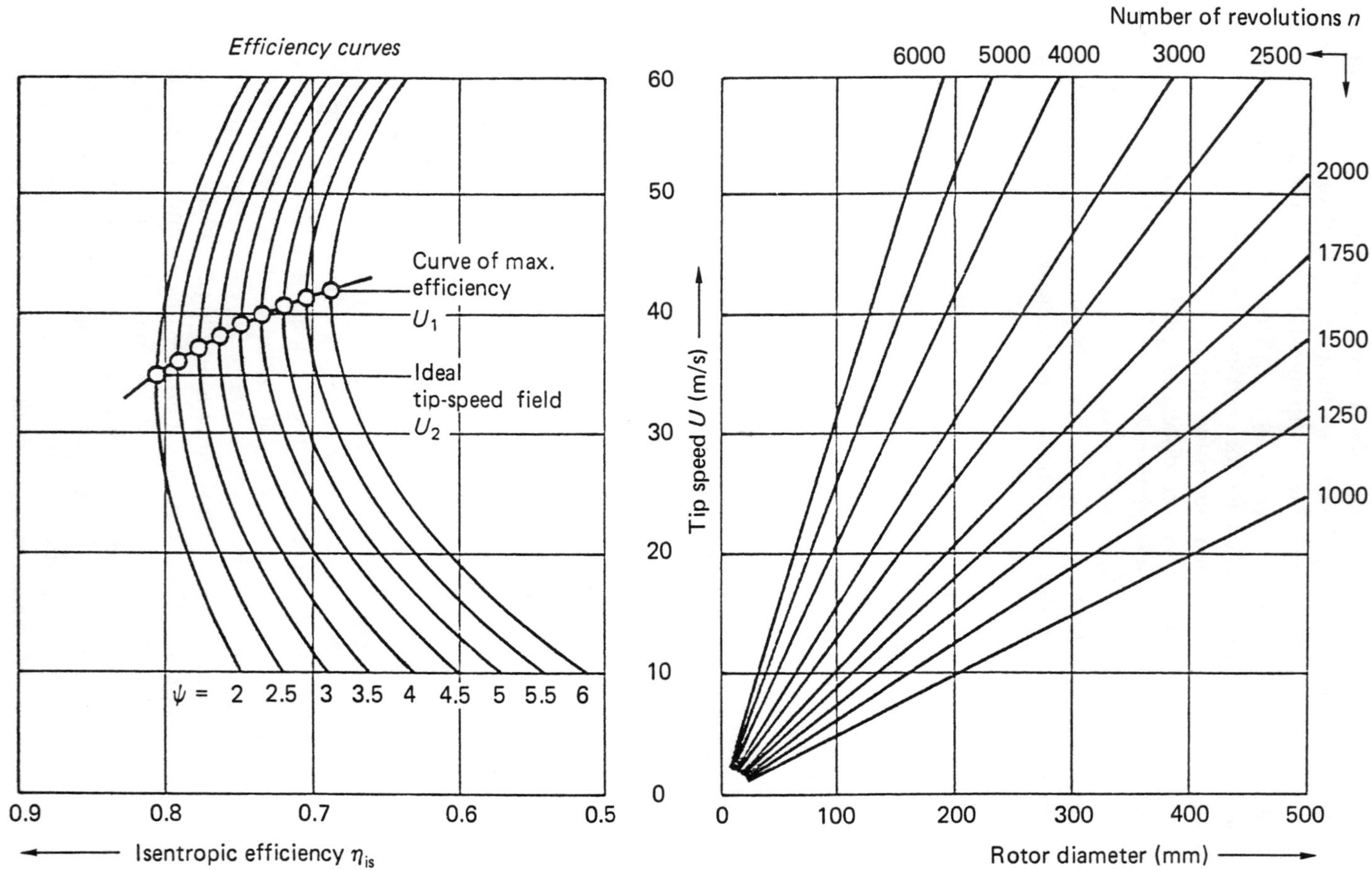

Figure 4.28 *Curves relating tip speed against rotor diameter and efficiency with different values of ψ and rotational speed for R717 screw compressors*

installation is lower than the pressure of the gas as it arrives at the discharge port then there is overcompression. In figures 4.29a and b we can see the conditions for over and undercompression shown on adjacent *pV* diagrams. These losses could go as high as 5%. When the end-compression pressure is lower than the condensing pressure then some gas that has already been compressed re-expands into the discharge port. If it were possible to continue the compression process up to the condensing pressure p_c, the energy amount represented by the black triangle shown on figure 4.29a would not be wasted, as it now is.

In figure 4.29b we see that by compressing up to a pressure p_i higher than p_c, we also waste energy as shown by the black triangle. This effect has been eliminated in modern screw compressors by equipping them with either an automatic or a manually adjustable in-built volume ratio. Figure 4.30 shows the principle of such an arrangement. By moving the slide stop, which is incorporated opposite the capacity control slide valve, we can change ψ by changing the radial outlet port (see below).

The efficiency of a screw compressor can only be optimized when the in-built volume ratio corresponds to the required compression ratio. Manufacturers provide a range of screw compressors with different in-built volume ratios to allow for a range of capacities at different compression ratios. For example 1:2.5 or 1:3.6 are typical values for one make of compressor. When we know the average pressure ratio at which the compressor will be operating, we can choose the most suitable in-built volume ratio. But even then during actual operation the pressure ratio will inevitably vary from time to time and there will be a deviation from the

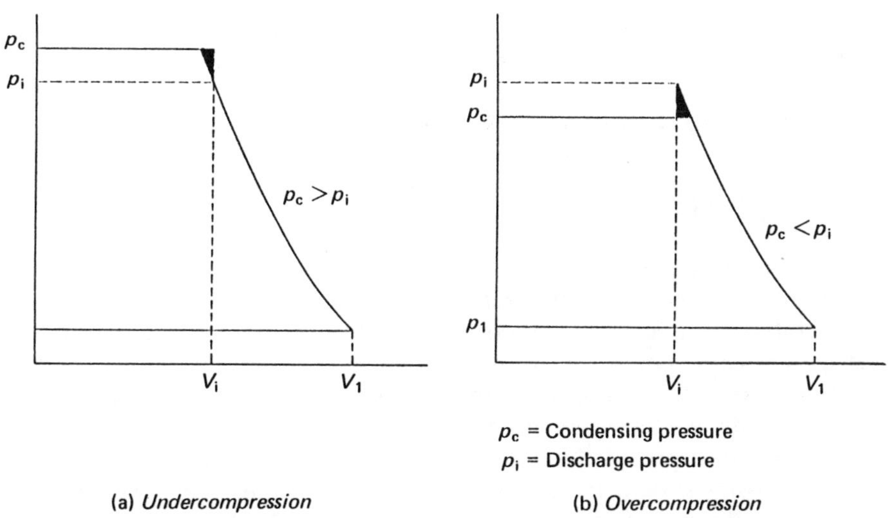

Figure 4.29 *Indicator diagrams showing undercompression and overcompression*

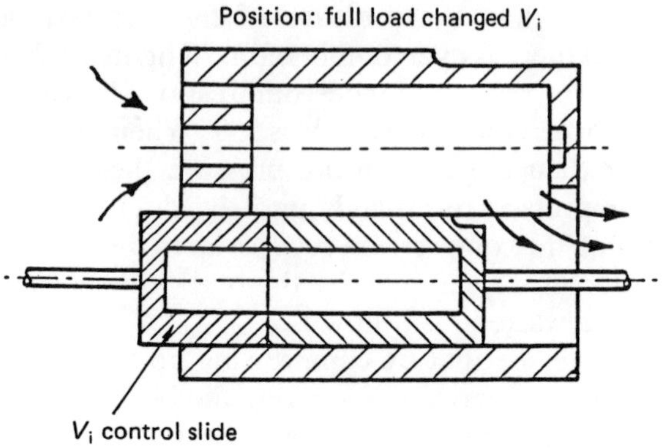

Figure 4.30 *An adjustable in-built volume ratio*

ideal isentropic efficiency. Note that the relation between in-built compression ratio and in-built volume ratio is given by $(p_d/p_e)^{1/k} = V_1/V_2$.

We can see that pressure ratio depends on the value of k, which depends on the particular refrigerant used. This must lead us to the conclusion that the pressure ratio also depends on the refrigerant used. But the volume ratio is independent of the refrigerant used. For that reason we prefer to use this ratio as the design characteristic for this type of compressor.

Now let us go back to the subject of factors affecting performance. Leakage through the rotor clearance is a more serious problem than leakage past the piston of a piston compressor.

As we said before, this leads to turbulent contra-flows of gas, resulting in a lower value of η_{is}. The amount of leakage depends on the clearance between the lobes and the tip speed. By injecting a sufficient amount of oil, a large part of the leakage and at the same time the heat-transfer losses are reduced. As there are no valves but rather large ports, the losses due to streaming of gas through those ports are also very small.

Discharge Temperature

Up to a compression ratio of 5 the discharge temperature will increase; after that, because of the greater influence of the cooling oil, the temperature remains almost constant. The oil is cooled either in a refrigerant-cooled or in a water-cooled oil cooler, and in the latter case the heat load on the condenser is decreased. In a refrigerant-cooled oil cooler, gas is immediately fed by gravity to the condenser, and there will be no loss of compressor capacity. See figure 4.31 for details of this arrangement.

In a booster system the low stage has no oil cooler, but a refrigerant injection system between the discharge port and the low-stage oil separator.

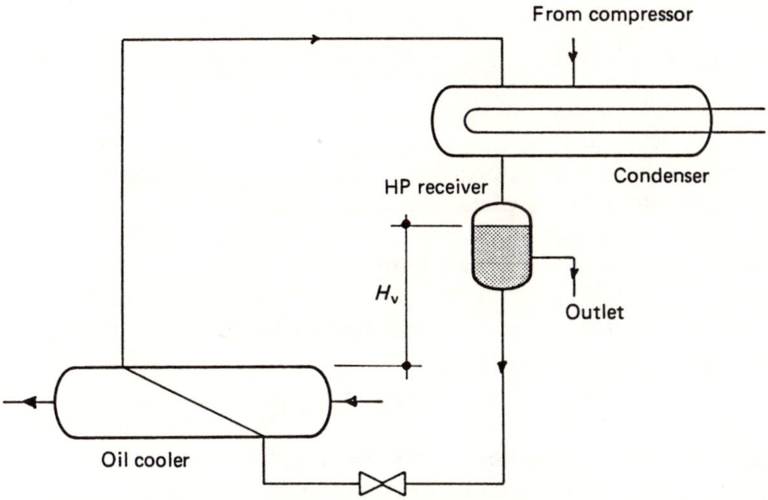

Figure 4.31 *Refrigerant-cooled oil cooler circuit*

A screw compressor is much more sensitive to high compression temperatures causing expansion problems, therefore 100°C (373K) is the safe operating limit. Single screws are even more sensitive in this respect.

Power Consumption

If the work for compression is P then we have the situation shown on figure 4.29a:

$$P_{th} = \int V \mathrm{d}p + (p_c - p_i)V_p$$

Note that the difference between this formula and that for piston compressor is the second term $(p_c - p_i)V_p$: the extra triangle shown on the indicator diagram, figure 4.29a. In these formula p_0 = initial compression pressure, p_i = final compression pressure and p_c = condensing pressure. For adiabatic compression we saw that

$$p/p_0 = (V_m/V)^k \quad \text{or}$$
$$V = V_m(p_0/p)^{1/k}$$

We can substitute V in the equation of the power consumption after integration:

$$P = V_m p_0{}^{1/k} k/(k-1) p_i{}^{(1-1/k)} - p_0{}^{(1-1/k)} + (p_c - p_i)V_p$$
$$p_i/p_0 = (V_m/V_p)^k \quad \text{and} \quad p_i/p_0 = \psi^k \quad \text{so}$$
$$P = V_m p_0 k/(k-1)(\psi^{k-1} - 1) + (p_c/p_0)V_p$$

In the ideal case:

$$p_0 = p_i \quad \text{so that} \quad P_1 d = V_m p_0 k/(k-1)[(p_c/p_0)^{1-1/k} - 1]$$

We can now see that the supplementary triangle has disappeared. For a specific application, a screw compressor must be chosen with an in-built compression ratio that approaches as closely as possible to the desired compression ratio of the plant.

Suction Volume and Capacity Control

The suction volume V_m of a double-screw compressor is calculated from the formula

$$V_m = a(A_m + A_f)ln$$

where a = number of lobes of the driven rotor
n = rotational speed of the driven rotor
l = working length of the rotor.

A_m and A_f are the areas of the male and female rotor spaces respectively, calculated from a cross-section of the compressor. A_m and A_f can be expressed as a function of the rotor diameter D when all the radii of the circles are known. Then $V_m = (A_m + A_f)l$.

The number of lobes a on the male rotor is usually 4, while the number of flutes on the female rotor is 6; this results in the rotational speed of the female rotor being two-thirds of that of the male. As we saw above, manufacturers are now producing rotors with asymmetric profiles in order to increase the volumetric efficiency. Design calculations of this nature are in fact very complicated and require processing on computers. The above treatment of the subject is only intended to serve as a guidance for better understanding rather than for practical calculation.

As we stated above, because of the absence of clearance volume, filling ratios of screw compressors are better than those of piston compressors. Against the advantage of low final compression temperature, brought about by oil injection between the rotors, must be offset the disadvantage of temperature sensitivity. This leads to a risk of contact between rotors owing to the expansion of the materials. On the other hand, screw compressors are less sensitive to liquid slugging.

Capacity control is obtained by means of a movable slide valve, creating a variable-sized bypass between lobe inter-space and the inlet port. A variable amount of suction gas already between the lobes is returned to the suction port before being compressed. Figure 4.32a shows this process.

This system of capacity control can be compared with a hypothetical method of changing the length of stroke of a piston compressor. This explains why moving the slide results in two different actions. First, the compression starts later, and second there is less compressed gas. Under this arrangement if the discharge port does not change, the gas would be

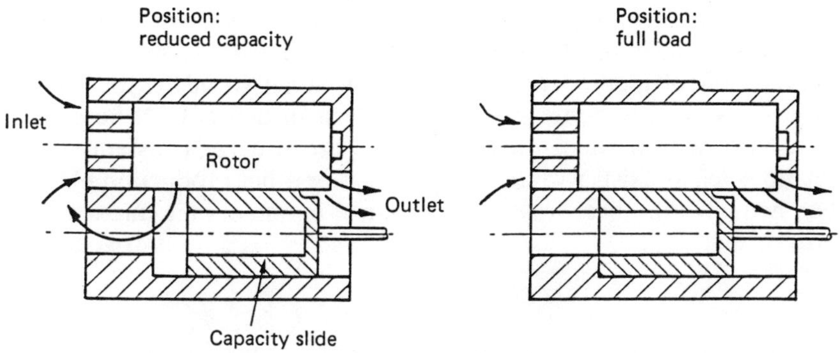

Figure 4.32 *The capacity control slide mechanism*

compressed to a lower end-pressure. The in-built pressure or volume ratio ψ should be reduced and vary from the desired ratio, and efficiency should decrease. But as we see in the figure, the end of the slide has an opening, which functions as a radial port of the discharge port.

Moving the slide means changing this radial part of the discharge port, in other words, an adjustment is made to the value of ψ. This is only true down to about 70% of the full capacity. This is the reason for the decrease of power consumption being more or less proportional down to a capacity reduction of 70%. Below this percentage, the efficiency rate decreases significantly. See figure 4.33 for details of this effect.

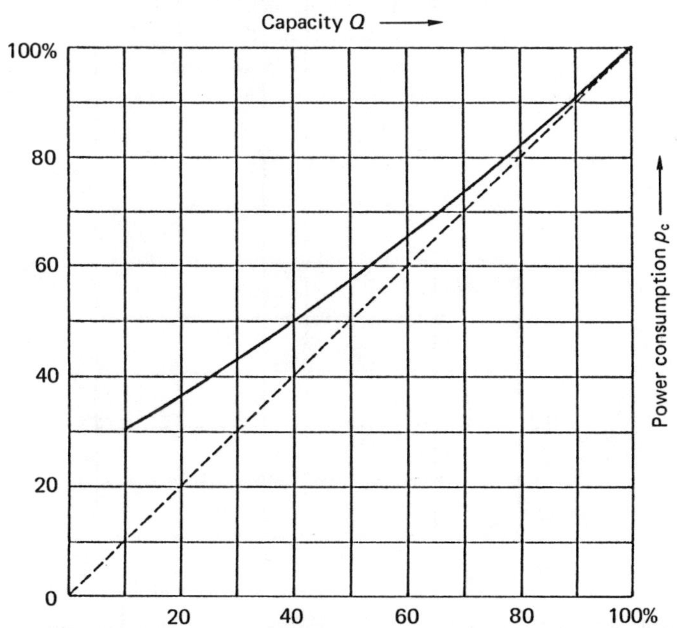

Figure 4.33 *Efficiency losses for an ammonia screw compressor operating with slide capacity control*

While practically it is possible to reduce the capacity down to 10%, this is not advisable since the efficiency will decrease disproportionately. Using a capacity control method in which the rotational speed of the rotors is varied gives slightly better results, as shown in figure 4.34. Screw compressors work best when used to satisfy a reasonably constant capacity demand, such as exists with freezing tunnels or other production plants, or when acting as the base-load compressor of a system handling a variable load.

Adjustable In-built Volume Ratio

As stated above, modern screw compressors can now be provided with an adjustable in-built volume ratio. In order to achieve this a second slide or slide stop is added to the design as shown in figure 4.30. By moving this slide stop to the right or the left we can see that the radial outlet is also changed at full load, when the recirculation slot is closed.

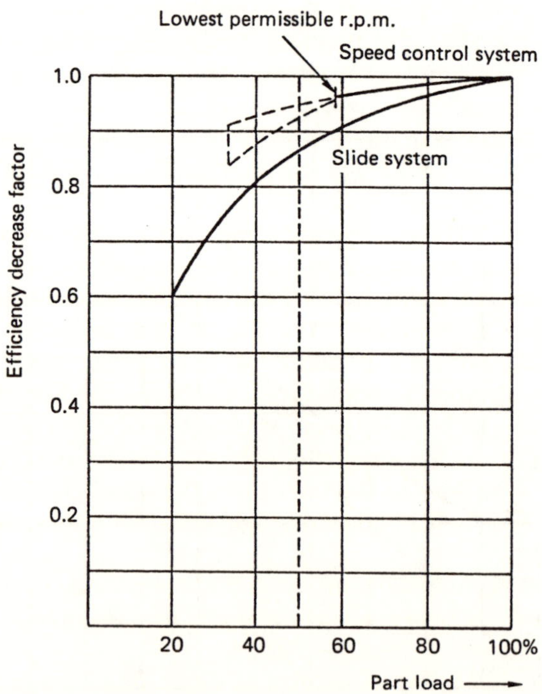

Figure 4.34 *Efficiency losses for a CFC screw compressor using slide regulation and speed-regulation methods of capacity control – manufacturer's data*

Super-feed or Economizer Operation

It will be recalled that the efficiency of a refrigeration cycle can be improved by subcooling the liquid refrigerant. In CFC plants this is achieved by heat exchange between the cold suction gases which become superheated, and the warm liquid coming from the condenser. In doing this, part of the efficiency gain is lost because of the following:

(a) superheat is increased, so increasing the specific volume of the gases on the inlet of the compressor
(b) pressure drop in the heat exchanger.

In two-stage plants, subcooling is achieved in the intercooler with a much improved efficiency, because of the two-stage expansion, which is partly realized in closed systems and fully realized in open or flash systems. Flash gas formed in this secondary expansion down to intermediate pressure is induced by the high stage only.

Because of the constructional design of screw compressors it is possible to obtain this advantage in single-stage compressors. See figure 4.35 for two arrangements illustrating this principle. The flash gas of the secondary expansion is actually channelled into the normal flow through the compressor. A special hole is drilled into the housing, through which the flash gas enters at a specific point during the compression process which is already taking place.

The circuit diagrams in figure 4.35 show arrangements with an intermediate cooler and a coil working at the intermediate evaporation temperature and pressure. Liquid refrigerant is injected over the coil, cooling down the warm liquid inside. The gas created is then fed into the supplementary port of the compressor described above. The manufacturers of screw compressors use several terms for this system, of which the best known are 'super-feed' and 'economizer'. The use of such a system can produce a capacity increase between 10 and 30%; even higher figures are obtained when the evaporating temperatures are lower and/or the condensing temperatures are higher. This also depends on the refrigerant used.

Some compressors are supplied with two of these ports or economizer-connections – one close to the inlet port, the other close to the discharge port. In this way a choice between a high and a low intermediate pressure is available.

Two types of economizers are available: open flash, normally used with ammonia, shown on figure 4.35a; and direct expansion heat exchangers for use with R22, shown on figure 4.35b.

The compression follows the form of single-stage compression until the special connection is reached. At that connection the intermediate pressure gas from the economizer or super-feed vessel joins the suction gas

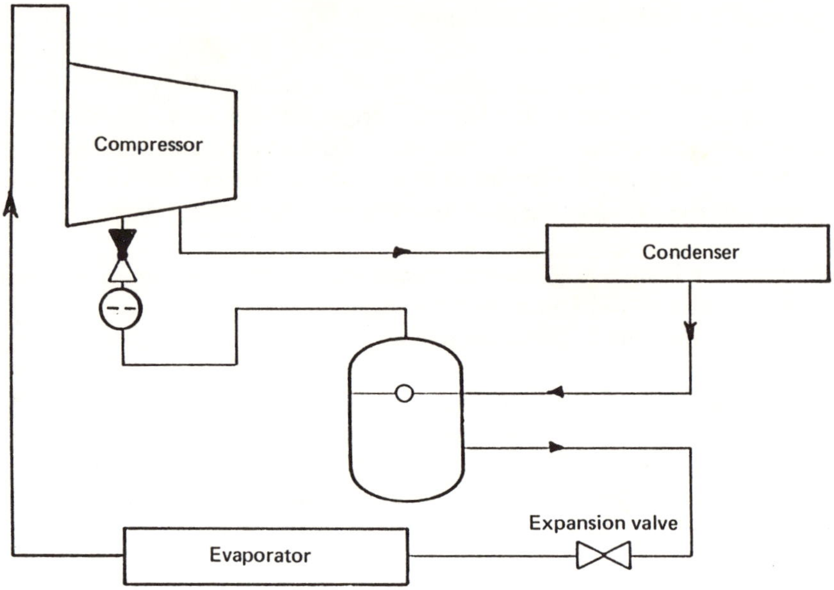

(a) *Open-type intercooler*

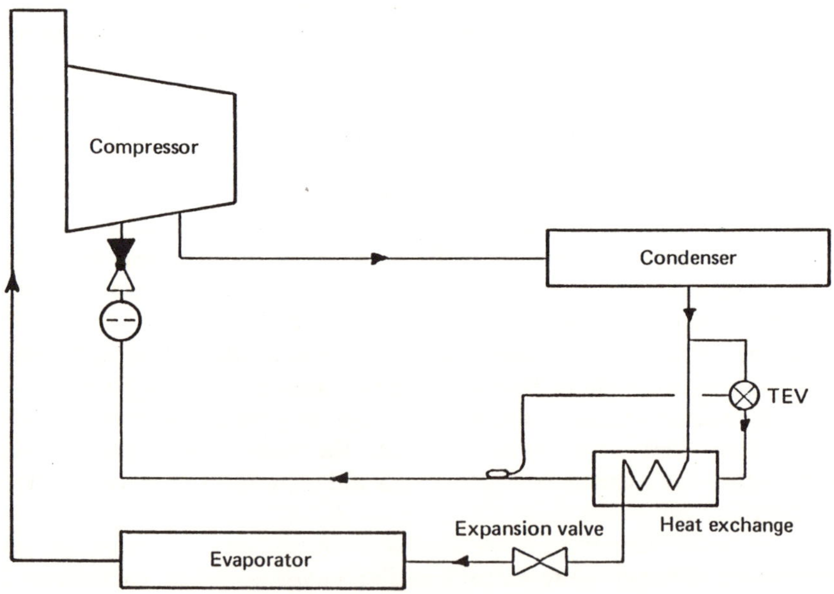

(b) *Direct expansion intercooler*

Figure 4.35 *Shell and coil and direct expansion variations of the superfeed system with* p/h *diagram*

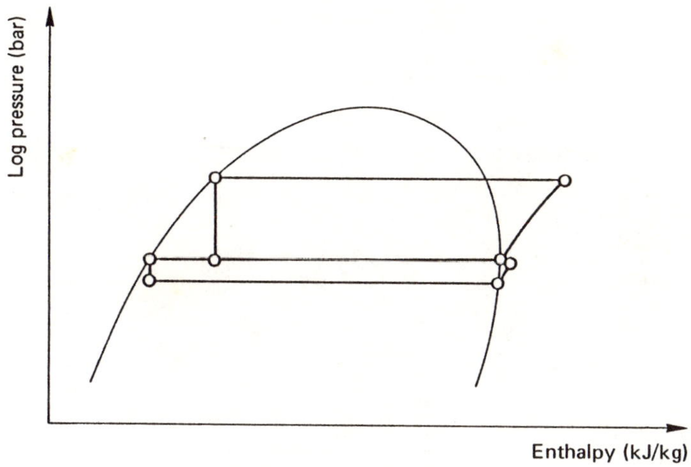

(c) *Open-type superfeed*

Figure 4.35 (*continued*)

already under some compression. The 'second step' of the compression handles a larger refrigerant mass flowrate than the 'first step', which only handles the mass flowrate evaporated in the plant evaporator. Because of the larger mass handled in the 'second step' of compression, the power input is increased according to the selection of the first or second port.

Single-screw Compressor

Figure 4.36 shows two views of a single-screw compressor. The fundamental operating principles of the single-screw are the same as those for the double-screw. The teeth of the stars can be considered as the pistons of a double-acting reciprocating compressor. If the volume of a lobe interspace filled by a tooth is V_1, and assuming that the wheel has 12 teeth, then the suction volume V_m is given by $V_m = 12V_1n$.

Capacity control is obtained by using a ring which has the same function as the slide of the double-screw. Moving the ring opens a bypass to the point where compression normally starts. A part of the gas flows back to the opening. When the ring moves as described at the same time the discharge port is somewhat obstructed, and this reduction of the discharge port automatically adjusts ψ.

Advantages of Screw Compressors

It is useful to have a summary of the advantages of screw compressors:

- Simple construction so far as the compressor body is concerned – no valves, crankshaft, connecting rods etc. The main parts which are

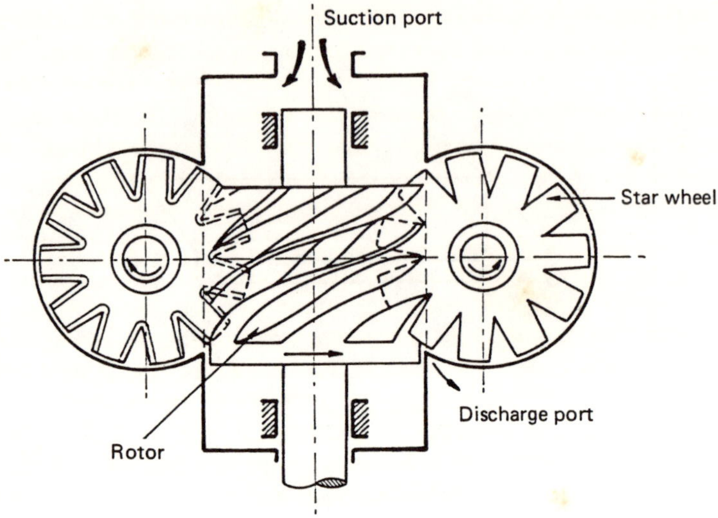

Figure 4.36 *Sectional view of a single-screw compressor*

subjected to wear-out or breakdown are the bearings and the oil-pump. However, the complete compressor-unit including oil-cooling system, controls etc. is rather a complex piece of equipment

- Low maintenance costs: 40 000 operating hours between maintenance. However, an inspection of bearings is advised every 20 000 hours in order to avoid emergencies or heavy overhaul costs, for example as a result of mechanical seizure between the rotors. Some single-screws need replacement of tooth wheels every 25 000 hours, since they are manufactured from plastic
- Light foundations, because rotary motion results in fewer out-of-balance forces than do reciprocating compressors of similar capacity
- Reliability
- Gradual capacity control down to 10%
- Low discharge temperatures
- Non-pulsating gas discharge
- High compression ratios possible
- Less sensitive to liquid slugging
- Available in large refrigeration capacities per single unit
- Able to utilize the economizer facility.

Comparison Between Screw and Piston Compressors

It is difficult to make a comparison between piston and screw compressors on the basis of efficiency rates. On the other hand it can be all-too-easy to draw general conclusions. The design parameters of both types of compressor vary greatly. It is always difficult to compare compressors of similar type and construction and even harder to make comparisons between screw and piston compressors. When comparing the two compressor forms by examining performance graphs, for example, the type of piston compressor as well as the type of screw compressor must be considered. Comparisons made only on the basis of equal suction volumes are very risky because a comparison is perhaps being made between a piston compressor of very good design and a screw of extremely low performance, and vice versa. It becomes even more complicated for compressors being compared under different working conditions or their behaviour at reduced capacities. Figure 4.28 shows the relationship between isentropic efficiency, tip speed, rotor speed and rotor diameter. From this it might be concluded that small screw compressors with small rotors need to run at a high speed of rotation in order to perform at optimum tip speed and highest efficiency. Figure 4.37 presents a graph in which the coefficient of performance or refrigerating ratio is given at different working conditions and varying rotor diameters at constant tip speed. It is interesting to see

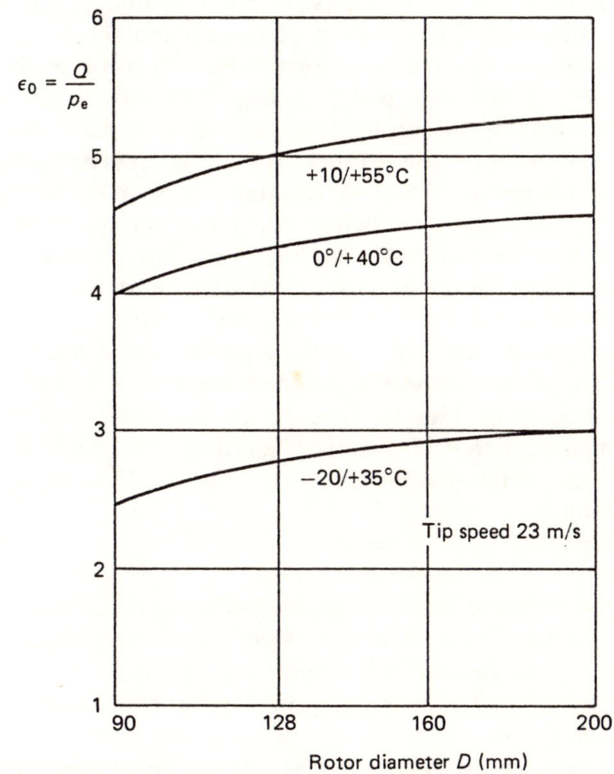

Figure 4.37 *Graphs showing the variation of the refrigeration ratio with rotor diameter at different working conditions at constant tip speed for an ammonia screw compressor*

that this ratio decreases with decreasing rotor diameter, but we cannot consider practical values since we do not know what operating conditions, such as superheat and subcooling, are involved. Small screws are of less interest from an efficiency point of view, because they require high velocities and even gearboxes, which results in extra mechanical losses. Nevertheless, sometimes small screw compressors exhibit a better η_m than is expected, the reason being that small screws often have specially developed rotor profiles. Improved profiles result in continuous contact between the rotors over a complete revolution. Traditional rotors lose contact at a point during a revolution. This point is called a 'blow-hole'. Blow-holes are not found in single-screw compressors. For these new profiles the optimum tip speed is reduced and may be as low as 12 m/s. When we examine the graph in figure 4.38a, in which piston and screw compressors are compared, we can conclude that the isentropic efficiencies of screw compressors in general lie below those of piston compressors. See also figure 4.39.

It is not always clear whether the authors of articles on this subject are considering P_i or P_e, in other words, whether or not η_m has been taken into account. Figure 4.40 supplied by another manufacturer shows graphs that lead partly to other conclusions. The effect of a fixed in-built compression or volume ratio is a handicap, especially in a system which undergoes frequent changes in the working conditions and when using the capacity control system below 70%. In some cases it is difficult to operate with optimum tip speed. When we examine figure 4.38b we see that the filling ratio is in each case better for large screws than for piston compressors. The total overall efficiency of a compressor in a plant is influenced by many different factors, so care must be exercised when giving an answer to the question: 'Will screw compressors replace piston compressors in the future?'. In fact the engineer must make a judgement and a choice for every new project based on the particular circumstances of the project concerned. Finally, here is an illustration of this treatment based on practical figures. It should be noted that these figures should not be used to draw general conclusions, for they have been selected very much at random.

Let us compare:

(a) a screw compressor with a large rotor diameter and 2950 1/min
(b) a screw compressor with a somewhat smaller rotor diameter at the same speed, but lower condensing temperature
(c) a screw compressor with a much smaller rotor diameter, but with the same rotor speed
(d) the same compressor as in (c) but at a higher rotor speed of 4425 1/min

Note that all the screw compressors are chosen so that the in-built compression or volume ratio corresponds to the actual working conditions. In this way the effects of under or overcompression on η_{is} are eliminated.

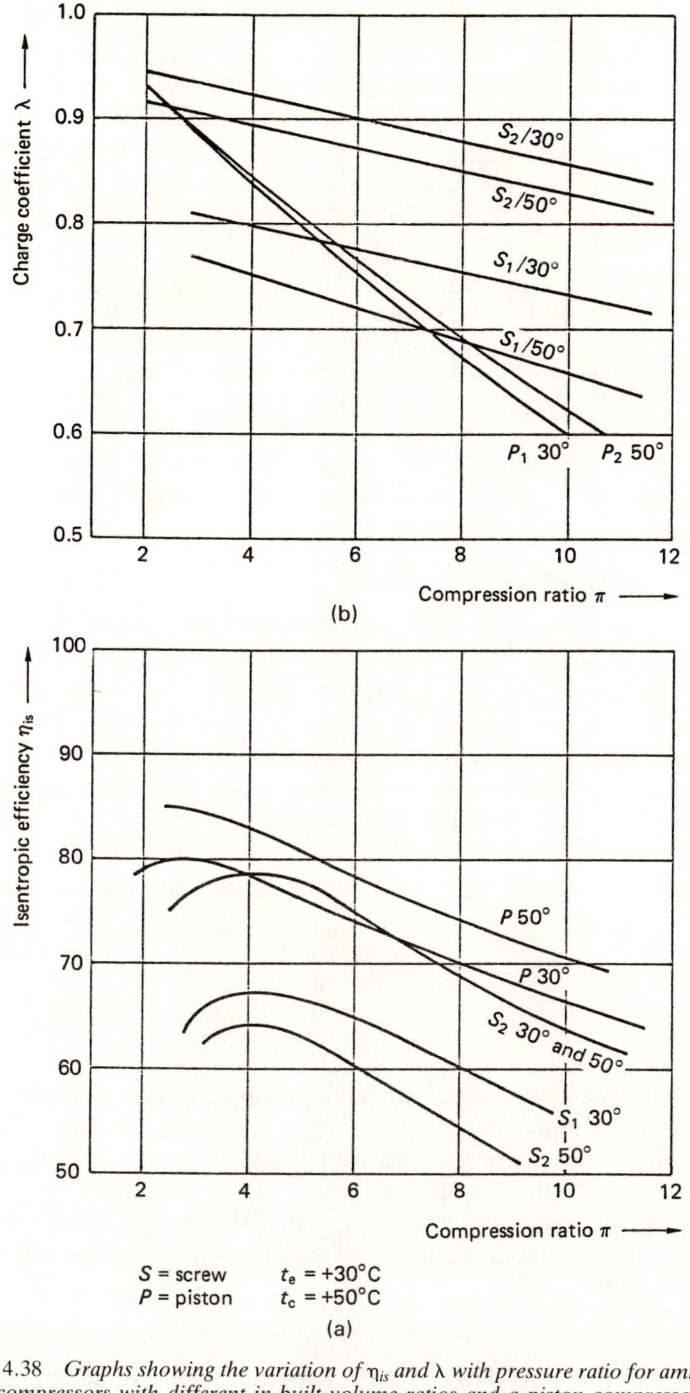

Figure 4.38 *Graphs showing the variation of η_{is} and λ with pressure ratio for ammonia screw compressors with different in-built volume ratios and a piston compressor with equal capacity*

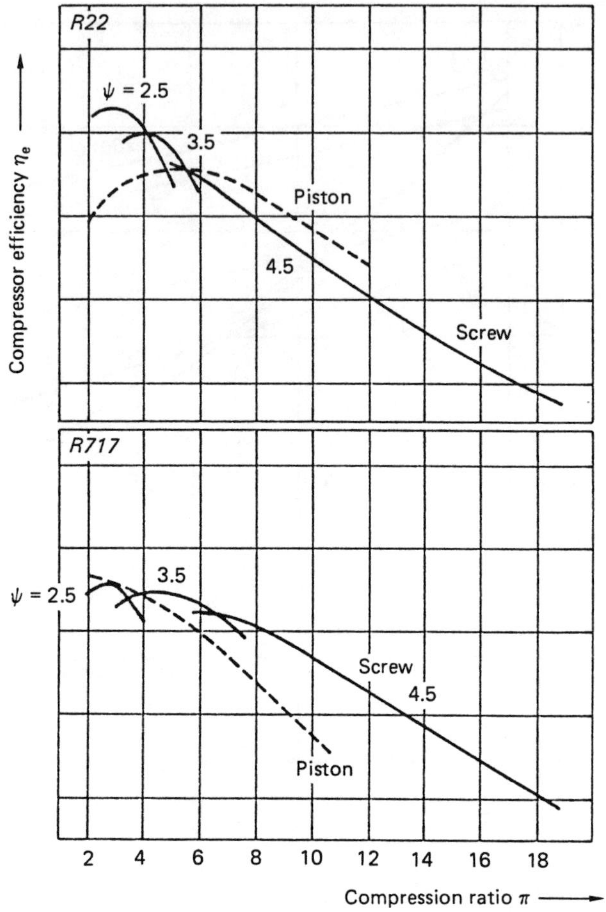

Figure 4.39 *Graphs comparing screw and piston compressors for η_e against pressure ratio for R22 and R717*

(e) a large piston compressor with a big piston section, with equal capacity to the screw under (b)
(f) a small piston compressor with a small piston section, corresponding to the capacity of that under (c)
(g) the same piston compressor type but with a bigger stroke than that under (f) and a somewhat lower speed, so the capacity corresponds that to under (c).

Note: All plants are working with 5K superheat and 5K subcooling.

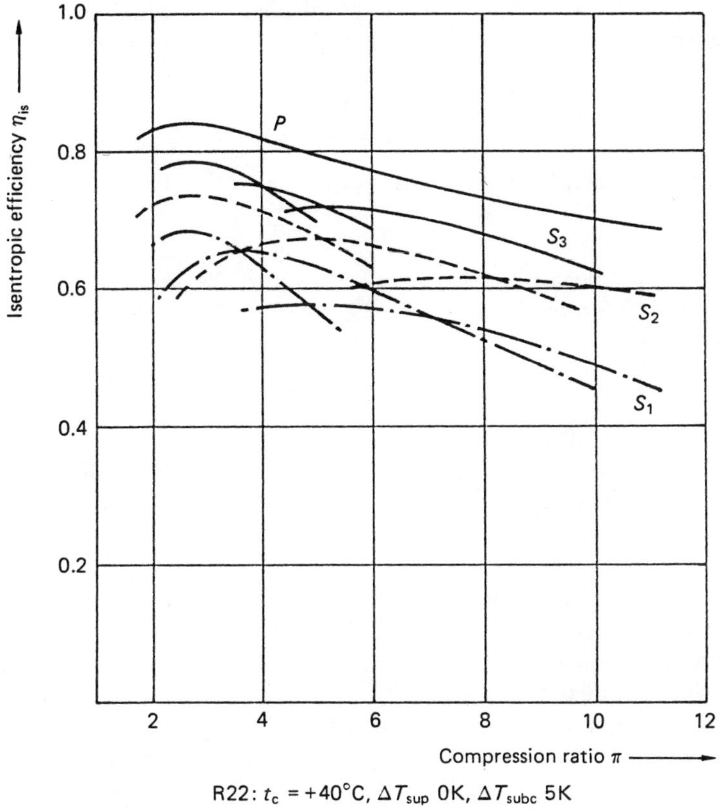

Figure 4.40 *Graph comparing η_{is} against pressure ratio for three screw (S) and one piston (P) compressor operating with R22*

(a) Rotor 345 mm, $L/D = 1.6$, $\psi = 3$, $n = 2950$ 1/min:

$$t_0 = -10°C$$

$$t_c = +33°C$$

This corresponds to a volume ratio ψ of 3.06.

$$\eta_{is} = 0.79$$

(b) Rotor 270 mm, $L/D = 1.6$, $\psi = 2.3$, $n = 2950$ 1/min:

$$t_0 = -10°C$$

$$t_c = +24°C$$

This corresponds to a volume ratio ψ of 2.59.

$$\eta_{is} = 0.87$$

η_{is} of (b) is lower than that of (a) because of the higher t_c.

(c) Rotor 127.5 mm, $L/D = 1.7$, $\psi = 2.6$, $n = 2950$ 1/min:

$$t_0 = -10°C$$

$$t_c = +25°C$$

This corresponds to a volume ratio ψ of 2.59.

$$\eta_{is} = 0.67$$

η_{is} is lower because of the smaller rotor at the same speed.

(d) Rotor 127.5 mm, $L/D = 1.7$, $n = 4425$ 1/min, working conditions as under (c):

$$\eta_{is} = 0.78$$

This is better than η_{is} in (c) because of the higher speed.

(e) 8-cylinder piston compressor, $D = 180$ mm, $S/D = 140/180 = 0.78$, $n = 970$ 1/min, working conditions as in (c):

$$\eta_{is} = 0.83$$

This is somewhat lower than η_{is} under (b), but remember that the stroke is short.
Compare with (f) and (g).

(f) 6-cylinder piston compressor, $D = 100$ mm, $S/D = 80/100 = 0.8$, $n = 1170$ 1/min, working conditions as in (c):

$$\eta_{is} = 0.81$$

This is much improved when compared with (c).

(g) 6-cylinder piston compressor, $D = 100$ mm, $S/D = 100/100 = 1$, $n = 1065$ 1/min, working conditions as in (c):

$$\eta_{is} = 0.86$$

This is better than in (c), and better than in (f), because of the longer stroke resulting in a 'square machine'.

4.16.3 Turbo-compressors

In refrigeration the term turbo-compressor usually denotes a centrifugal compressor. In this form of compressor the discharge pressure is limited by the maximum permitted tip speed. For high compression pressures a set of impellers must be arranged serially. The suction volume and other characteristics are calculated in the same way as for centrifugal fans and pumps. The isothermal and adiabatic efficiency are low. The two common capacity control methods operate by changing the blade angle of the movable blades fitted to the diffuser or the guide vanes at entry to the impeller.

The main applications of turbo-compressors are found in situations requiring high suction volumes at high suction pressures, such as for air-conditioning and other systems utilizing large quantities of chilled water or brine solutions. Although in industrial refrigeration and in mine-cooling plants with such requirements they are seldom used.

Turbo-compressors experience a very strong negative value of η_{is} when operating conditions change. Turbo-compressors are built in open and semi-hermetic units. For further information the reader is recommended to refer to specialist literature dealing with air-conditioning applications and theory.

References

Virendra Charan and Ram Lal, Optimum Intermediate Pressures in Compound Refrigeration System, *Climate Control*, Feb., 1973.

E. Fornasieri (Instituto Fisica Tecnica, University of Padua, Italy), Analisi del processo di compressione dei fluidi frigorigeni mediante compressoni alternative, *Il Fredo*, 3 (1983).

R. Klein (GHH, Oberhausen, Germany), *Schrauben Verdichter fur den Einsatz in Wärmepump Anlagen.*

K. Mötz (Man, Nurenberg, Germany), KI Klima, Kälte and Heizung, 10 (1983).

Joachim Paul, Integral Technology, Schrauben und Kolbenverdichter in Vergleich, *Die Kälte und Klimatechniek*, 12 (1981); Kolben und Schraubenverdichter in der Kälte, Wärme und Klimatechnik, *KI Klima, Kälte und Heizung* (1983), Flensburg, Germany.

V. Villadsen and F. V. Boldvig (Sabroe Ltd, Denmark), *Oil in Refrigeration Plants*, IIAR Annual Meeting, 19–22 February, 1984.

5

Evaporators and Condensers

5.1 Heat Transfer

Before treating the subject of evaporators and condensers it is necessary to recall those theoretical principles which relate to heat transfer. Only certain basic subjects directly related to refrigeration techniques are treated here. For a more detailed treatment of heat-exchange processes we refer the reader to the specialized literature on this subject.

In refrigeration techniques we deal mainly with heat transfer from one material or system to another, such as from food products to air, from air to insulated walls, from air to the air cooler or to the air-cooled condenser, through tubes, over fins, from refrigerant to a liquid or to the air etc.

The reader should remember that 'cold' does not exist, nor is there such a thing as 'cold' exchange. Coldness could be considered to be the absence of heat, or negative heat. Heat is a relative term, which we explained in the earlier chapters and suggested that a better name for heat is thermal energy.

Three distinct modes of heat transfer are identified in thermodynamics: conduction, convection and radiation. Heat transfer can only take place from a system or body at a higher temperature than the system or body to which the heat transfer is taking place. The science of heat transfer not only explains how it takes place but also the rate at which it takes place – information that is very important in the design and rating of evaporators and condensers.

(a) *Thermal conduction* is the transfer of thermal energy from one part of a substance to another part, or from one substance to another in physical contact with it, by means of vibrational impact between adjacent molecules, without any significant displacement of the molecules. The energy transfer takes place from atoms of high energy to those of lower energy.

(b) *Thermal convection* is subdivided into two forms:
(i) *Natural convection*. In this form the temperature difference causes a difference in the density of the fluid which results in different buoyancy forces causing mass movement of the fluid and a transfer of energy. This is sometimes referred to as 'thermosiphoning' in heating systems using water.
(ii) *Forced convection*. In this form the mass movement of the fluid is caused by the action of a pump or fan, with a temperature difference between thc fluid and the body to which the energy is being transferred.
(c) *Thermal radiation*. Here the transfer of energy is by electromagnetic radiation in the range of wavelengths between 10^{-6} and 10^{-3} m. All substances emit radiant energy but the net heat flow is from the hotter to the colder body. In other words the cooler body emits less heat than it absorbs. The equation for this type of energy transfer is:

$$Q = \epsilon\sigma A \left(\frac{T_1{}^4}{100} - \frac{T_2{}^4}{100}\right) \text{ W}$$

where ϵ is the emissivity of the surfaces
and σ is the Stefan–Boltzmann constant, 5.67×10^{-8} W/(m^2 K^4).

Generally, this kind of heat transfer is of little importance in refrigeration techniques. There are some applications of cooling by radiation, for example the cooling of chocolate, in which convection must be avoided. The shell of a water-cooled condenser loses some thermal energy by radiation. The most important thermal radiation to be dealt with is the solar energy which is transferred to the walls and especially the roofs of cold stores. In solar power calculations this kind of heat transfer is of course extremely important.

5.1.1 Heat Transfer by Conduction

Steady-state Conduction

Steady-state conduction is a thermal energy flow which does not change during the process. This is the simplest form of heat transfer, which is assumed when considering heat transfer by conduction through a flat plate, as in the case of building insulation, tanks, in plate freezers or in plate heat exchangers.

In heat exchangers such as evaporators and condensers, heat transfer by both conduction and convection takes place. When we are dealing with problems such as the calculation of chilling, freezing or defrosting times, however, introduction of the time factor complicates the situation. Time-dependent heat transfer by conduction is called transient conduction, in contrast to stationary conduction where the energy flow does not change throughout the process.

The starting point for the treatment of thermal conduction is Fourier's law, which states that the quantity of thermal energy that flows per unit of time Q is proportional to the surface area A of the material concerned and to the temperature gradient dT, and is inversely proportional to the layer thickness dx of the material. Then applying Fourier's law we have:

$$\text{Rate of heat flow } Q \propto -A(dT/dx)$$
$$\text{or } Q = -kA(dT/dx)$$

where k, the constant of proportionality, is called the coefficient of thermal conductivity. The coefficient of thermal conductivity can be defined as the thermal energy flow per unit area per unit time with a temperature decrease of one degree in unit distance.

When the equation $Q = -kA(dT/dx)$ is integrated then

$$Qx = -A \int_{T_1}^{T_2} k dT$$

Most materials we use exhibit a reasonably constant value of k with temperature so this can be solved as

$$Q = -kA(T_2 - T_1)/x$$

Since k can be easily confused with the overall heat-transfer coefficient k as used in Continental Europe, λ is now used, and δ instead of x for thickness.

The units of λ are W/(m K) where Q is in W, A in m^2, $(T_2 - T_1)$ in K and δ in m. The inverse of thermal conductivity is resistivity r which equals $1/\lambda$.

An important notion is the heat flow density q expressed in W/m^2. Thus $q = Q/A = (\lambda/\delta)(T_2 - T_1)$ in the case of heat flow through a single layer of material of thickness δ. We can also say that $\lambda/\delta = q/dT$, that is heat-flow density per temperature difference.

The treatment of heat flow through composite structures will be dealt with later in the chapter.

In shell and tube heat exchangers we deal with the curved surface of the tubes, as well as with insulation techniques of pipework.

The effects of the curved surface with inner radius r_1 and outer radius r_2 must be introduced in the equations:

$$Q = -\lambda A dT/dr = -\lambda 2\Pi r dT/dr$$

After integration between T_2 and T_1 and r_2 and r_1:

$$Q \ln(r_2/r_1) = -2\pi\lambda(T_2 - T_1) = 2\pi\lambda(T_1 - T_2)$$

Hence $Q = 2\pi\lambda(T_1 - T_2)/\ln(r_2/r_1)$.

We see that the temperature curve is logarithmic. λ depends on the temperature level of the matter, since at high temperatures the molecules are more widely spaced and λ will consequently decrease. λ is largest for solid materials and smaller for liquids and gases. It decreases most in liquids when the temperature nears the saturation temperature. When the pressure on a liquid increases, λ increases also, because the molecules are more closely spaced. Under 50 bar this influence, however, can be neglected. λ for liquids and gases also decreases if the molecular mass increases. When a heat exchange surface is fouled by scale or other impurities the effect of the added thermal resistance must be added to the original thermal resistance.

Transient Heat Transfer by Conduction

In this form of heat transfer not only is λ of importance but also the density ρ and the specific heat c_p.

If a = the thermal diffusivity by conduction through the material, then $a = \lambda/\rho c_p$ expressed in m^2/s, and

$$d^2T/dx^2 = (\rho c_p/\lambda)(dT/dt) \quad \text{or} \quad (1/a)(dT/dt)$$

a is a measure of the velocity with which the change in temperature penetrates the material.

The increase in temperature in a material, for a given amount of thermal energy stored in it, is greater when c_p and ρ are smaller. However, it also depends on the amount of thermal energy supply and on λ. As this supply is proportional to λ, then the temperature will penetrate faster if λ is larger and c_p and ρ are smaller.

5.1.2 Heat Transfer by Convection

Convection occurs between a solid and a fluid (liquid or gas). Energy is transferred by an increase in the molecular vibrations of the fluid, which results in the molecules near the surface being continuously pushed away and replaced by others lying deeper in the current layer and being in turbulent motion. This replacement or pushing-away creates friction and so causes resistance or pressure drop.

From fluid mechanics we know that there are two kinds of flow: laminar and turbulent. Laminar flow exists at low current speed, turbulent flow at high speed. In laminar flow the current speed is low, the viscosity sufficiently high and the tube section sufficiently small that the layers of matter slide over each other. The velocity in each layer is the same in value and direction. See figure 5.1.

The pressure drop, owing to the friction of the molecules, depends directly on the velocity.

When the fluid velocity U, tube diameter d and kinematic viscosity are in a certain relationship to each other, the type of flow changes. Random

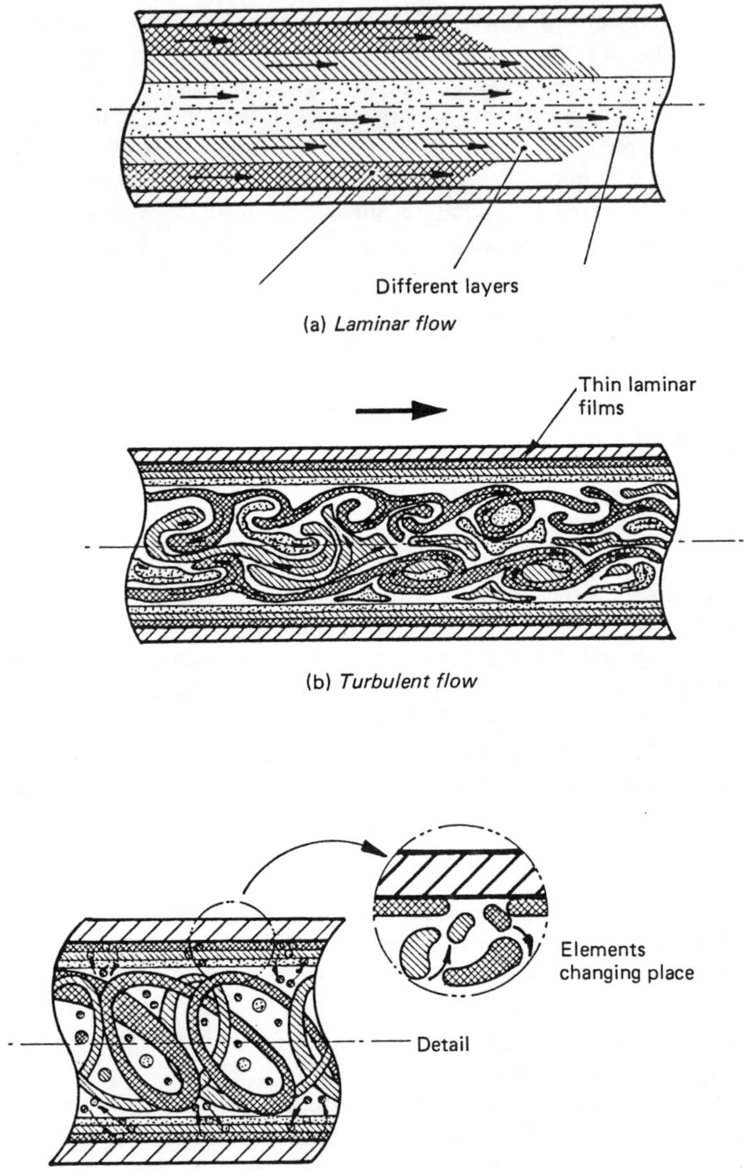

Figure 5.1 *Laminar and turbulent flow sources*

motion, often in the opposite direction to the flow, occurs. The various currents no longer have the same value and direction. Pressure drop now depends on the square of the velocity.

Reynolds found that by changing from laminar to turbulent flow, velocity and section increase and viscosity decreases. As we saw, these are the conditions required to change the type of flow. He characterized this by a dimensionless number, called the Reynolds number *Re*.

$Re = Ud/\nu$ or if we use the dynamic viscosity $\eta = \rho/\nu$, *Re* becomes $Ud\rho/\eta$. Here η is expressed in kg/m s and U in m^2/s. If the value of *Re* is bigger than 2320 the flow is turbulent. *Re* is a dimensionless number or similarity factor, and could be viewed as the slowness force (velocity), $\nu U^2/d$, divided by the friction force (viscosity), $\eta U/d^2$, that is a ratio between two parameters which characterize the type of flow.

When *Re* is large, the velocity dominates the viscosity, resulting in an 'uncontrolled' behaviour of the flow, that is, turbulent flow.

Other important similarity factors are:

$$\text{Péclet number } (Pe) = Ud/a$$

This can be viewed as the ratio of the heat transferred under the influence of the velocity, Uc_p/ν, divided by the thermal diffusion flux created by conduction, λ/d.

$$\text{Grashof number } (Gr) = fd^3\beta dT/\nu^2$$

in which β = volumetric expansion coefficient v/k. This can be viewed as the ratio pushing power per unit volume/slowness power per unit volume. A combination of *Pe* and *Re* is also used:

$$\text{Prandtl number } (Pr) = Pe/Re = \eta c_p/a$$

or the ratio kinematic viscosity/thermal diffusion flux created by conduction.

The above-mentioned dimensionless numbers are named after the following physicists;

O. Reynolds	(1842–1912)
J. Péclet	(1793–1857)
F. Grashof	(1826–1893)
L. Prandtl	(1875–1933)

The heat transfer coefficient for convection is α W/(m^2 K). Its value depends on many factors, some of which we have discussed in connection with the above-mentioned numbers: characteristics of the fluid, the type of motion of the fluid, temperature T, pressure p, velocity U, density ρ, heat transfer coefficient by conduction λ, thermal diffusion flow by conduction a, specific heat c_p, viscosity ν or D, and shape, state and roughness of the surface of the objects involved in the heat transfer.

In a laminar stream, heat transfer takes place only by conduction. In a turbulent stream this is true only for a thin boundary layer at the surface of the wall, the thickness of the layer depending on the viscosity. Further into the stream, heat transfer occurs by the above-mentioned uncontrolled motions of the molecules, typical for turbulent flow. When motions increase, heat transfer increases too, because more molecules will be pushed away from the boundary layer and be replaced constantly by others.

Gas molecules are in more violent motion than those in a liquid, and they are at a greater distance from one another. Therefore in a gas it will be more difficult to remove molecules and replace them by others.

During very violent processes, such as evaporation and condensation, there is considerable motion and good heat transfer by convection.

When there is a liquid on both sides of the heat-exchanger wall, the overall heat-transfer rate is very good; if there is a vapour on one or on both sides, the overall heat-transfer rate is less. A water-cooled condenser has a better heat-transfer coefficient than an air-cooled condenser. A flooded evaporator, the tubes of which are filled with liquid, gives better heat transfer than a dry evaporator with tubes filled with little liquid and much vapour.

5.1.3 Nusselt Number (and Other Numbers)

The well-known German physicist W. Nusselt (1882–1957) gave us a practical method for the calculation of heat transfers. This was greatly needed, as the results obtained up to that point using differential equations were unacceptable.

Nusselt applied the laws of hydrodynamics to thermodynamics, and used the basic equations about continuity of flows, motions of fluids and energy. The congruence theory states that physical processes depend on dimensionless standard numbers – numbers such as those discussed above. If those numbers are known, then one can generalize the results of experiments and even predict the results of processes.

The dimensionless heat transfer Nusselt number for tubes is:

For forced flow: $Nu = \alpha d/\lambda$

For free flow: $Nu = \alpha d/\lambda = f(Re, Pr)$

We can say that: $Nu = \alpha d/\lambda = f(Gr, Pr)$

Nu is the ratio (real heat flow determined by α)/(heat flow by conduction through a thickness d).

When talking about a flow through a tube the influence of the ratio length/diameter, l/d, of the tube must be introduced, so $Nu = f(Re, Pr, l/d)$ or $Nu = f(Re, Pr, Gr, l/d)$.

Remarks

(1) Velocity in a tube is not the same over the whole section – it is zero at the surface and a maximum in the middle, with an average velocity of 0.82 times the maximum velocity.

(2) For the Reynolds number *Re*, the velocity used is the highest local velocity in the middle of a bundle of tubes.
(3) The parameters of the fluid must be considered at the average temperature in the boundary layer.

Example

For heat transfer with turbulent flow in a tube, Hausen found:

$$Nu = \alpha d/\lambda = 0.0235(Re^{0.8} - 230)(1.8Pr^{0.3} - 0.8)[1 + (d/l)^{2/3}](\eta_f/\eta_v)^{0.14}$$

This is valid between the limits $2320 \leqslant Re \leqslant 10^6$ and $0.6 \leqslant Pr \leqslant 500$. *Re* can be calculated if *c*, *d* and ϕ are known. *Pr* is found in tables in *Wärme Atlas* or in Perry (see References at end of book). So α can be calculated.

There are similar equations given by Hausen for various cases, such as heat transfer in a straight plate, a cylinder, a sphere, a cylindrical tube, with external and internal flow, tube bundles etc., and for laminar as well as turbulent flow, the latter being for free as well as forced circulation.

For further details on this subject, refer to the specialized literature on heat transfer, in particular to the works by R. H. Perry, D. Green and General Electric Company which can be found in the list of References at the end of the book. Also refer to the information contained in section 5.5.

5.1.4 Biot Number

The Biot number *Bi* is given by $Bi = \alpha r/\lambda$ in which *r* is the distance from the core of a product to the surface. *Bi* is the relation between the external heat transfer by convection and the internal thermal conductivity. Sometimes this number is used in combination with the Fourier number *Fo*, where $Fo = at/r^2$, in which t = cooling time in seconds and a = the thermal diffusivity, which was defined earlier.

In order to calculate the pull-down time of a product, instead of using Plank's equations described in chapter 8, we can calculate *Bi* and θ where:

$$\theta = (T_{\text{final}} - T_{\text{fluid}})/(T_{\text{initial}} - T_{\text{fluid}})$$

Graphs are available that give *Fo* by plotting *Bi* on the base and θ. With *a* calculated, the time *t* can now be found.

5.2 Evaporators: Technology, Design, Selection and Applications

After compressors, evaporators are the most important components of the refrigeration plant. In fact one could say that they are the most important, since they transmit the 'cold' directly to the material that we want to cool.

They are the components which finally determine the success of the conditioning process. On top of that, they are the most expensive part of the installation so far as the investment cost is concerned and, therefore, also the most likely subject for price manipulation.

It is necessary that we pay attention to factors such as:

1. definition of 'dry' and 'flooded' evaporation
2. definition of specific heat load
3. definition of d*T*
4. definition of *U*-value
5. concept of matter transfer
6. hoar-frost and defrosting of coolers
7. internal pressure drop
8. positioning of fans
9. selection of manufacturers' data sheets
10. air flow in the coldstore
11. wet-air cooler
12. liquid chillers
13. ice accumulators.

Great attention must be given to the problems concerning air coolers, since this area is the most complicated. Originally, these coolers consisted of a heat-exchange surface built up from coils of steel tubes, through which the refrigerant passed. In the most primitive applications, they were placed in one or two layers against ceiling and/or walls of the cold room.

These were the so-called natural convection evaporators or coolers. In low temperature rooms it was a problem to defrost them efficiently, so they were mostly covered up with a layer of ice, which had of course a negative influence on their performance. Energy consumption became very high. Ice could only be removed if the coldstore was partially emptied, which increased costs considerably.

Since the *U*-values were low, between 10 and 15 $W/(m^2 K)$, large surfaces were needed. Because of the natural air flow, it was however possible to use a large temperature difference between coil and room air, which partially compensated for the low *U*-value. Because of this, coils with forced air flow were constructed by placing a set of coils in a steel casing and connecting a fan. Using air velocities of 5 m/s, *U*-values were increased to between 20 and 25 $W/(m^2 K)$ for ammonia, with cross-flow and a twisted tube pattern. The 'tubes' were placed at 110 mm (or 55 mm) centre distances. When using CFCs the *U*-values were 20–25% lower. It was possible to use high air velocities as there was little resistance over the coils.

Next it became possible to install drain pans and to defrost regularly, mostly by using water-spray systems. The coolers still occupied much space and were mostly placed outside the cold room itself, and they were still expensive. Subsequently, heat-exchange surfaces were built more economically by putting fins on the coils. In fact, a cheap secondary surface made from thin metal sheet was added to the expensive primary surface of smooth tubes. This construction was also more compact. Pressure drop on the air side increased however, and air velocities were limited, in order to reduce heat loss and energy consumption of the fans. Generally, a fan power of not more than 3% of the total heat load of the cooler is allowed. For special applications however, fan power can be about 10%.

Another reason for controlling velocities was to limit the velocity of the outlet air, since false ceilings or air-ducts were no longer used and long air jets were not yet in fashion. Figure 5.2 shows a modern air cooler.

It should be remembered that the gravity-type evaporator still has its uses – with finned coils, built as compact units with drip trays. These are now used when forced air flow is impractical or only partly allowed, as for instance in display cabinets. In order to stimulate some natural air flow, the distances betwen coil-set drip pans and air deflectors must be in a prescribed relationship with each other, as is shown in figure 5.3.

Air inlet and outlet dimensions, *W* and *K*, are about 1/7 of *D*. The drip pan has a slope of 1 in 6. Both the drip pan and the vertical casing are insulated in coldstore applications. Air inlet is on the side facing the door opening, so that incoming air moves directly to the coils.

We now return again to the subject of coolers with forced air flow. A cooler with a big secondary surface compared with the primary one is

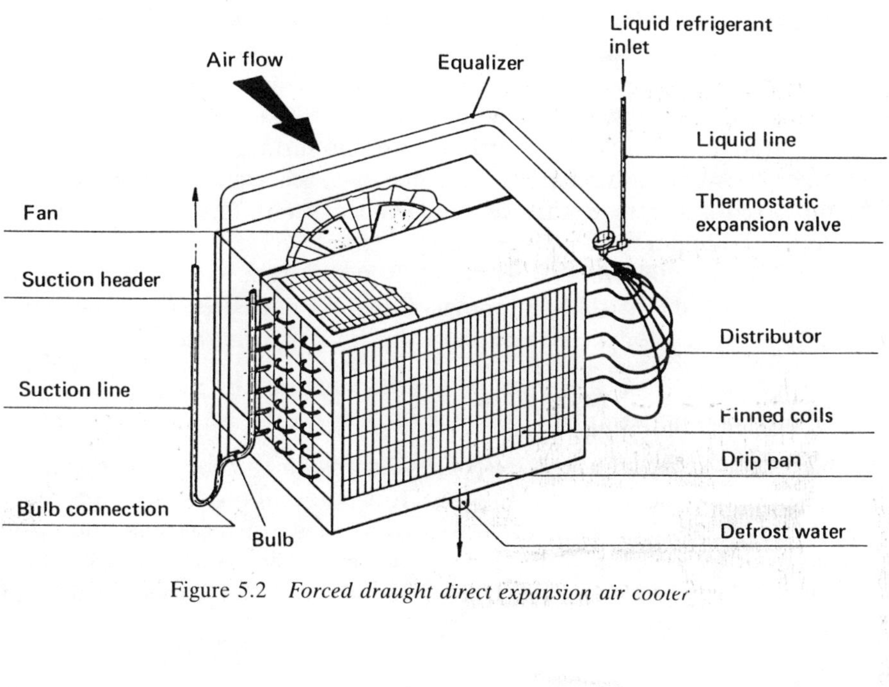

Figure 5.2 *Forced draught direct expansion air cooler*

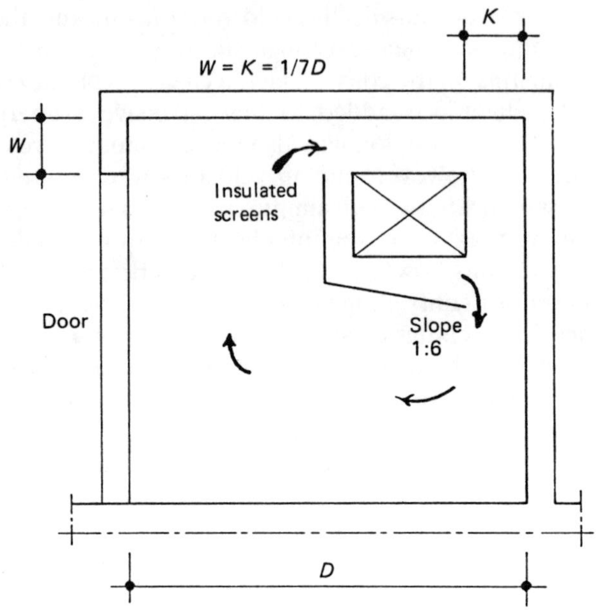

Figure 5.3 *Recommended location for gravity-type air cooler showing airflow pattern*

cheap to build. When the fin spacing is narrow, ice will quickly close the fin space and frequent defrosting becomes necessary. Coolers using CFCs or brines usually have narrow fin spacing.

Judging a cooler on its total surface only is not adequate. Attention should be paid to the heat transfer over the fins. Fins are pressed mechanically or galvanized onto the tubes, and the heat transfer properties depend greatly on how this is done; in effect on the efficiency of contact between the tube and the fin. Furthermore, the surface temperature on the fin is not constant, and certainly different from the surface temperature of the tube itself. Of course, surface temperature also differs greatly from the evaporating temperature or brine temperature.

Therefore a certain fin efficiency, η_f, is introduced, which enables us to calculate using dT at the beginning of the fin and later on correct the calculated surface for the deviation of the fin temperature with η_f:

η_f = the actual heat transfer of the fin/the ideal heat transfer of the fin if the temperature was equal all over the fin surface

Shape, shape ratios, the relation fin dimension to tube dimension, fin spacing and thickness – all these are factors determining the fin efficiency.

There is a big difference in construction between ammonia and CFC evaporators. The former are built from steel coils and fins, then hot-galvanized after manufacture. The latter are built of copper coils and aluminium. The fins are also made from copper when they are to operate in an aggressive environment, and sometimes the whole coil-set is coated with tin or an epoxy finish.

As the reader will have noted, we sometimes use the word 'evaporator' and sometimes the words 'cooler' or 'air cooler'. This is common practice. It would be more accurate to use the word 'air-cooler' when we speak about an evaporator used to cool down an air stream. An air cooler is not always an evaporator, as sometimes brine or ice-water is used as a refrigerant; there is no evaporation but only the exchange of sensible heat of the brine or the ice-water.

Definition of 'Dry' and 'Flooded' Evaporators

As happens many times in the refrigeration industry, the commonly used names are not the most appropriate. A 'dry' evaporator suggests a process without liquid, which of course is impossible. What is really meant is that the gases in the final part of the coils are dry or (almost) free of liquid drops. In contrast, in the 'wet' or so-called 'flooded' evaporator, the outlet gases are saturated with a mist of fine liquid drops. It would be more accurate to speak of coolers with 'complete evaporation' and 'incomplete evaporation'. 'Flooded' evaporators are fed with an overdose of liquid, from which only a part (20% or 25%) is evaporated when the refrigerant leaves the coils, and only the part that must be evaporated to satisfy the heat load. The other liquid refrigerant serves to keep the inner surface of the tubes wet, increasing the internal heat transfer, and at the same time washes the oil away.

The over-flooded operation is obtained by using forced circulation by pump or by gravity feed. This will be considered in more detail when we discuss refrigerant pumps.

If a static pressure is used, the separator is placed at a level above the evaporator which compensates for the pressure drop over the evaporator.

The refrigerant level in the separator is maintained by feeding the separator via a floating valve. For 'complete evaporation' the supply of refrigerant to the evaporator is controlled by a throttling valve, usually a thermostatic expansion valve. This valve lets an exact amount of refrigerant pass, which must be evaporated because of the given capacity demand. The end tubes of the evaporator are indeed almost 'dry' and serve as a superheater compartment. So without the use of a big liquid separator, the compressor will receive dry gases and no liquid. As we will see later on, superheat is necessary to activate the thermostatic expansion valve.

Superheat can be divided into 'used' superheat and 'wasted' superheat. The first category refers to that part which absorbs the thermal energy of the fluid we want to cool, in the evaporator and in the adjacent pipework.

The second category refers to the heat lost outside the evaporator and coldroom, for instance in the suction line leading to the compressor or in the motor windings of semi-hermetic or hermetic compressors. The temperature of superheat varies between 4 and 8K; wasted superheat can go up to 20K, for example at low evaporation temperatures, in long suction lines or in gas-cooled electric motor windings.

It is clear that the 'dry' system will result in a lower heat transmission coefficient than that of the 'flooded' system. The 'dry' system is mostly used in CFC plants, although the same principle applies to ammonia plants. When using CFC refrigerants, which are rather expensive, it is an advantage if this system were to use a relatively small amount of refrigerant.

Speaking of 'dry' and 'wet' evaporators can lead to confusion with the so-called 'wet-air coolers', described later, which transfer the thermal energy by intermediate action over cooled water.

Definition of Specific Heat Load q (W/m^2)

$$q = U\mathrm{d}T(\mathrm{W/m^2})$$

The specific heat load q per unit of a heat-transferring surface is a measure of the intensity of the supplied or transferred thermal energy per time unit. In daily practice, q is used for preference. It is in fact a combination of the U-value and the temperature difference between the refrigerant and the medium to be cooled. The U-value is difficult to use as a characteristic of an evaporator, as it depends not only on the construction, but also on the air velocity over the surface and the condition and behaviour of the refrigerant in the tubes, depending itself on the pressure drop in the coils; it also depends on the positioning of the thermostatic expansion valve or setting of the super heat.

The value of q is determined by the value of the internal heat-transfer coefficient α_i. With increasing α_i and mass stream ϕ, q increases too. See figure 5.4.

In figures such as 5.4, which have been established for only one set of experimental conditions, such as a certain tube section, refrigerant and evaporating temperature, q values are given for various mass streams of refrigerant in the tube and a range of α_i values. Figure 5.5 shows, by means of a correction factor, the influence of different tube sections on α_i from the previous figure 5.4.

The situation is very complicated, mainly because of the phenomenon of 'nucleate boiling' about which little is known at present. During evaporation or 'boiling' of a liquid in a tube, the behaviour of the fluid varies with different temperature differences dT and q values.

Nukijama established figure 5.6 in which we see three different sectors with different behaviour patterns during 'boiling'. Below a certain value of

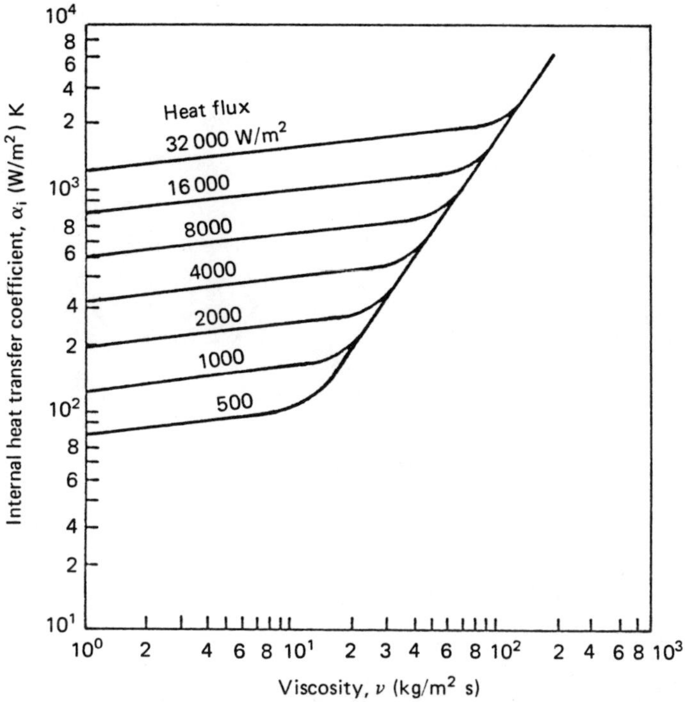

Figure 5.4 *Effects of viscosity and heat flux on internal heat transfer coefficient for R22 evaporating in a tube at −30°C*

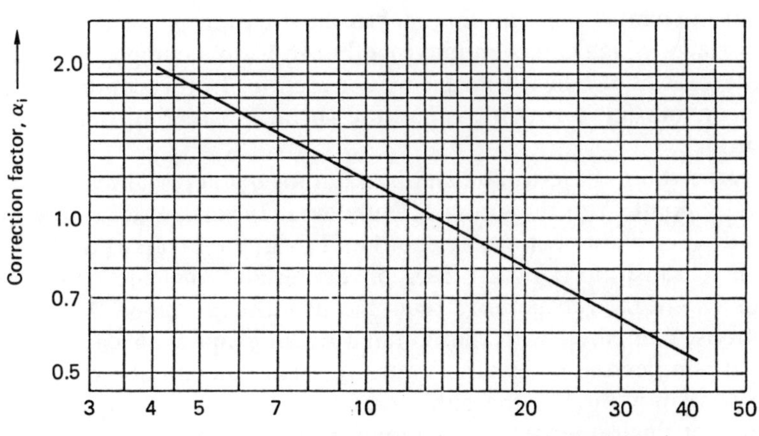

Figure 5.5 *Effect of tube diameter on internal heat transfer coefficient,* α_i

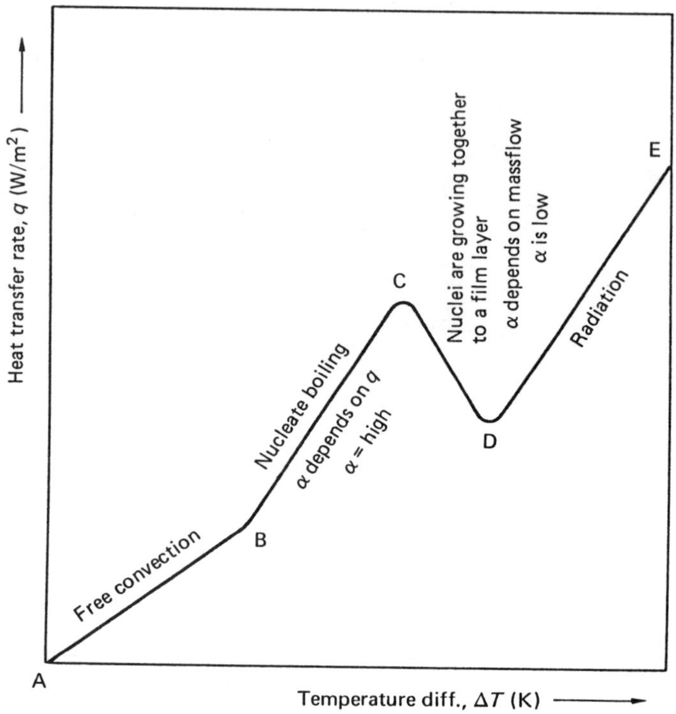

Figure 5.6 *Heat transfer from a heated surface towards a saturated liquid*

dT between A and B there is free convection and no gas bubbles are created. From B to C we come to the section where nuclei begin to form, the so-called 'nucleate boiling', and this creates turbulence which increases the heat transfer. The turbulence is increased (see explanations about heat transfer in chapter 11), and that means the creation of gas pockets on the tube surface, which can even join together to form a complete gas layer or film, decreasing the heat transfer by convection. In this condition there is only heat transfer between gas and the surface instead of between liquid and surface.

So somewhere there will be an optimum for q at a certain value of dT as shown by point C on the graph of Nukijama. In refrigeration technology, dT always lies in the sector between B and C. In the first part of this sector, there is a transition between free convection and nucleate boiling where the velocity of the refrigerant stream has its influence on α_i. At smaller dT conditions, compared with the nominal dT in evaporator catalogues, the mass stream decreases, as does the refrigerant velocity as well as α_i.

We shall not discuss the subject of how to calculate α_i any further; instead, for this rather theoretical matter, we refer the reader to the VDI *Wärme Atlas* or other specialized literature on heat transfer.

What we would like the reader to retain from this theoretical review is: when dT increases, q increases too, and when α_i (or U-value) increases, q increases more than proportionally with dT. When reading manufacturers' catalogues it is advisable to take care in the selection of evaporators with dT values other than those of the basic dT of the tables, because the actual capacity may not be proportional to the dT used.

Definition of dT

There is much misunderstanding concerning this simple, but important concept; for this reason we shall provide further explanation. The misunderstandings are created because of the fact that there are different definitions of dT.

First there is the logarithmic mean temperature (log dT):

$$\text{Log d}T = \{(T_{a1} - T_e) - (T_{a2} - T_e)\}/\ln\{(T_{a1} - T_e)/(T_{a2} - T_e)\}$$

where T_e = evaporation temperature
T_{a1} = air-inlet temperature
T_{a2} = air-outlet temperature.

First of all we have to agree on the definition of T_e, because about this, different points of view exist. Some call it the saturated gas temperature. This is not correct, or at least not quite complete. It should be defined as the saturation temperature at the inlet of the compressor.

It is important to note that this is not the temperature of the inlet gases of the compressor, as the gases at this point are superheated. The real evaporation temperature is the one where evaporation really takes place, that is, in the evaporator, and this temperature is not constant there. In the distributor of the expansion valve and in this valve itself there is a pressure drop, and an even greater one in the tubes of the evaporator. When the pressure varies, the evaporation temperature varies too. We cannot do calculations on the basis of a temperature that can vary all the time, therefore the above proposed definition was chosen.

This temperature is easy to check and the plant design engineer does not need to take into account the pressure drop over the evaporator.

Some manufacturers use another definition of dT, the arithmetic mean temperature difference (dT_{mr}):

$$\text{d}T_{mr} = \{(T_{a1} - T_e) + (T_{a2} - T_e)\}/2$$

The following definition is also used:

$$\text{d}T_1 = T_{a1} - T_e$$

and called the air-in temperature difference. In this case T_{a1} is considered to be the coldroom temperature, but this is not quite correct. There is no such thing as 'a' room temperature, because the temperature differs from place to place in the room.

Some manufacturers use as room temperature $(T_{a1} + T_{a2})/2$ and define dT as $\{(T_{a1} + T_{a2})/2\} - T_e$, which is another presentation of dT_{mr}.

We recommend considering T_{a1} as the room temperature.

All this should make it clear that, before using a selection of table for evaporators, make absolutely sure which definitions are being used by the manufacturers under consideration.

Now here is an example using the following temperatures: $T_e = -25$, $T_{a1} = -18$ and $T_{a2} = -21$°C. Then log dT = 4.6°C, dT_1 = 5.5°C, dT_{mr} = 7°C and dT = 5.5°C.

The temperature difference between the outlet air temperature and the evaporation temperature must not be chosen outside the range 2–4°C for coldstores used as storage rooms. Larger differences give a smaller air volume, meaning smaller fans, but certainly also a bigger log dT, which means more weight-loss from the goods stored and a lower evaporation temperature, resulting in more power consumption.

Normally, the difference between air inlet and outlet temperatures is chosen between 2 and 3°C, which results in negative temperatures in an air volume equal to the refrigeration capacity. An air cooler with a large air volume and a small dT has a large outlet area and a small depth, and vice versa.

Definition of the Overall Heat-transfer Coefficient

As mentioned above, the thermal conductivity of a single layer of material has the symbol λ in Europe and k or λ in the United Kingdom and k in the United States. The overall heat-transfer coefficient has the symbol k in Europe and U in both the United Kingdom and the United States. U has therefore been chosen to reduce any confusion in the subject.

The inverses of heat-transfer coefficients can be considered as individual resistances against heat transfer, and, as in the analogy with electrical resistances, they can be added to produce the total resistance $R_{total} = R_1 + R_2 + R_3 + \ldots$. This means that the overall heat-transfer coefficient U of a wall, taking into account the heat-transfer coefficients α on both sides of a wall and λ/δ of the wall, can be determined by the equation:

$$1/U = 1/\alpha_{external} + \delta/\lambda + 1/\alpha_{internal} \ (\mathrm{m}^2\ \mathrm{K})/\mathrm{W}$$

The determination of $\alpha_{external}$ is as difficult as for $\alpha_{internal}$. It depends very much on the construction of the evaporator, tube distance, tube pattern, tube section, fin space, fin shape, air velocity in the cooler, state of hoar-frost on the tubes and the relation between the latent and sensible heat of the air; all these play their part. Calculations are based on empirically determined data.

Finally we can calculate U per metre length for a curved surface such as a tube from:

$$1/U = 1/\pi d_e\alpha_e + (\ln[d_e/d_i])/2\pi\lambda + 1/\pi d_i\alpha_i \text{ (m K)/W}$$

For α_i we have to consult existing tables or graphs in the specialist literature (see above)

and d_i = inside tube diameter

d_e = outside tube diameter

λ = coefficient of thermal conductivity of the tube material.

Note that a simplified equation for the U-value of finned evaporator tubes is given by:

$$1/U = A_r/A_i\alpha_i + 1/(\alpha_e[1 - \{A_r/A_i\}(1 - \eta_f)])$$

in which A_i and A_r are the inside tube surface and finned surface areas respectively.

Calculation of the surface area for natural convective heat transfer

Newton's empirical law of cooling states that

$$Q = \alpha A \mathrm{d}T \text{(W)}$$

which is similar to Fourier's law. A can be calculated from

$$A = Q/\{\alpha(T_2 - T_1)\} \text{ m}^2$$

Concept of Mass Transfer

When speaking about air coolers and cooling towers we are not only dealing with heat transfer but also with mass transfer. The mass here is water. The air used to cool contains a certain amount of water vapour, which will condense on the outside of the heat-exchanging surface when the surface temperature is below the dew point. In most cases in the refrigeration process, we will have to deal with such a wet heat-exchange surface.

By analogy with the theory about heat transfer in which we use a heat-transfer coefficient α, we use a mass transfer coefficient σ. Mass transfer takes place between the air in a boundary layer just above the film of condensation on the tube or fin. σ = water vapour transmission factor, expressed in m/s at 1 kg difference in absolute moisture contents, dx. So

$$\mathrm{d}m = \sigma(x - x_k)\mathrm{d}A\rho$$

where x = moisture content of the air in kg/kg at a certain air temperature t

x_k = saturated moisture content in the boundary layer
A = surface area.

Lewis's law says:

$$\sigma = 0.622\alpha/\{c_p\rho(p_t - p_w)\}$$

in which p_t = atmospheric pressure and p_w = water vapour pressure. Lewis stated that $\sigma = \alpha/c_p\rho$ when there is turbulent streaming of humid air. This is only approximately correct at air temperatures above 0°C.

In most of the literature you will find β used instead of α when dealing with this subject. β is the rate of water vapour transmission in kg/m² s at a pressure difference of 1 Pa. $\sigma = \beta/\rho$. Lewis stated that $\beta = \alpha/c_p Le$ and that above 0°C, $Le = 1$, so $\beta = \alpha/c_p$ at that condition, or $\sigma = \alpha/c_p\rho$ ($\sigma\rho$ = m/s. kg/m³ = kg/m²/s).

α and β can also be defined as the quantities of air that must flow over a humid surface to saturate it. Rather than considering these as transmission values, they can be considered as a vapour deficiency. This vapour deficit is proportional to the existing vapour pressure deficiency. So instead of calculating with the difference in moisture content and $m = \beta A\mathrm{d}x$, we can also calculate with the difference in vapour pressure dp. The transferred volume of water $m = \beta A\mathrm{d}p$. This consideration is easier to understand by analogy with heat transmission.

From the above described values and equations, Merkel derived an equation for the total heat transmission Q_t that included vapour transmission, latent heat Q_l and sensible heat Q_s:

$$Q_t = \beta(h - h_k)A$$

We know that the cooling line can by approximation be considered as being a straight line between the air inlet and the air outlet condition in the Mollier graph. If the line was completely vertical, it would mean that only sensible heat would be taken out. We could then calculate simply $Q_s = \alpha(t - t_k)A$, with a so-called 'dry' α. β does not come into consideration. See figure 5.7, line A–B.

As the line A–B deviates further from the vertical and the angle β becomes progressively less than 90°, so α will be more affected by moisture exchange. The cotangent of the angle, $(x - x_k)/(t - t_k)$ is a measure for the deviation of α from α_{dry}. α_{wet} differs from α_{dry} by the ratio $(x - x_k)/(t - t_k)$. This ratio determines the relation between sensible and latent heat Q_l/Q_s. Since $Q_l = LA(x - x_k)\beta$ then:

$$Q_l/Q_s = \{LA(x - x_k)\beta\}/\{\alpha A(t - t_k)\} = (L\beta/\alpha)\{(x - x_k)/(t - t_k)\}$$

and since $\beta = \alpha/c_p$ as we saw before, we can write:

$$Q_l/Q_s = (L/c_p)\{(x - x_k)/(t - t_k)\}$$

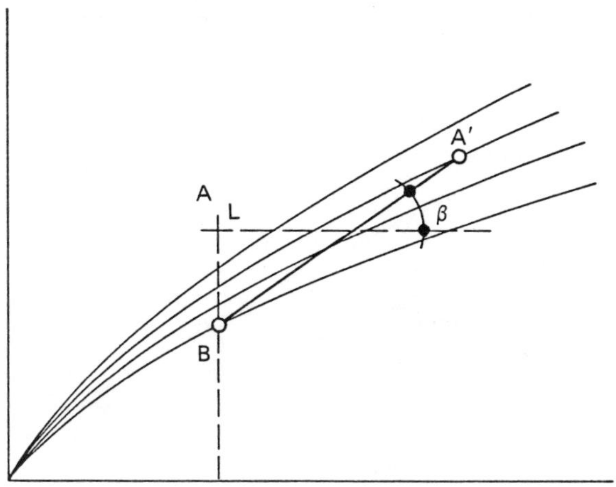

Figure 5.7 *Slope of cooling line on Mollier chart*

so

$$\alpha_{wet} = \alpha_{dry}(L/c_p)\{(x - x_k)/(t - t_k)\}$$

and

$$\alpha_{total} = \alpha_{dry} + \alpha_{dry}(L/c_p)\{(x - x_k)/(t - t_k)\}$$

The subject, however, does not end with this complication. In many cases the condensed water will be changed into ice, when the theories no longer hold. When only a small amount of hoar-frost forms we can make an approximate calculation by adding $\lambda_{hoar\text{-}frost}$ to $\delta_{hoar\text{-}frost}$. However with thicker layers this approximation is not valid. In this case, experimentally established values must be used for hoar-frost.

In the section below dealing with hoar-frost and defrosting problems of evaporators, we will discuss this subject in detail.

This theory as explained so far will indicate to the reader how careful one must be when judging the refrigeration capacities of air coolers given in manufacturers' data sheets. We must know whether the given Q values are based on a dry, a wet or a hoar-frost situation, and in the last case with little, medium or heavy hoar-frost, as shown on figure 5.8.

Q can be any of the following:

Q^* = 'dry' capacity.
Q_0 = nominal capacity, meaning gross capacity half-an-hour after the start of the cooling period. Gross capacity is a capacity that includes the (wasted) capacity for the fan motor cooling.
Q_k = The net average capacity during the cooling period.

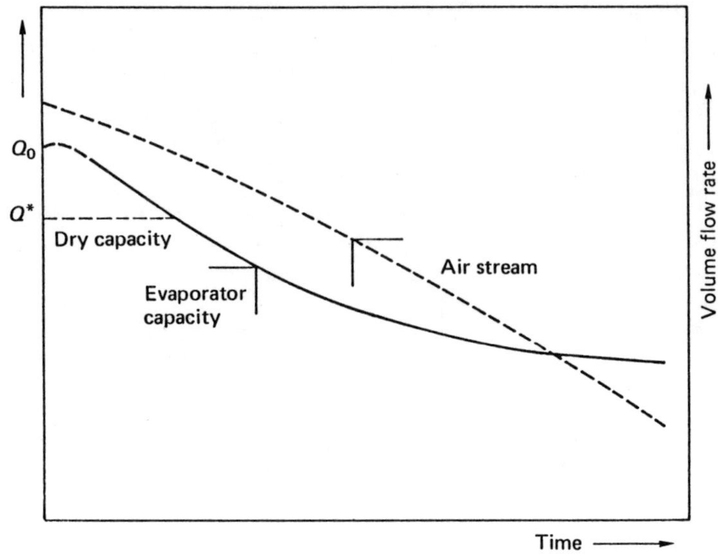

Figure 5.8 *Graphs of capacity and air volume flowrate dependence on hoar-frost thickness*

Q_e = The effective average capacity during the cooling period and the following defrosting period together.

In The Netherlands, an attempt is being made to standardize all these definitions and include them in one specification (NEN 1976).

Hoar-frost and the Defrosting of Air Coolers

At surface temperatures below 273K (0 C) the water vapour will freeze on the coils. We get hoar-frost or even an ice layer. The result will be:

(a) $\alpha_{external}$ will decrease.
(b) The effective free air-passage through the coils and fins will decrease, so the pressure drop of the air increases and the volume flow rate decreases.

For both the above reasons, power consumption of the plant will increase, making regular defrosting necessary. Energy consumption will still go up, though how much will depend on the method used. We can apply, for instance, a method of defrosting that injects hot gases into the coils. These are gases diverted from the hot gas line after they have left the compressor. See figure 5.10 for details of such a system. This defrosting method, acting on the inside of the tubes, is very effective and is a kind of energy recovery. The heat is available in the plant itself. A part of this energy gain will however be lost under certain conditions such as:

(a) Whenever the compressor is obliged to run in order to produce the necessary hot gases
(b) In systems which do not have enough pressure available in winter periods, necessitating the installation of a throttling valve in the hot gas line of the compressor. That means that in winter we cannot take advantage of the lower condensation pressure, which would normally be the case with colder air flowing through air-cooled or evaporative condensers.
    We must however remember that this drawback might also be caused by the need for a sufficiently high pressure in cases where we have to use thermostatic expansion valves in the plant, their capacities depending on the pressure difference before and after the valve. When this difference becomes too low, problems with liquid feeding to the evaporators will occur
(c) In systems which waste defrosting energy, the thermal energy transferred to the air of the coldstore; so this heat is not only wasted for defrosting but also imposes an extra heat load on the plant. This wasted energy can be as much as 10–20%.

Defrosting from the outside to the inside with electrical resistors always costs, of course, more energy. This electrical energy must be supplied from the outside. In special cases, if the coldstore temperature is above +5°C, defrosting can be done partly or completely with store air, which of course is interesting from an energy point of view. Systems that defrost by means of water or by continuously spraying glycol solutions have practically disappeared because of the maintenance problems involved.

Systems with a reverse cycle are not interesting from an energy point of view. In order to defrost, the compressor has to run just to produce hot gases, and energy is used and wasted. For freezing tunnels there is a system that avoids the build-up of hoar-frost. In this system, compressed air is blown by a nozzle system which moves backwards and forwards in front of the cooler, and the ice is blown away as soon as it is formed.

The old fashioned 'wet' cooler, which is now in use again for cooling vegetables and flowers, is free of defrosting problems, as well as 'dry' coolers working at a temperature above 5°C (272K).

Altogether, defrosting is a considerable inconvenience, because:

- It disturbs the temperature and humidity in the coldstore, if not by adding heat and moisture, then in each case by interruption of the cooling cycle.
- It brings wasted defrost-energy into the cold store.

- A chimney effect blows freezing mist on to the walls and ceiling.
- The fans require a delayed start after a defrosting period, otherwise they would blow moisture into the room.
- The heat of the warm air expands the air in the cold store in such a way that intolerable pressure occurs on the walls and ceiling; a problem that is only solved by taking special precautions, as described in chapter 10.

To determine the period and frequency for optimum defrosting is very difficult. Experiments done at the Univesity of Delft in The Netherlands showed that frequent defrosting gives an increasing amount of wasted defrost heat, but when defrosting takes place less frequently the thickness of hoar-frost will increase. Figure 5.9 shows results of this work, which attempted to establish an optimum frequency for defrosting. The figure shows that with a lower defrost capacity, the Q_e is reduced and the optimum defrost cycle becomes longer. The sharply curved shape of the lines on the left side shows that the influence of the wasted defrost energy is proportionally larger for frequent defrosting.

The best frequency found seems to be at intervals resulting in 6–8 defrosting cycles per 24 hours. Defrosting periods with hot gases were determined at between 15 and 20 minutes, while for electrical defrosting the figure was about 30 minutes.

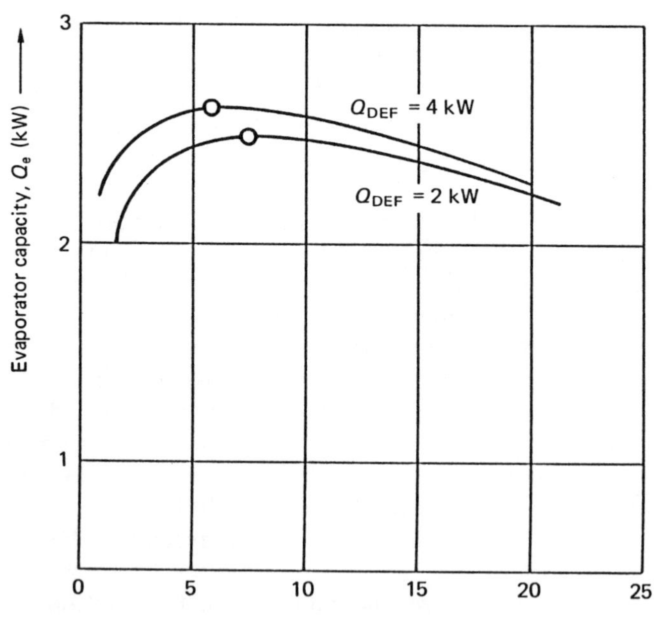

Figure 5.9 *Variation of evaporator capacity* Q_c *with total cycle time (total cycle time =* $t_{cooling} + t_{defrost}$ *(Courtesy Van Male, The Netherlands)*

How does one determine the right moment to initiate the process? Today, time-regulation with a clock is the most widely used, the defrosting being terminated by means of a suitable thermostat – although this system is unattractive from the energy point of view. This method is usually backed up by an additional fixed time-sequence on the clock.

Energy-wise, more efficient systems are based on the measurement of the ice thickness, the increase in pressure drop of the air or the decrease of that of the air stream, or measurement of the increase of dT between air and fin temperatures; even weighing the cooler has been used.

In computerized plants the so-called may–must system can also be used. With this system the use of hot gases for defrosting can be maximized. The computer receives a signal from coolers which need to be defrosted. This signal is sent by a sensor on the fin surface. These coolers have priority in the program, but if there is excess hot gas available it is used to start a defrosting cycle on other coolers; the latter are coolers which could benefit from defrosting but do not need it urgently. The sensor of the cooler is also used to control the hot-gas stream and to give the order to terminate the defrosting.

The may–must system could be used elsewhere in the plant, for example, to maximize the use of the compressor capacity in controlling the actual room temperatures.

Finally, to complete this subject, we show in figure 5.10 an example of a hot-gas defrost system, and the adjacent pipework in a typical refrigeration plant.

Internal Pressure Drop

Figure 5.11 shows the effects of temperature drop of the air and refrigerant through a direct expansion air cooler. Because of the flow of refrigerant inside the evaporator tubes, friction occurs and the pressure drops. At the end of the evaporator tube or coil, there will be a lower saturation temperature compared with the beginning.

When using a thermostatic expansion valve with a distributor, which directs the flow of the refrigerant equally into the different circuits, a pressure and corresponding temperature drop from t'_2 to t_2 occurs. From this point the pressure decreases further to point 3, where superheating starts and the temperature rises to t'_3. In this last part of the circuit the pressure drop is negligible and it is generally considered that the evaporation temperature is equivalent to the temperature t_{r2}. This is the evaporating temperature that was discussed earlier in the book.

The reader should be aware that not all cooler manufacturers use this definition for the evaporating temperature, and not all of them give in their data sheets the amount of superheat applicable.

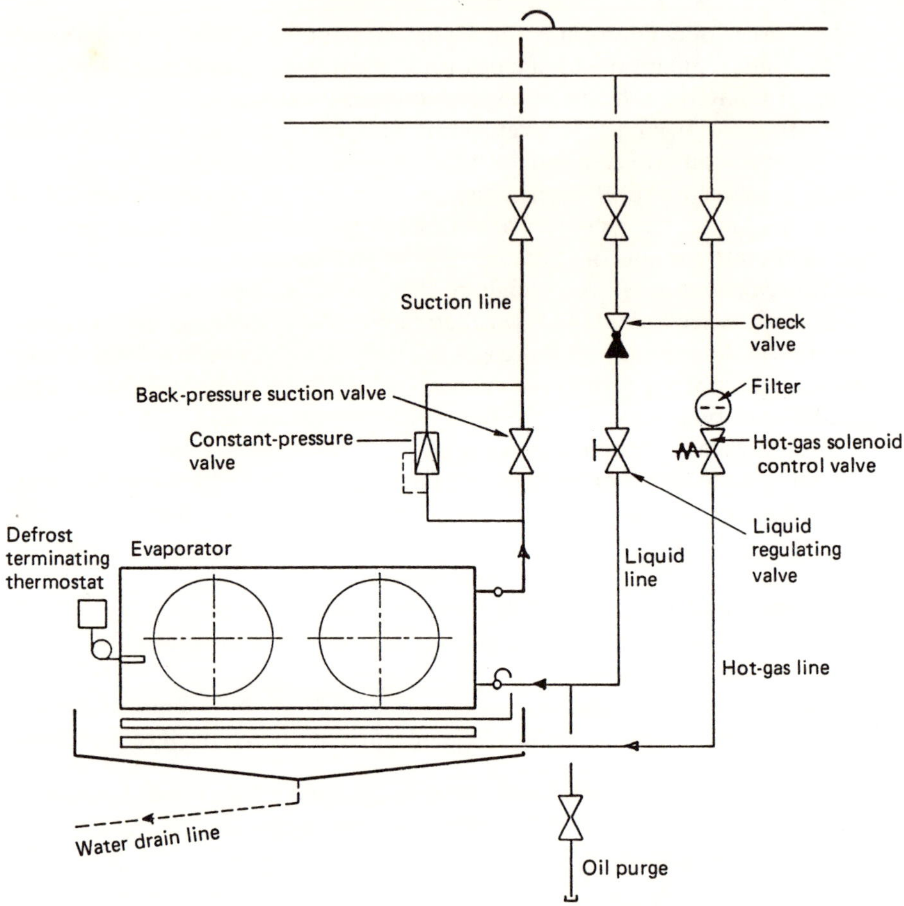

Figure 5.10 *Ammonia hot-gas defrost system*

The differences that might be encountered can be seen in figure 5.11, where it is apparent that the hatched surface (A) reflects the actual cooler capacity. This surface can differ much depending on the position of S, since lower superheat means a bigger surface.

In daily practice, we are always bound to work with approximately 6K superheat. Also remember that pressure drop increases with increasing refrigerant velocity, although this situation does increase α_{inside}; but somewhere there is an optimum.

The value of pressure drop in R717 plants is the equivalent of between 1 and 3K, while for R12 and R22 it is 4 and even up to 6K. The value of the pressure drop is determined by the length of the tube per circuit, and by the construction of the cooler.

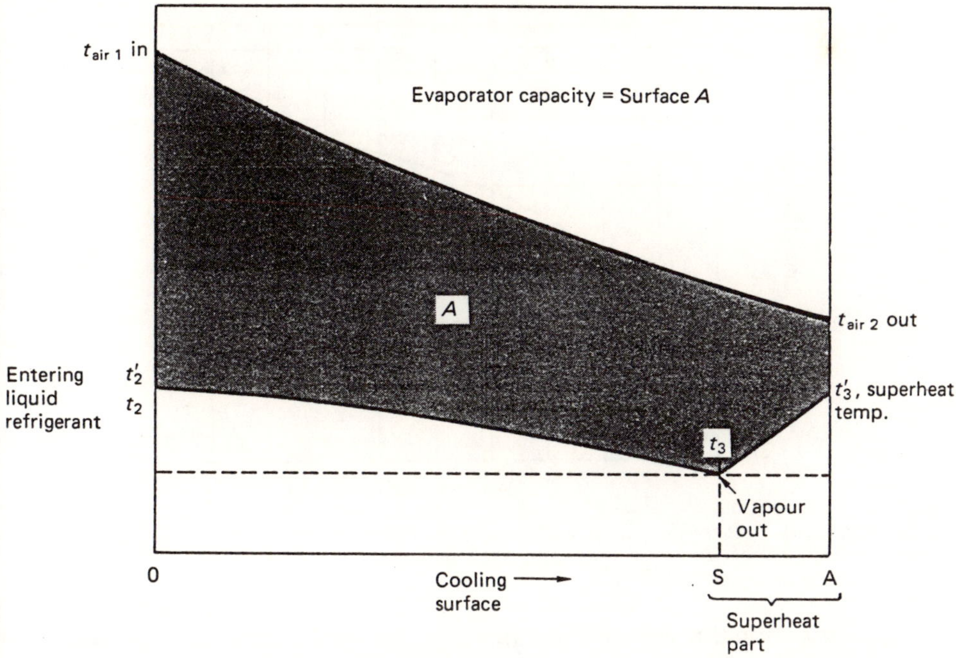

Figure 5.11 *Effects of refrigerant pressure drop through an evaporator on air temperature and refrigerant temperature*

There is a maximum admissible tube length per circuit, which is different for the various diameters and materials of the tubes, their roughness and the refrigerant specific mass. All this has been established by Bernoulli, and expressed as the law of Bernoulli. For instance, for R22 and R502 the maximum length of a $\frac{5}{8}$-inch tube per circuit is between 30 and 50 m.

We must know the value of the pressure drop when selecting a refrigerant pump. Bäckström gives the following simple equation:

$$dp = 0.04k(l/D)(1/\rho)(m/A)^2 \ (N/m^2)$$

in which k is a correction factor available from figure 5.12, and

where ΔT = the ratio of dT at entrance/dT at outlet of the evaporator (normally > 2 in refrigeration practice)
D = tube diameter in m
l = tube length in m
A = tube section in m^2
ρ = specific mass of the refrigerant in kg/m^3.

The manufacturer must ensure that the circuits are well balanced and that the right refrigerant distributor is chosen, to maintain equal refriger-

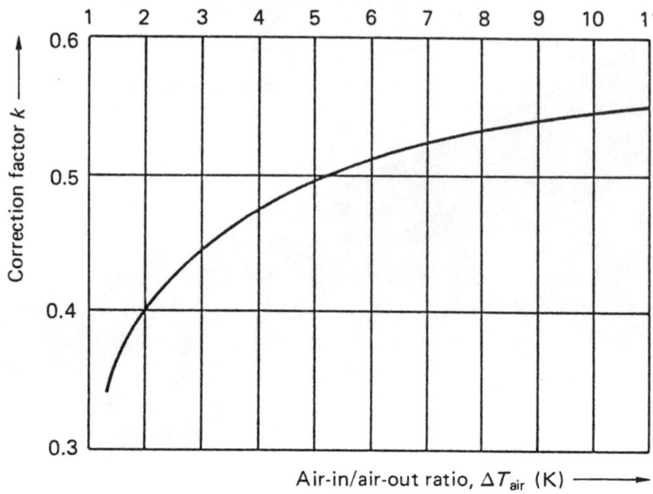

Figure 5.12 *Correction factor* k *for use in the Bäckström equation as a function of air-in/air-out temperature ratio*

ant distribution throughout the cooler. An oversized distributor will lead to an unequal refrigerant distribution and too much subcooling.

The suction manifold must be of such size that the gas flows are equalized before they enter the compressor. The pressure must be equalized, otherwise some circuits will be better fed than others. When the heat load differs from circuit to circuit, for instance because of unequal air flow somewhere in the system, incomplete evaporation may occur and there might be incorrect superheating.

Normally, coolers are bottom-fed, which will given an acceptable α_{inside}. The suction manifold is placed on the air-entering side, which will thus give adequate superheat.

Positioning of the Fans

Figure 5.13 shows the difference that results from putting the fans in front or behind the battery of cooling coils. Coolers with fans in front of the coils, and blowing air through them, have to absorb the fan motor heat, resulting in a larger logarithmic dT. The cooling capacity will be bigger than that in a design where the fans are placed behind the battery of coils, the air being sucked through them.

Sometimes this latter design is preferred, for instance when a high outlet velocity is required to produce a long jet. This also has the advantage of drying the air by passing it over the fan motor at temperatures above zero degrees, and there will be less opportunity for water-drop distribution.

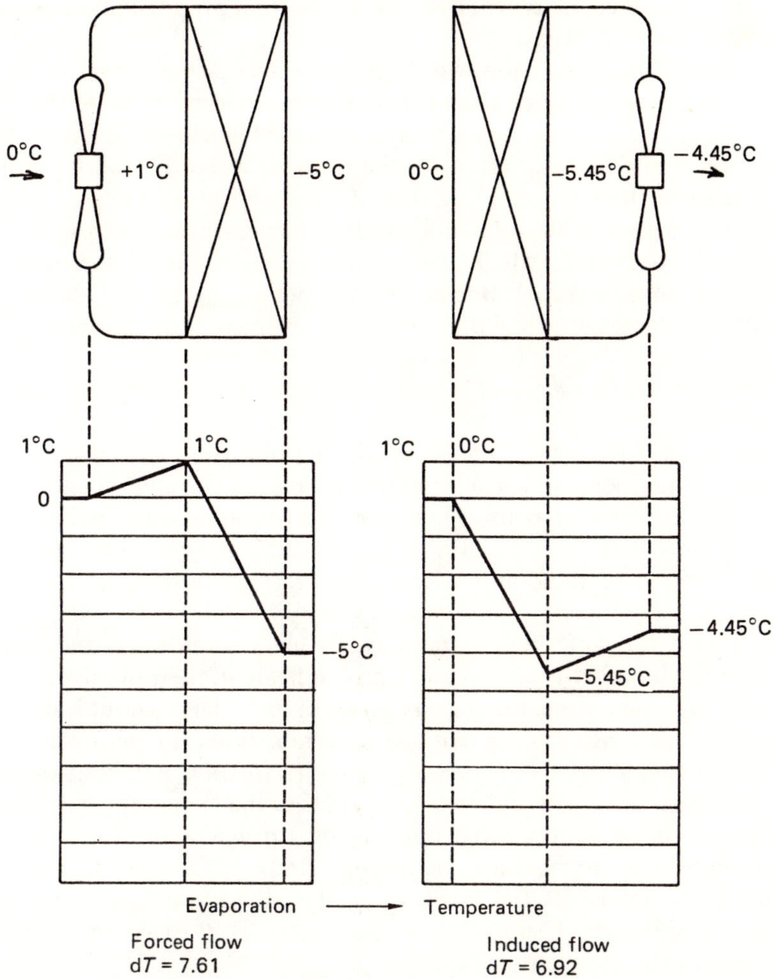

Figure 5.13 *Influence of the position of the fans on the temperature of the air passing through the evaporator*

Selection from Manufacturers' Data Sheets

By now the reader will have learned to look first at the definitions of dT and evaporation temperature used in manufacturers' data sheets. Frequently, the capacity of the coolers is given only at one dT and evaporation temperature. For other conditions, one is obliged to use correction factors taken from tables. It is quite certain that capacity does not change in direct proportion to changing temperature difference. On decreasing dT, α_{inside}

decreases too, so capacity will decrease more than proportionally with dT. The U-value also decreases.

The internal pressure drop will decrease at decreasing capacity, and the picture of the variation of the evaporating temperature inside the evaporator shown in figure 5.11 will change also; the logarithmic dT is inversely proportional to dT. At a dT of 10, the ratio between dT and logarithmic dT, which was 0.89 degree at dT 7, will increase to 0.915 degree, but the differences between air inlet and outlet temperatures and evaporation temperature decrease, which will make the SHF factor increase (less latent, more sensible heat). Because of this fact, α_{inside}, the U-value and the capacity will decrease; when more hoar-frost appears, the influence will be less important.

Extrapolation, by means of dT, can only be done by using experimental correction factors.

Capacity also does not change in direct proportion to changes in the evaporating temperature. The manufacturer must specify which sensible heat factor (SHF) he has used and give correction factors for other SHF values. Often data is based on a RH of 85% for the inlet air and a superheat of between 2.5 and 5K; in daily practice, however, superheat is at least 6K.

The figure for superheat is not always given, neither is the one for subcooling. Manufacturers should clearly indicate in their published literature whether the stated capacity is gross or net. They should also state which hoar-frost situation is being considered. Some manufacturers give more than one table, with correction factors for low, medium or heavy frost. Furthermore, it should be stated whether the capacities are based on the use of thermostatic expansion valves ('dry evaporation') or 'wet' evaporation using refrigerant circulation pumps; and, last but not least, which refrigerant is used. Correction tables should be available for the different refrigerants. Using secondary refrigerants, the capacities will vary at different brine concentrations.

Practical example We wish to select a cooler which can achieve a room temperature of +6.5°C with an evaporating temperature of 0°C and a capacity of 45 000 W.

Manufacturer 1 makes his selection based on dT_1 and RH 80%. His data sheet gives a nominal capacity of 52 600 W. For the evaporation temperature and the dT_1 concerned, the table shows a correction factor of 1.21. The data sheet does not mention hoar-frost conditions.

The corrected capacity is 52 600/1.21 = 43 471 W.

The amount of air over the cooler is 44 200 m^3/h or 12.28 m^3/s. This results in dH = 43 471/(12.28 × 1.26) = 2810 J/kg.

When we put this dh on the cooling line in a Mollier chart, we will find the outlet condition of the air to be +4.1°C and 81% RH.

The second manufacturer uses a dT_{mr} of $(6.5 + 4.1)/2 = 5.3$°C. So $dh = 43\,000/(8.08 \times 1.26) = 4324$ J/kg.

Put on the cooling line from $t_{l_1} = 8$°C and 70% RH, this gives $T_2 = 4.8$°C and RH = 80%.

This is more or less the condition we assumed.

Although both manufacturers may think that they have fulfilled the wishes of their clients, the results so far as inlet and outlet air temperatures and relativc humidities are concerned are very different. In this particular case, the cooling surfaces have different sizes.

Manufacturer 1 proposed a smaller surface, partly explained by the fact that $\alpha_{external}$ was superior because of the larger amount of air and the better fin-efficiency resulting from the use of a twisted tube pattern. For α_{inside} this is less relevant: the tube section was smaller but because of the smaller log dT, heat-flux or specific heat-load must be lower.

Air Flow in the Coldstore

We would now like to draw the reader's attention to the behaviour of the air leaving the cooler. A knowledge of this behaviour will help in choosing the correct position of the air coolers in the coldstore, and we will meet the so-called Coanda effect. Coanda discovered that an air stream, delivered with a high velocity by a fan, has a tendency to create an underpressure above this air stream. So, when the air cooler is hung from the ceiling, an underpressure exists between the air stream and the ceiling.

In this situation, the air stream is held against the ceiling and a laminar streaming pattern results which is influenced by the condition of the friction surface (see figure 5.14).

This Coanda effect is beneficial when air has to be transported over a long distance, so that a high velocity has to be created at the outlet of the air cooler. Furthermore, a high velocity helps to lengthen the jet of the air stream (gravity law), and for the same reason it is better to 'shoot' the air from a high position. See figure 5.15. The contrary also is true, so care must be taken with long coldstores having a low ceiling.

Putting the fans in front of the cooler, thus drawing the air through the coils, the air leaves the cooler with a turbulent air-stream pattern as shown on figure 5.16. This has a negative influence on the Coanda effect, which is best served by a laminar stream-pattern obtained by putting the fans

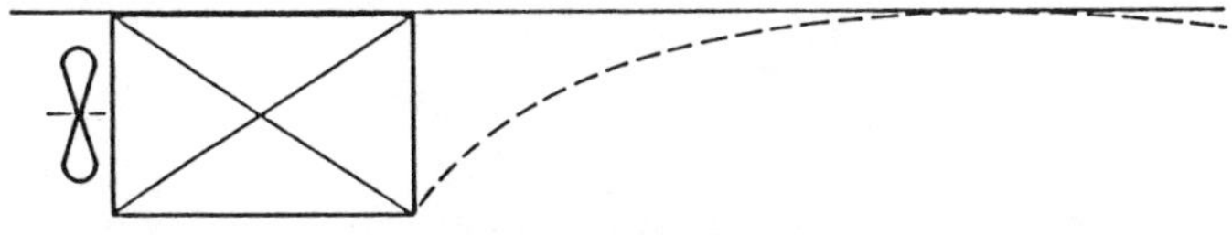

Figure 5.14 *Coanda effect*

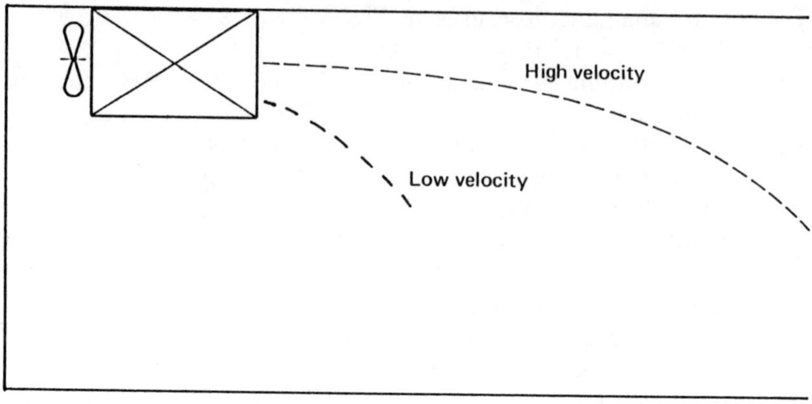

Figure 5.15 *Influence of air velocity on jet length*

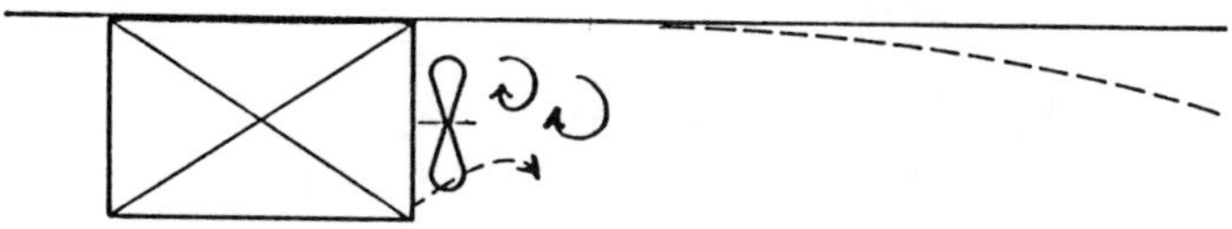

Figure 5.16 *Influence of turbulence on Coanda effect*

behind the cooler. However, this lowers the air velocity, as a big part of the energy is converted into thermal energy by friction and collision. Another disadvantage of placing the fans behind the cooler and pushing air through the cooler is the fact that the air is not evenly spread over the cooler area, especially when the distance between fans and coils is short; even the fins prevent the air from spreading, resulting in uneven charging of the cooler and uneven frost layers. Opinions on this subject differ and the final choice will depend on the total design of the cooler. The power of the fan motors and their efficiency depend on the shape and angle of the fan blades.

In order to intensify the Coanda effect it is interesting to direct the air stream in an upward direction by using a short diffuser or a simple air-screen as shown on figure 5.17. This is very important in long rooms with a low ceiling (see figure 5.18).

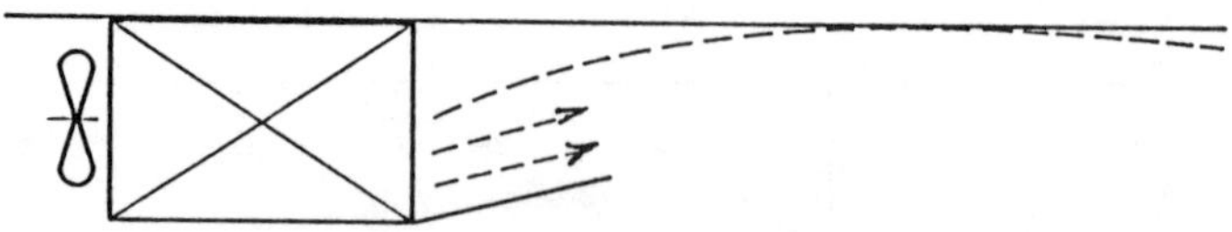

Figure 5.17 *Enhancement of Coanda effect by the use of an air screen*

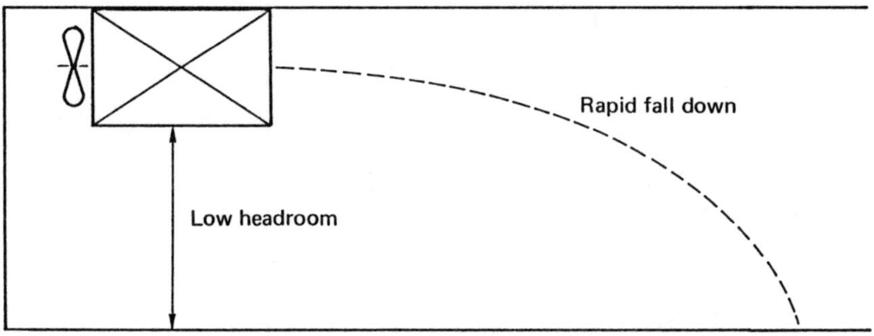

Figure 5.18 *Low ceiling effect on air jet length*

It is not always possible to realize a high air velocity at the outlet, for instance in cases where positive temperatures exist, when there is a danger of water-drop distribution. In those cases, air velocities no higher than 2.5 m/s may be used. Some manufacturers like to work with high air velocities in order to build cheap coolers, with increased $\alpha_{outside}$ and smaller surface, and try to eliminate the water-drop distribution by using water eliminators which increase the pressure drop and give turbulent streaming patterns.

As we saw before, a short diffuser can be added, having a secondary advantage: the outlet section can be optimized, so giving us a high air velocity at the outlet. A cylindrical diffuser should give the best results, though this is difficult to realize from a construction point of view, because of the rectangular shape of the coolers. Therefore, an attempt is made to give it an almost square shape.

In order to create a long air-jet it is also good practice to work with small differences between air-outlet temperature and the temperature of the room, which means a small dT, because cold air has a high density and thus a tendency to fall.

Now have a look at the horizontal dimension of the air stream shown in figures 5.19 and 5.20. When the velocity is low, at the point where the air leaves the cooler, the air stream is very large and it soon spreads out sidewards. The stream has a tendency to stay spread out over a short distance. When the velocity is high, the air stream stays compact over a longer distance and the spreading extends far into the room; this, however, means the sphere of influence in small. The spreading out is influenced by the velocity and the difference in viscosity. In the core of the stream, where the temperature and the viscosity are lowest and where there is no friction with the surrounding air, the pattern is different from that on the sides where the influence of the ambient air on the stream is large. This ambient

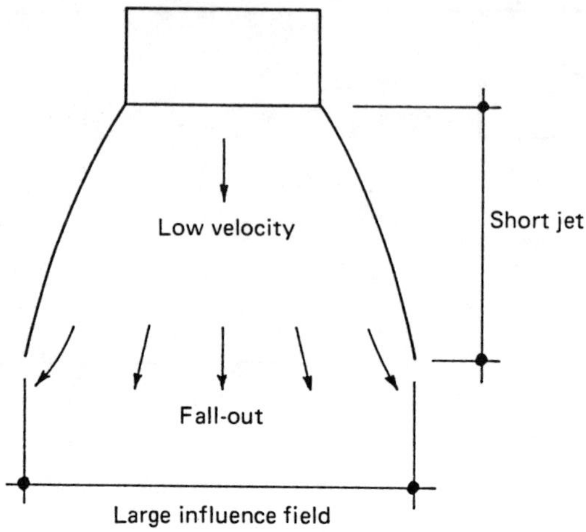

Figure 5.19 *Wide and short jet at low air velocity*

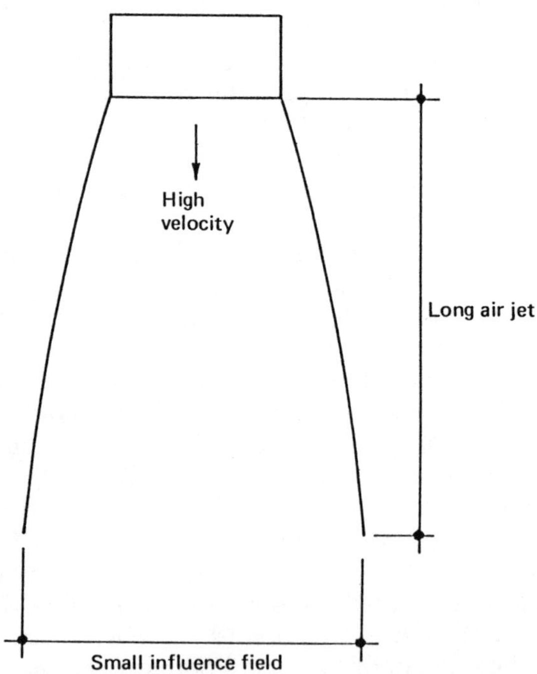

Figure 5.20 *Narrow and long jet length at high air velocity*

air slows down the outside layers and makes the air spread out. The position and degree of this spreading-out depends on the air velocity.

Induction An air stream causes induction streams too. Calm air in the surroundings is dragged in by the air stream which passes with high velocity. The more this induction stream increases the volume and density of the original stream of air, the more the velocity of the stream decreases. The induction streams also have a negative influence on the length of the air jet. When the influence fields in the horizontal level are almost touching one another, there is little space left between the jets for induction streams.

Conclusions In order to create long air jets the following rules must be followed:

(a) High velocities of the air coming out of the cooler or diffuser.
(b) Small dT of the air over the cooler.
(c) In the case of evaporating temperatures above freezing point, the maximum leaving velocity should be 2.5 m/s and a diffuser must be used.
(d) Support the Coando effect with an inclined sheet or diffuser with an upward-directed outlet opening.
(e) A standard modern air cooler has a jet with a length of 15–20 m. When longer jets are needed in the range 25–30 m, special fans with high outlet velocities should be used. When the height of the room is less than 5 m, air-ducts, false ceilings or at least diffusers must be used.
(f) Place the coolers close together. Standard air coolers have an effective width of between 6 and 8 m depending on their front length. In the case of very long jets, leave less space or apply (e).
(g) Sometimes long jets are asked for, but at the same time high air-velocities are not permitted. In that case only ducts or false ceilings can be used. Air velocities in ducts may not exceed 6 m/s.

Wet-air Coolers

Wet-air coolers are air coolers in which the air is in direct contact with a cooling medium, normally water; in the past, brine solutions were also used. Two typical wet-air coolers are illustrated in figure 5.21. Today, the popular medium is chilled water. For this reason, applications are restricted to positive room temperature. Chilled water is sprayed by means of a sparge pipe over a filling mass inside a casing and fans draw the air in counterflow through this wet mass.

The principle is analogous to a cooling tower but with the aim reversed. In a cooling tower the air stream cools down the water; in a wet-air cooler the water stream cools down the air. Wet coolers that have been used in

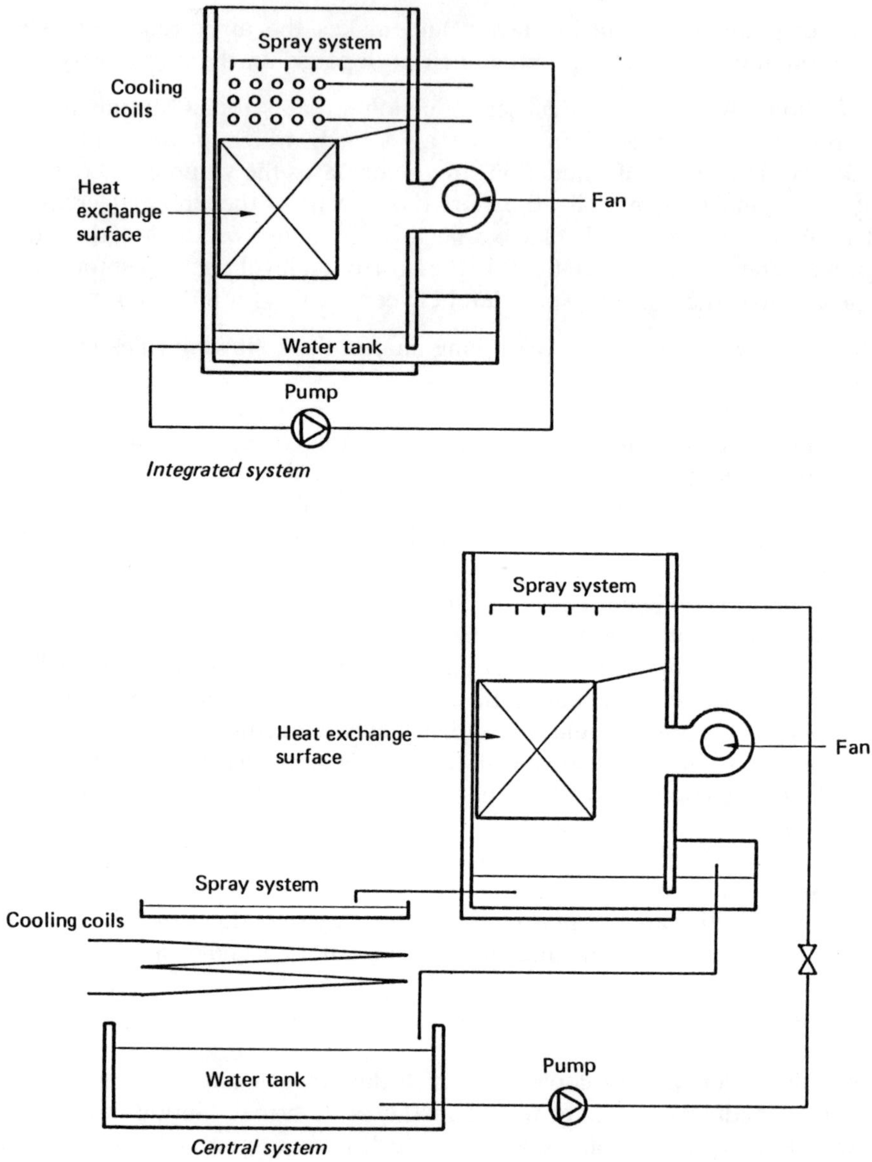

Figure 5.21 *Wet air coolers*

the past few years are in fact very old-fashioned. They were in use 50 years ago but disappeared because of their disadvantages:

(a) they are expensive to buy and maintain, and require a high investment and upkeep against corrosion;

(b) they require considerable space;
(c) they use more energy than dry-air coolers.

There are of course advantages, otherwise they would not have made a comeback. Their main advantage might be that higher relative humidity can be obtained than with a dry system. As the air is not in contact with a dry surface, but with water, it leaves the cooler at almost 100% RH. The product does not necessarily deliver the moisture, but the cooling water does. The use of large air volumes is possible without a great drying effect on the product and variations in air volumes are possible without disturbing the condition of the air, and since there is no frost, no defrosting is necessary.

All these advantages can be achieved when we use dry coolers with:

(a) a very high evaporating temperature
(b) a surface temperature above freezing point
(c) a small dT
(d) a wet surface because of condensation
(e) a large air volume by using secondary fans outside the coolers to create the same large air velocity over the product.

The energy costs of wet-air coolers are higher because of the higher power consumption for the water pump; sometimes a secondary pump is required in ice-bank applications.

Another reason for higher energy consumption is the fact that the system works with an intermediate secondary refrigerant – water. Energy saving can be achieved by using ice-banks in order to accumulate cold in the ice around evaporator coils. The cold production can then take place during hours when electricity is cheaper. This 'storage' also gives the possibility of having a large quantity of refrigeration capacity available when a load of freshly introduced warm product is being cooled down. Altogether it is quite a good system if rapid chilling at relatively high temperatures and short storage times is needed, and it is used mostly for milk, vegetables and flowers that are about to be disposed of at auction.

Up to this point we have described centralized systems, involving different air coolers connected to a central ice-water tank or ice-bank. There is also an integrated system in which evaporator coils are placed in the cooler top, under the spraying system, so the water is cooled by flowing over the coils inside the air cooler itself. In the coils the temperature is about −6°C and a thin layer of ice forms on the tubes, controlled by a thermostat which stops the refrigeration installation when a certain predetermined ice thickness has been reached. The water reaches an even temperature just a little above the freezing point. The air-circulation rate in a storage room is normally about 40 times per hour and then the air stream over the cooler can be determined. The temperature difference between air and water may not be above 0.5K. In order to achieve RH values of 90%, temperature differences between 0.1 and 0.2K must be applied. In order to achieve the best results, air must be blown directly over the product and the fan must be placed in front of the filling mass. The ratio air:water mass stream lies in the range 1:1 to 1:1.5.

Example

$\rho_{air} = 1.3$ kg/m^3.
Air mass stream = 36 000 m^3/h or 46 800 kg/h.
$\rho_{water} = 1000$ kg/m^3. With a ratio 1:1, the water stream is 46.8 m^3/h.

The section of wet-air coolers is made by a trial and error method using a computer. Selection is based on air mass stream, conditions of the air, dimensions of the exchange surface, air velocity and heat and mass transmission coefficients, water temperatures, and the variables of the cold room and product characteristics.

The cooling time is checked to make sure that it corresponds to the refrigeration capacity of the cooler. By varying the water mass stream, air velocity or exchange surface, different results can be obtained.

The following controls are used:

1. The above-mentioned thermostat or other device to control the ice thickness
2. A thermostat to control the compressor capacity
3. A timer to keep the installation out of action for periods of about 10 minutes in order to melt part of the ice regularly and maintain a consistent ice layer.

Liquid Chillers

Liquids such as water, brine but also milk, beer and other products with low viscosities, can be cooled down by pumping them through multi-tubular so-called shell and tube coolers or chillers. See figure 5.22 for a typical liquid chiller.

Although there are some vertical chillers, the horizontal system is the standard chosen. A special type is the inclined chiller used to obtain very low water temperatures near the freezing point. This version is in fact a combination of a 'dry' and a 'flooded' system.

Another type of liquid cooler is the so-called submerged evaporator, examples of which are shown in figure 5.23. In this system, the liquid to be chilled is stored in a tank in which the refrigerant cooling coils are placed. These evaporators are often used to produce chilled water, and to accumulate refrigeration capacity in an ice layer on the coils or in brine

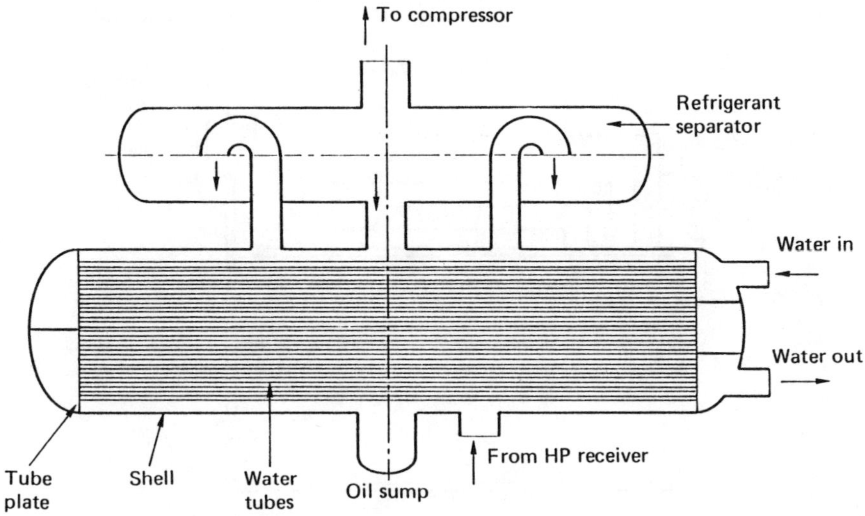

Figure 5.22 *Shell and tube flooded-type liquid chiller*

stored in the tank. In the chemical and metallurgical industries they are used to cool process tanks. In breweries they can be placed inside fermentation tanks.

Another design involves winding the coil around a drum in which the liquid is cooled. Alternatively, instead of coils, double-plate constructions can be used in all these applications, in which the refrigerant evaporates in the space between two steel plates.

The horizontal shell and tube chillers have steel tubes when using R717; usually they have an anti-corrosion coating on the water side when a corrosive liquid is chilled. Special bi-metal tubes are used for marine purposes – sea water chilling. For the other refrigerants, mostly outside finned or even inner finned copper tubes are used.

Two groups of shell and tube chillers are distinguishable:

(a) The refrigerant flows inside the tubes – these are evaporators with a dry system when used with CFC refrigerants.
(b) The refrigerant flows around the tubes in the shell. These are the 'flooded' evaporators mostly used for R717.

In case (b), maintenance is simple. By removing the water lids, access is gained to the water tubes, for inspection and mechanical cleaning.

In case (a), maintenance is more complicated. The side where the water is can only be cleaned using a chemical treatment.

Sometimes the tubes are placed in a coil in a hair-pin shape, so there is only a tube-plate on one side of the chiller; the chiller is built in such a way

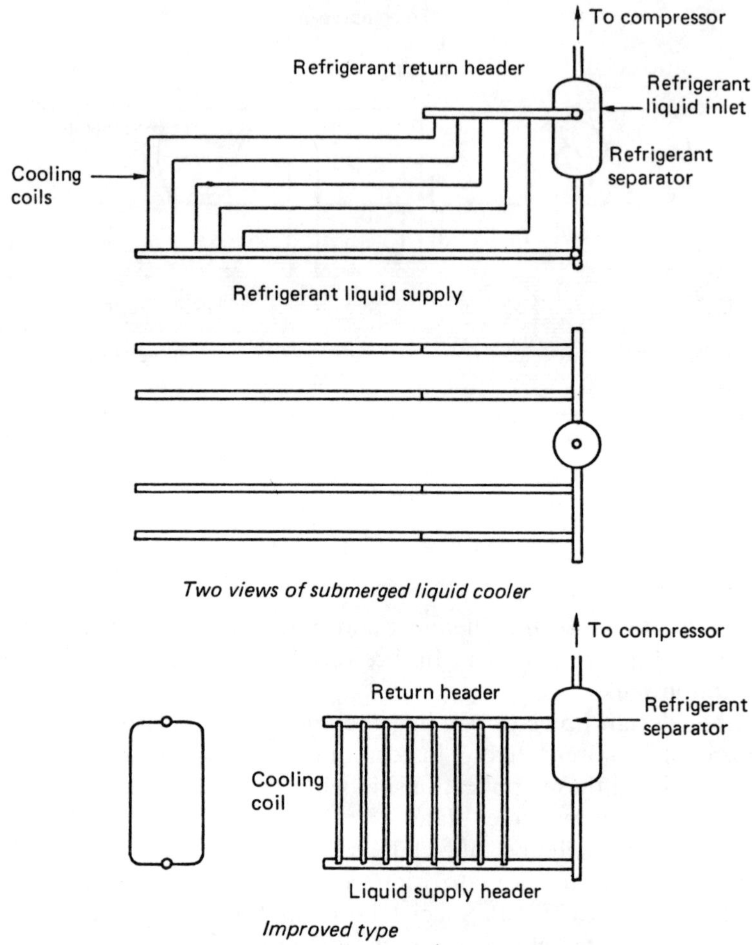

Figure 5.23 *Submerged evaporator coils*

that the coil with the tube-plate can be taken out for inspection and cleaning. Finned tubes are more difficult to clean than smooth tubes.

When using system (b) it is obviously much easier to build a refrigerant-tight system, as the gaskets are placed on the water side, than in case (a), where the gasket must seal the refrigerant side.

In case (b), the tubes are normally placed up to 2 rows above the middle line of the chiller. The rest of the shell acts as a liquid separator.

To build a shell with a smaller section, tubes can be placed over the whole section, but then it becomes necessary to build an external liquid separator above the chiller.

The refrigerant flow in case (a) is regulated by thermostatic expansion valves; frequently, two coils are placed in the shell, each one with its own expansion valve. Now with the capacity reduced, one coil can be disconnected to guarantee constant refrigerant flow and velocity in the tubes, which is very important in CFC plants, as we will see in the chapter dealing with pipework.

Generally, the refrigerant flow in this kind of liquid chiller is very irregular, especially when the temperature difference dT is small. The distribution of the refrigerant over the tubes is not adequate because in every turn in the curves the gas and the liquid must be re-distributed, and not every tube receives the necessary quantity. The instability of the flow can be improved by using long circuits and high refrigerant velocities, but this again increases the pressure drop.

The chillers can be used only for relatively high water temperatures, high dT, for instance not lower than 4–6K, and evaporating temperatures above the freezing point of the water.

In case (b), the refrigerant flow is mostly regulated by a float-valve device, keeping the refrigerant level constant, just above the top row of tubes. The refrigerant used is normally R717.

When water at a low temperature is required, a constant-pressure valve is placed on the suction connection in order to prevent the evaporation temperature from falling below 0°C.

A safety thermostat is placed with the bulb in good thermal contact with the refrigerant by means of an oil-filled tube, to safeguard against freezing the water.

Most manufacturers only permit a minimum chilled water temperature of +4°C. For lower water temperatures the inclined chiller or an immersed evaporator must be used. These are described later. Another way to prevent freezing of the liquid is to use, instead of water, a brine solution such as $CaCl_2$, NaCl, ethylene glycol, propylene glycol, acetone or alcohol.

The water velocity in the tubes may normally not exceed 2 m/s for reasons of pressure drop and cavitation corrosion damage.

Sometimes construction (b) is used for R22 plants, especially when low water temperatures between 2 and 4K must be achieved. In these flooded type of chillers there is no limit on dT. A difference of 1K between evaporation temperature and water temperature can even be achieved. In the case of R22 chillers, special care must be taken to 'bleed' the separated oil floating on top of the refrigerant liquid level. In the case of R717, the separated oil is collected in an oil-sump built under the shell, which can easily be drained by means of a hand-cock or be led to an oil-collecting drum.

One of the special water-chiller designs, the inclined shell and tube chiller, is shown in figure 5.24. This is a flooded evaporator but with the refrigerant inside the tubes. In order to evacuate the gas bubbles in a sufficient and easy way, the chiller is put in an inclined position. This chiller is used when CFC compounds must be used as refrigerant and the

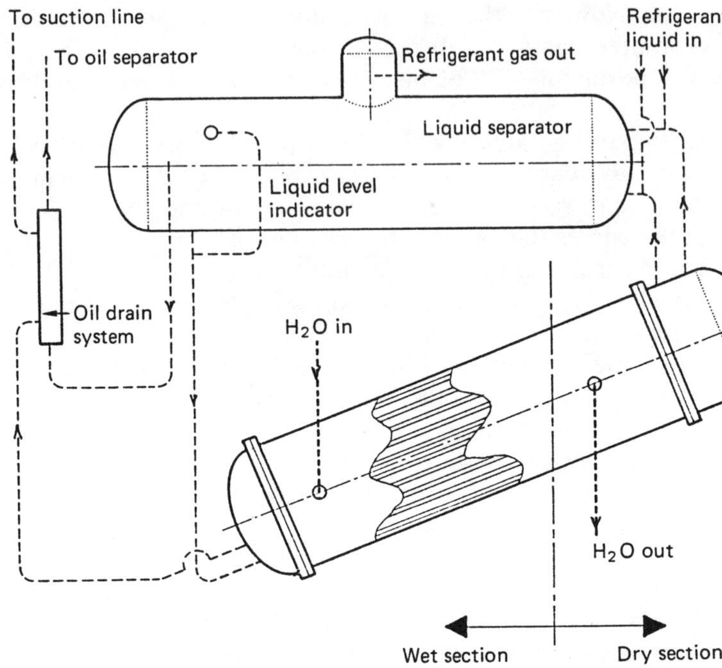

Figure 5.24 *Inclined shell and tube chiller*

outlet temperature of the chilled water, however, must be very low. A dT of about 2K can easily be obtained, a figure that would be impossible when using a dry system.

In flooded chillers the refrigerant is simply controlled by a float-valve system, not by means of a thermostatic expansion valve such as those used for dry systems. The chillers are used for special applications such as in central water-chilling plants serving coal-mine air conditioning.

It is even possible to obtain water at about 0.5K with a special type of closed chiller; in that design the chiller is built as a plate heat exchanger, the type we know from milk or beer cooling. For that purpose, those plate heat exchangers consist of a series of plates between which alternatively the milk or beer to be cooled and the cooling chilled or glycol water pass in counterflow. See figure 5.25.

In particular cases where we would like to produce chilled water with a temperature of 0.5K, the water flows through the spaces where the milk or beer flows, and refrigerant evaporates between the plates where the chilled water or glycol solution flows. The U-values of these plate evaporators are extremely high: water/ammonia, 2500–4500 W/(m^2 K); water/R22, 1500–3000 W/(m^2 K). Since the refrigerant charge is 10% or less than shell and tube chillers and the overall dimensions are much smaller, this type of

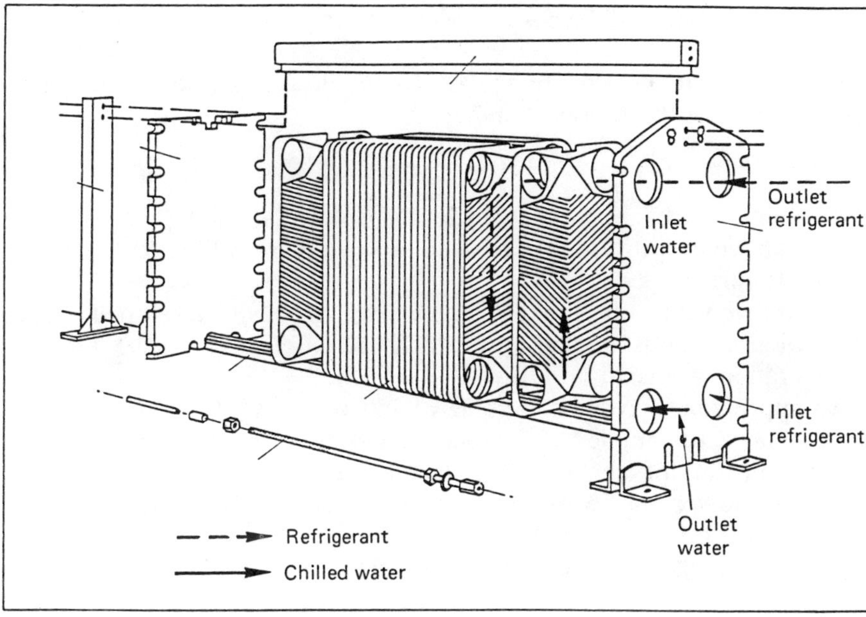

Figure 5.25 *Plate-type liquid chiller*

evaporator is being used much more, not just for its low chilled water temperature.

When open chillers are allowed, water of +0.5K can be obtained by letting the water flow in a thin film over plates in which the refrigerant evaporates, such as the Baudelot cooler.

Overall heat-transfer coefficient for shell and tube liquid chillers This U-value is obtained by adding the following thermal resistances:

$$1/U = 1/\alpha_i + \delta/\lambda + 1/\alpha_o$$

where i denotes inside and o the outside.

For round tubes this becomes:

$$1/U = 1/(2\pi r_i\alpha_i) + \ln(r_o/r_i)/(2\pi\lambda) + 1/2\pi r_o\alpha_o$$

where r stands for radius.

As the values of U for CFC refrigerants are so low, the surface has to be increased by using fins on the tubes, which results in compact constructions. Also, in order to increase the heat-transfer efficiency, the surface in contact with the refrigerant is provided with rough points, screw-shaped fins, porous coating and other elements that create roughness of the surface. By doing this the heat transfer by convection and nucleate boiling are increased and stratification effects are neutralized.

α Values for 'Flooded' Types

The α value on the side of the chilled liquid inside the tubes can be calculated using the Nusselt number (*Nu*):

$$Nu = \alpha_i d_i/\lambda = 0.012(Re^{0.87} - 280)Pr^{0.4}(1 + \{d_i/L\}^{2/3})$$

This equation is by V. Gnielinsky and is only valid for 1.5 $Pr < 500$. A correction factor must be applied as in practice there will be fouling of the waterside surface.

For the α value on the refrigerant side, Slipcevic developed simple calculation methods and graphs which are published by the German Refrigeration Association, DKV.

An approximate average *U*-value of 760 W/(m^2 K) for 'flooded' R717 chillers may be applied. The actual *U*-values lie between 500 and 1000 W/(m^2 K) for clean water, depending on the construction, tube diameter and spacing, water velocity and density of liquid to be chilled. When using brine, the values are lower and can go down by 50% depending on the brine concentration. For CFCs, the values are 25% lower.

In the case of a refrigerant inside the tubes and the water around the tubes α_o can be calculated from a formula by Donohue:

$$Nu = \alpha_o d_o/\lambda = 0.25Re^{0.6}Pr^{0.33}(\eta_{f1}/\eta_w)^{0.14}$$

η are viscosity coefficients of liquid in the middle of the mass and in the layer near the tube surface. The last part of this equation, $(\eta_{f1}/\eta_w)^{0.14}$, can be considered as 1. More details about the type of calculation can be found in the VDI *Wärme Atlas* (GR–I).

The surface coefficient on the side of the refrigerant, α_i, can be found with the aid of graphs produced by J. M. Chawla of the University of Karlsruhe, Germany. They are available for R11, R12, R21 and R22. Figure 5.26 shows the graphs for R22 and R717.

The method of Slipcevic can be used when dealing with other refrigerants. For example in the case of nucleate boiling:

$$\alpha_i = B\phi q_i^{0.7}/d_i^{0.5}$$

where B is a value that depends on the material characteristics, and is shown in tables by Slipcevic,
ϕ = specific mass stream,
q_i = heat flux on the outside of the tube,
d_i = inner diameter of the tube.

Note that this equation is valuable for smooth tubes only; for finned tubes, correction factors have been published.

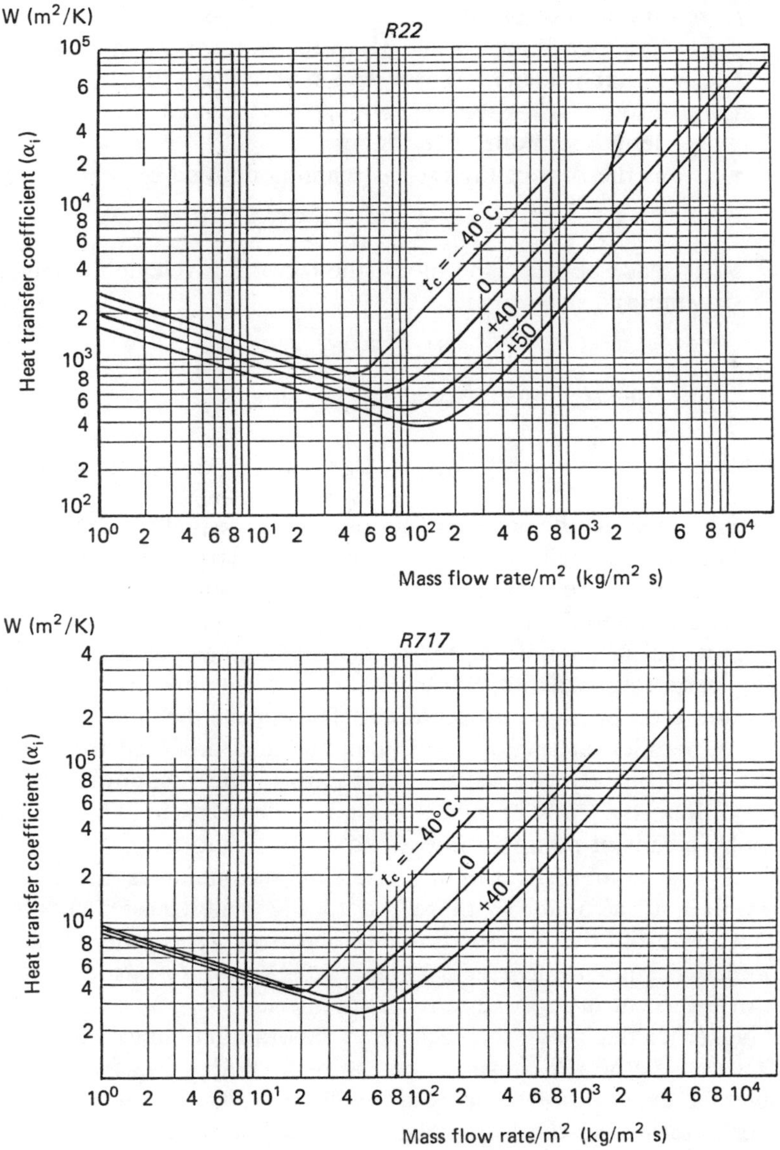

Figure 5.26 *Values of heat transfer coefficients for R22 and R717 condenser tubes at different refrigerant massflow rate densities*

In those chillers in which the water flows over the tubes, partitions must be placed to create the desired water velocity.

The pressure drop on the water side is calculated as follows:

$$dp_{tot} = \rho U_{dw}^2/2\tau_{tot} = \rho U_{dw}^2\{(n + 1) + \tau_{dw}Z_{dw} + n\tau_1\}$$

where n = number of partitions,

Z_{dw} = number of tubes over which the water flows in cross-stream,

U_{dw} = water velocity in cross-stream across the tubes,

τ_1 = $2(S_{dw}/S_1)^2$, where S_{dw} and S_1 are the areas of the cross and length sections of the tubes,

τ_{dw} is the part of the cross-streaming defined by

$$\tau_{dw} = \{3/(\eta_{fl}/\eta_w)^{0.14}\}/[\{(ST - d_o\}^{0.2} Re_{dw}{}^{0.2}]$$

ST is space between the tubes (the factor 3 is relevant to turbulent streaming)

$$Re_{dw} = U_{dw} d_o / \nu$$

where U_{dw} is the stream velocity across the tubes.

Ice Banks

As stated above, ice banks are evaporator coils or plates submerged in a tank of water. The coils or plates are placed at a larger distance than they would normally be from each other in order to permit a reasonable layer of ice to build up. See figure 5.23.

Without ice accumulation the normal U-value, when used for water chilling purposes, is about 500 W/(m^2 K) for ammonia/water when designed and used as 'rapid evaporators'. This means the coils or plates are placed in parallel on suction, and liquid headers are used to enable the gases to flow quickly and have a short path to the separator. The water velocity along the coils or plates must be adequate and the circulation forced by means of pumps, compressed air or agitators.

When the submerged coils are used as accumulators the necessary surface is not calculated with regard to the heat transfer in the first instance, but on the ice load that is needed. Once the capacity to accumulate is known, the necessary ice load is known too. Normally an ice layer of 3–3.5 cm is considered as economical.

At bigger ice thickness, the necessary evaporation temperature becomes too low and the power consumption too high. With a 3–3.5 cm ice layer, the average evaporation temperature will be −5°C.

After determining the required plate or coil surface, it is necessary to check if the ice surface is large enough to deliver the quantity of refrigeration capacity per hour which the product demands. The temperature difference between ice and water is about 1.5K and the heat transmission coefficient using coils about 1200 W/(m^2 K). So $Q = 1.5A \times 1200$ W, A being the ice surface area. If necessary, A can be adjusted to cover the maximum capacity expected.

$\alpha_{outside}$ using plates can go as high as 1800 W/(m^2 K) at a water velocity of about 0.23 m/s. α depends on ice thickness, heat flux, water velocity and construction method.

In conclusion, those readers who wish to study the subject of heat transfer applied to evaporation and condensation more intensively are advised to consult the 'Arbeitsblatter' worksheets of Deutsche Kälte Verein (DKV) and the *Wärme Atlas* (Heat Atlas) published by the Association of German Engineers, Verein Deutscher Ingenieure and the books of Perry and General Electric (see References at the end of the book).

In particular, the publications of the following authors should be consulted: Chawla, Schulenberg, Schmidt, Henrici, Hirschberg, Slipcevic, Mann, Bäckström, Merkerl, Nusselt, Dhar and Jain, and Donohue.

5.3 Condensers

5.3.1 Introduction

Liquefaction of the refrigerant gases sucked out of the evaporator by the compressor is done in the condenser. The compressor must therefore raise the pressure of the gases to the necessary saturation pressure at which the transformation from gas to liquid can take place by heat transfer to the available cooling medium, usually air or water, taking into account the temperature of the cooling medium.

When the gas enters the condenser, the superheat is first removed. Next, condensation takes place and, finally, the liquid may be subcooled. In all stages the heat-transfer coefficient of course is different. In the first stage it is very low, because on the refrigerant side there is only gas; in fact, in air-cooled condensers there is gas on both sides – refrigerant and air.

This is also the case in some evaporative condensers, where water consumption is economized by desuperheating using air only. In the second stage, the heat transfer is very good because of the condensation process. In the third stage of subcooling, the heat-transfer coefficients are higher as there is liquid on both sides; except of course in the case of air-cooled condensers.

For the general theory about heat transfer, the reader is referred to the chapter on insulation techniques for more details.

For specific theoretical treatment of α_i and α_o during the condensation process, refer to the specialized literature on the subject. Here we will only give some general principles of the condensing process.

During the condensing process a liquid film is formed, either inside or outside the tubes, depending on the design. The thickness of this film affects the rate of heat transfer. When condensation takes place on the inside of the tubes, the thickness of film increases with the tube length. When condensation takes place on the outside of the tubes, the thickness of film increases according to how low the tubes are positioned in the coil. When condensation takes place inside the tubes the gas velocity will decrease as it progresses along the tube or coil.

Calculating the α values is very complicated as there is both laminar as well as turbulent streaming, and also two-phase streaming. α depends on the mass stream density and the temperature, and decreases with increasing mass stream density. The stream pattern in the middle of the refrigerant path is very different from that further on in the path where more liquid is present. At the beginning of the condensing process the liquid film is thin; it then increases, which results in a lower heat transfer coefficient. At a certain point on the heat exchange surface there is so much liquid that α will increase again as the flow turns from laminar to turbulent.

5.3.2 Water-cooled Condensers

The simplest condenser is the water-cooled condenser, of which several types have been used in the past. The first was the atmospheric condenser. A spray system divided the water flow over a number of vertical parallel coils in which the refrigerant was condensed. The α_o was very high; the cooling effect relied partly on the water and partly on the evaporation of some of the circulating water – the cooling tower effect which depended on the atmospheric conditions.

The disadvantages of these condensers were that they were expensive and occupied too much space, so they had to be placed outside the building where they suffered from corrosion and the growth of algae.

Another older type, which is still in use in very small installations, is the double pipe condenser. It is very compact with a high *U*-value. The refrigerant condenses in the space between inner tube and outer tube, while the cooling water flows in the inner tube. The complicated construction makes them too expensive for normal applications.

Nowadays horizontal shell and tube condensers, constructed like liquid chillers, are normally used (see figure 5.27). The vertical shell and tube condenser shown in figure 5.28 has almost disappeared from the market; however, for large capacities they are not so much more expensive than the horizontal ones. They have the advantage of taking up little ground space, are easy to maintain even during operation and have a high *U*-value, partly because of the chimney effect. Figure 5.29 shows graphs based on experimental data which enable the necessary surface area to be calculated when we know the application condensing temperature and the temperature of the available cooling water. Normally the water flow is 9 l/minute. See also tables 5.1 and 5.2.

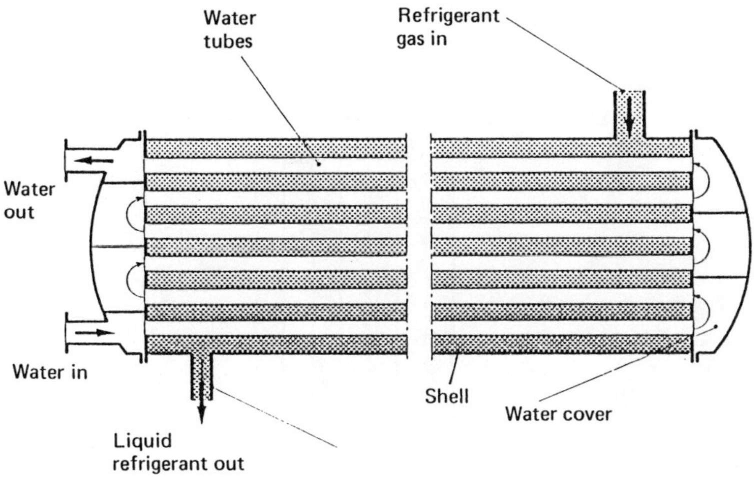

Figure 5.27 *Horizontal shell and tube water-cooled condenser*

Table 5.1 Surface correction factors (vertical condensers)

Water flow per tube (litre/min)	*Correction factor for condensers of length*			
	3 m	*4 m*	*5 m*	*6.5 m*
1.9	3.483	3.636	3.752	
3.79	1.914	2.045	2.149	
5.7	1.353	1.460	1.548	
7.5	1.053	1.142	1.216	
8.85	0.921	1.000	1.068	1.1475
9			1.051	1.0923
9.4	0.865	0.041	1.005	1.047
10.2	0.816	0.889	0.951	
11.4	0.738	0.806	0.863	
13.3	0.657	0.720	0.774	
15	0.605	0.666	0.719	

Table 5.2 Suction temperature correction factors (vertical condensers)

Suction temperature (°C)	*Correction factor*
−28	1.08
−22	1.05
−14	1.00
−11	0.98
−8	0.97
−3	0.95
+1	0.93

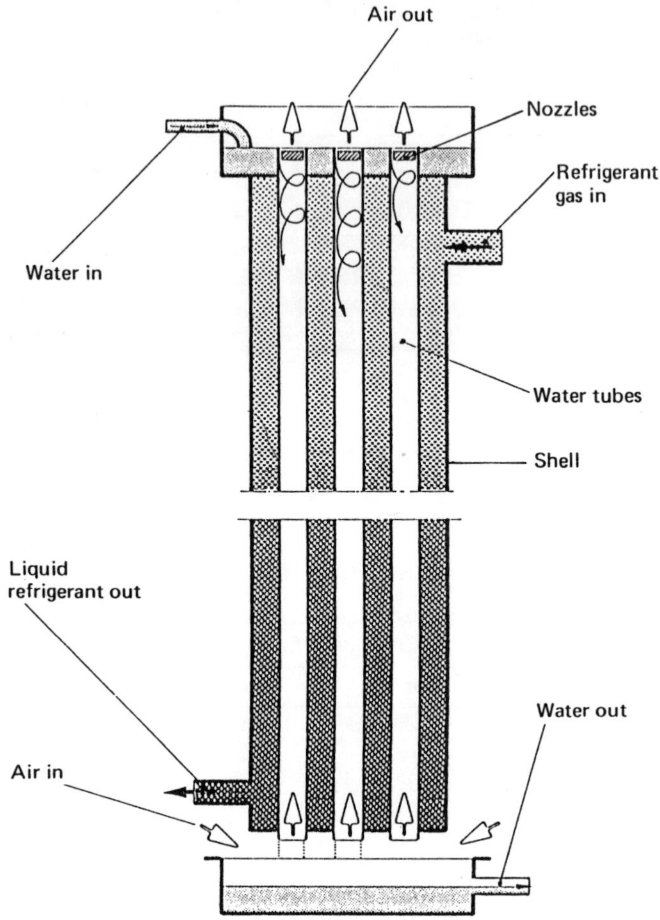

Figure 5.28 *Vertical shell and tube water-cooled condenser*

The horizontal shell and tube condenser is still in use today, although mainly in combination with a cooling tower owing to the shortage of clean water. On board ship or near the coast, water is available in sufficient quantity; in this environment the tubes must be coated with plastic or be made of special alloys to prevent corrosion from sea-water. In marine plant, galvanic plugs are placed in the water covers to protect against corrosion.

In ammonia installations, water flows inside the tubes and condensation takes place on the outside of the tubes. The same advantages as stated for liquid chillers of this construction type are valid. The covers only need to be water-tight and that is easier than making them refrigerant-tight. However, this is necessary when the refrigerant condenses inside the tubes,

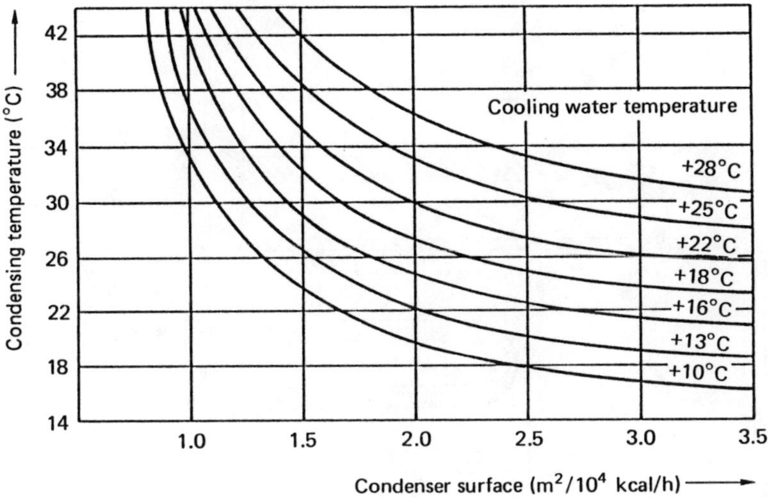

Figure 5.29 *Graphs of condenser surface against condensing temperature and cooling water temperature*

as is the case of some CFC installations. Cleaning problems are also encountered with this type of equipment, which sometimes can be solved in the same way as for liquid chillers.

Finned tubes are used to increase the surface, both inside and outside the tubes.

5.3.3 Evaporative Condensers

The evaporative condenser is a combination of condenser and cooling tower. It is an atmospheric condenser with a forced water and air flow as shown in figure 5.30. As in the case of an atmospheric condenser, water is sprayed over the rows of coils, which are placed parallel to each other and through which the refrigerant flows. The water is collected in a tank to which the circulation pump is connected, the coils are placed inside a casing, and air is blown or sucked in counterflow with the water stream by means of a fan.

Today, centrifugal fans are normally used, placed in the bottom of the tower. This is the best way to avoid maintenance and noise-level problems.

These condensers are normally placed in the open air and must therefore be protected against freezing. To do this a heater is placed in the water tank. Sometimes the water is drained to a water receiver placed in a frost-free place. The circulation pump is then placed inside the building. When outside temperatures are low enough, it is sufficient to use only the fans and not the water flow. The condenser then acts as an air-cooled condenser and the water is stored in a frost-free tank.

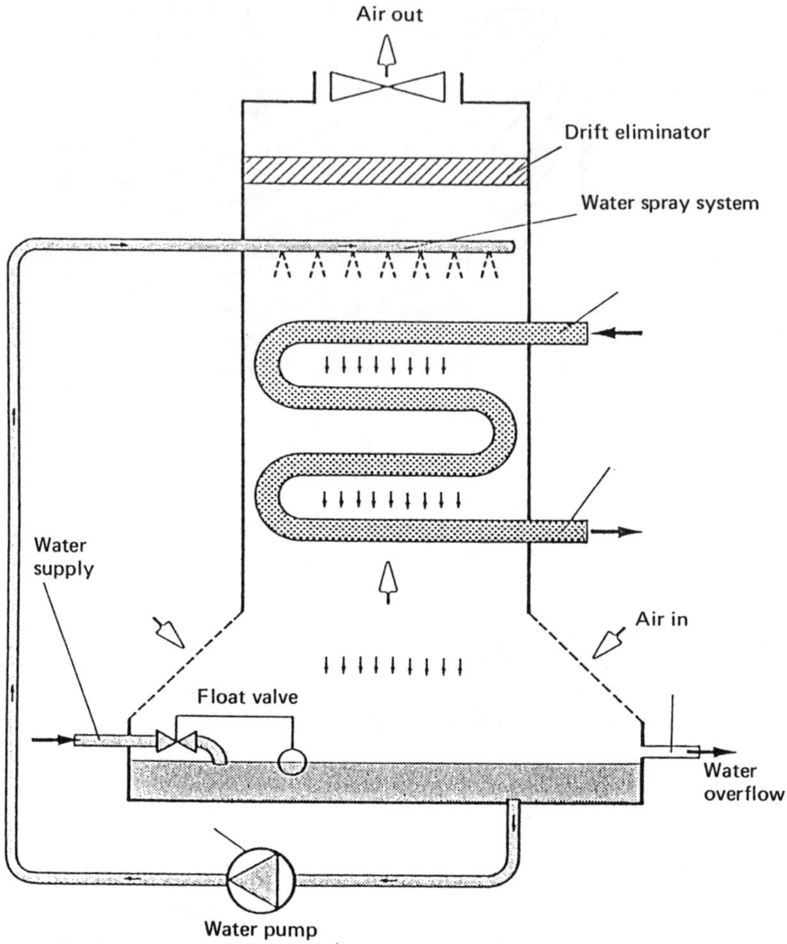

Figure 5.30 *Schematic diagram of an evaporative condenser*

Such condensers are used when water is difficult to obtain or is of inferior quality. The water is used in a closed circuit and a small part of the water evaporates, cooling down the rest of the water in the same way as is done in a cooling tower. The evaporated water is replaced by adding water using a float-valve built into the water tank. Water consumption is very low: about 5% of a water-cooled condenser. It can be calculated by reading the value of dx from the Mollier graph, dx being the difference in water content of the air between the inlet and the outlet, and multiplying this value by the mass of the airstream. The resulting value is then multiplied by 3 to allow for regular replenishment of water in the storage tank.

Compared with air-cooled condensers they have the advantage of working with lower condensing temperatures. Normally the condensing temperatures lie between 11 and 13K above the wet-bulb temperature, which means for European conditions +295K + 13K = +308K.

In addition to the type with an open water tank, called open circuit, there is a type with a completely sealed circuit, which is used when pollution is expected.

When two or more condensers are connected to one refrigeration circuit, care must be taken to follow exactly the erection instructions of the manufacturers in order to prevent unequal pressure drops which will block up the refrigerant in one of the condensers, which could then lead to a lower condenser efficiency. See figure 5.31 for a typical parallel pipework arrangement for two evaporative condensers and a liquid receiver.

The ratio air volume:refrigerant capacity has been determined by experiments at between 100 and 150 m^3/h per kW.

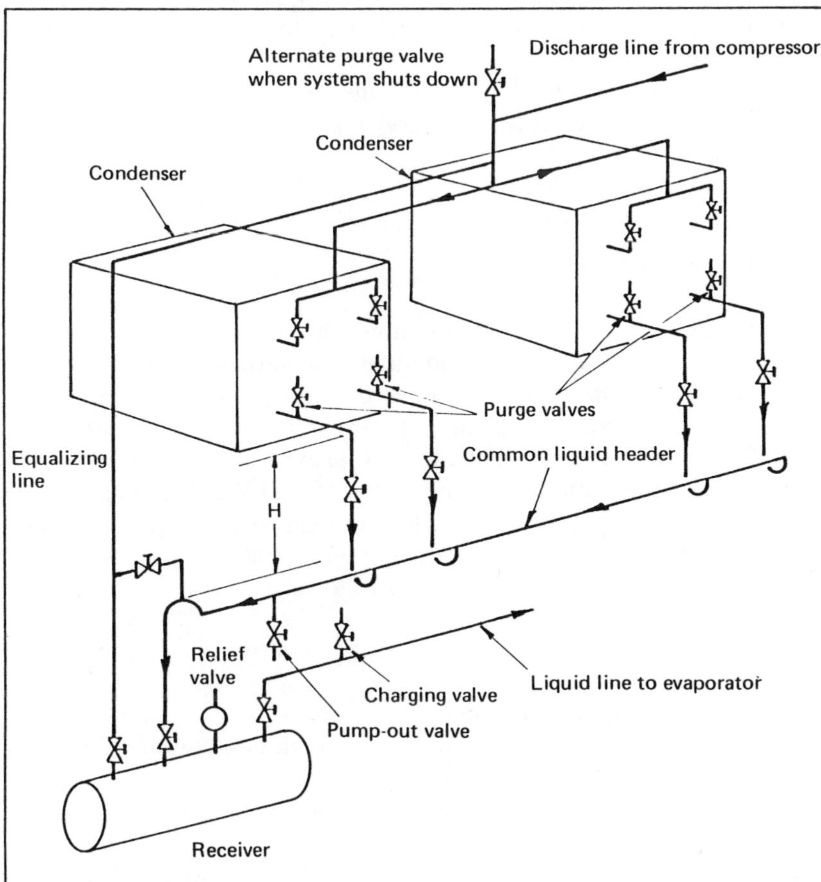

Figure 5.31 *Parallel evaporative condensers with common liquid header above receiver level (Courtesy Baltimore Air Coil Co., Inc.)*

In the calculation procedure for the U-value, not only heat-transfer coefficients but also the mass transfer coefficients of the water vapour are to be taken into account. As for the wet air coolers there are a lot of variable factors, so that only calculations by trial and error using a computer program can provide the U-value. Schmidt, however, established graphs which are published in the DKV worksheet, Arbeitsblatter 3–03.

The enthalpy difference between the inlet and outlet air can be calculated by

$$dh = 1000/150 \times 1.25 = 19.2 \text{ kJ/kg}$$

The air inlet conditions are the local dry and wet bulb temperatures, so with dh and the use of a Mollier graph for humid air, the outlet conditions of the air can be established. With the temperatures of the air and the condensing temperature, one can then calculate the logarithmic temperature difference.

With the condenser heat load Q_c, the U-value and the logarithmic temperature difference, the necessary heat-exchange surface A m^2 can now be calculated from

$$A = Q_c/U \log dT \text{ m}^2$$

5.3.4 Air-cooled Condensers

These condensers are the most commonly used type today owing to the difficulty in obtaining water of good quality, in sufficient quantity and at a reasonable price. Originally, they were only used as natural draught condensers for very small capacities, such as household refrigerators. Some condensers with forced draught air flow using fans were developed for commercial refrigeration up to capacities of 30 kW. A typical model is shown in figure 5.32. Today, air-cooled condensers are in use in industrial installations, even for those plants using ammonia.

Their disadvantages can be listed as follows:

- High condensing temperatures and high energy consumption of the compressors
- Power consumption of the condenser fans
- High condensing temperatures also mean high end-compression temperatures and high oil temperatures
- In some locations, noise can be a problem.

Normally, the condensing temperature is between 10 and 15K above the maximum ambient temperature. Under European conditions, this means condensing temperatures of between +40 and +45°C; and under tropical

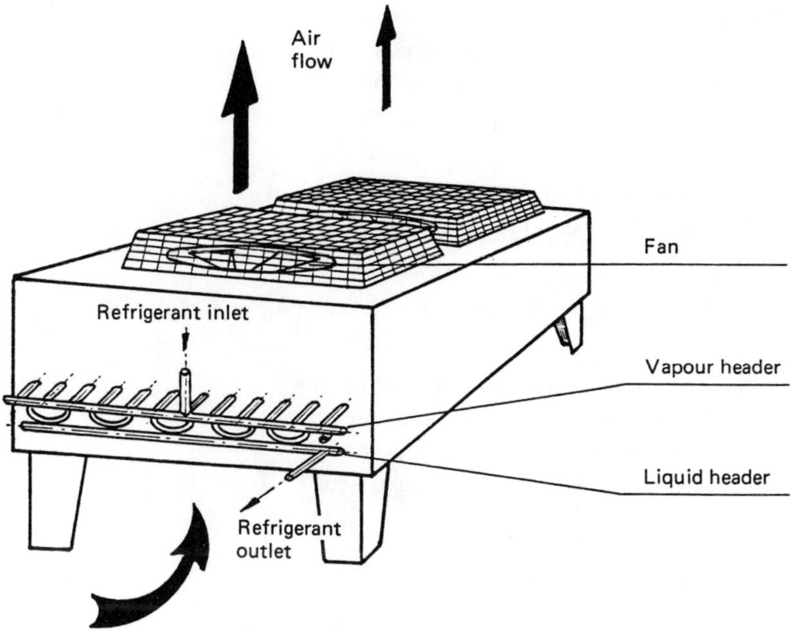

Figure 5.32 *Horizontal air-cooled condenser*

conditions between +50 and +60°C, which for most of the industrial refrigeration compressors is the absolute limit.

Compared with water-cooled or evaporative condensers, the *U*-value is low because the heat-exchange surface has gas on both sides for much of its effective area. The surface is increased by adding fins on the tubes. Since the fin spacing is small, 1–2 mm, and becomes quickly choked, the result is a higher energy consumption and the need for frequent cleaning of the condenser fins.

The condenser must always be located outside, or alternatively, be in contact with the outside air by means of air ducts. Small condenser types can only be positioned inside in a place where there is a large free volume of air available. Tables are available to guide such a choice. Summer and winter working conditions are very different because operation is directly related to the climatic conditions, in particular the dry-bulb temperature of the air. Using CFC refrigerants with thermostatic expansion valves it is not possible to take advantage of the low ambient temperature and low condensation temperatures. In order to maintain the efficiency of the thermostatic valves, a minimum pressure drop through the valves must always be available, in other words during periods of low ambient dry-bulb temperature the condensing pressure must be kept artificially at a high level. This is done by means of:

(a) disconnection of fans
(b) filling up a part of the condenser tubes with a refrigerant, thereby decreasing the effective condensing surface
(c) it is also possible to close some coils or to reduce the air supply.

A combination of (a) and (b) is normally used.

The U-value of these condensers can be calculated from the equation of Hofman:

$$U = \alpha_r(A_o + \phi_f A_2)/\{(A_o + A_2)\alpha_r(A_o + \phi_f A_2)/\alpha_i A_i\}$$

where $\alpha_r = \alpha_{fin}, \alpha_{outside}$
$\alpha_i = \alpha$ on the smooth side of the tube, α_{inside}
A_o = outside primary surface area
A_i = tube inside surface area
A_2 = outside secondary surface area
ϕ_f = fin efficiency.

A simplified equation gives:

$$1/U = A_o/(A_i\alpha_i) + 1/\alpha_2 \ (m^2 \ K)/W$$

The air velocities normally are between 3 and 10 m/s.

To calculate values, the reader is referred to the *Wärme Atlas* (see References at end of book), the theories of Chawla (*Kältetechnik-Klimatisierung*, 24 (1972) 33), Schulenberg (*Kältetechnik-Klimatisierung*, 22 (1970) 75) and the Arbeitsblätter' of DKV (2–02).

There are selection problems when choosing air-cooled condensers with the help of manufacturers' data sheets, and it is wise to take into account certain important facts. Many manufacturers do not allow for:

(a) high ambient air temperatures in tropical countries
(b) the type of CFC refrigerant that is used
(c) the altitude of the plant – the density of the air differs at higher altitudes.

In addition data sheets only give a capacity correction proportional to dT, which as we saw earlier in the chapter is not correct.

Example (a)

At ambient temperature +25°C the correction factor is 1, while at ambient temperature +40°C the correction factor is 1.063.

Example (b)

For R22 the correction factor is 1, whereas for R12 the correction factor is 1.05 and for R502 it is 1.025.

Example (c)

At sea level the correction factor is 1, whereas at an altitude of 1000 m the correction factor is 1.073 and at 2000 m it is 1.158.

Furthermore an allowance must be made for condensers connected to non-suction gas-cooled semi-hermetic compressor units and suction gas-cooled hermetic or semi-hermetic compressor units. The capacity correction factors for both evaporating temperatures and dT values are not the same, since the latter groups of compressors gain additional superheat from the motor windings which will have to be removed in the condenser.

When using long suction lines, a large quantity of superheat may need to be removed as well.

Whenever there are any noise limitations then it is advisable to use centrifugal fans. In extreme cases, natural draught air cooled condensers can be used. The air circulation then relies solely on the chimney effect. The heat-exchange surface in that case is built up from tubes with a secondary surface made from a special wired construction or fins.

When connecting condenser and the high-pressure liquid receiver, care must be taken to ensure that the condensate line is chosen large enough to guarantee that a liquid velocity of 0.8 m/s at full capacity is not exceeded.

It is recommended that the highest point of the condenser and the receiver be connected with an equalizing line to prevent liquid from blocking in the lower condenser tubes, so preventing unnecessary high condensing temperatures and pressures.

5.4 *U*-values in Daily Practice

In conclusion, some practical *U*-values will be given. These values are approximate and do depend on the design details and the local and working conditions. All the values below are in units of $W/(m^2\ K)$.
Horizontal shell and tube condenser:

Ammonia 800–1400.

For CFCs (finned tubes) 40% lower values

Fouling factors have to be taken into account.
Air-cooled condensers:

CFC refrigerants
- with fans, about 30
- natural convection, 6–15

Air coolers:
The lowest values are applicable for CFC refrigerants, the highest for ammonia:

15–40, depending on the refrigerant and the injection method used

Liquid coolers:

cold water, 500–1000
brine, 250–600

Immersed evaporators:

300–500
150 (ammonia coil immersed in a water bath without flow)

Intermediate coolers:

Ammonia, 250
CFCs, 200

Plate evaporators:

Ammonia/water, 4000
R22/water, 2500

U-values depend also on the kind of brine used and its concentration.

5.5 Modern Developments in Enhanced Heat Transfer for Evaporators and Condensers

The boiling nuclei are mainly created at the heat-exchange surface and less so in the body of the liquid stream. For that reason there is a tendency to produce a 'ring-streaming' pattern. Under the effects of this, the circulation ratios are decreased so that values of between 1 and 2 rather than 3 or 4 are now being used. The fact that nuclei are more easily created on a rough surface is related to the effects of surface tension σ on the heat-transfer process.

A porous heat-exchange surface gives a better heat-transfer coefficient than a smooth tube surface, as shown in figure 5.33. It is more able to create nuclei during the boiling process which results in a lower surface

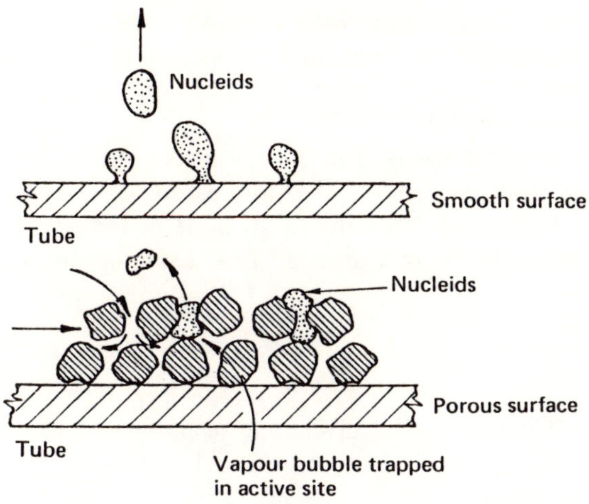

Figure 5.33 *Boiling at different heat-exchange surfaces*

temperature requirement for a given saturated temperature of the refrigerant. The porous surface also has a larger contact area.

Today, several techniques are available to obtain such porous surfaces. Manufacturers supplying tubes with porous surfaces are Union Carbide, Hitachi and Metofin. Figure 5.33 shows comparisons between high flux and normal surfaces.

Another modern design consists of providing the inside of the tubes with spiralling grooves. The grooves spread the refrigerant film over the heat-exchange surface by capillary action. This technique is used for dry evaporators by such companies as Daikin, Hitachi and Wieland.

The same technique can be used for condensers. The 'Thermoexcell-C' tube used in condensers is provided with 'teeth'. Capillary action promotes the drain of the refrigerant film towards the feet of the tooth.

In order to promote the heat transfer on the 'air side' of air coolers, fins with slits can be used. They can only be used on air coolers without frost. The slits increase the pressure drop over the cooler. This fin design is only used at present for small air-conditioning units.

Much research is yet to be done on the heat-transfer characteristics of evaporators. When we compare the graphs in the *Wärme Atlas*, 74 edition, with graphs that can be drawn from equations in the 84 edition, we note a large difference. Much has also been done recently by Dhar and Jain. Their equations have given more importance to characteristics previously not always used, such as the surface tension σ and the latent heat L_v. It appears that the effects of surface tension will gain in importance in the future. Unfortunately, different sources give different values for σ for the same refrigerant, a fact that does not allow reliable use of the new formulae, especially where ammonia is concerned. From this we can see that optimization of heat-transfer surfaces for ammonia is still in its infancy. Better heat-transfer characteristics will make it possible to reduce the amount of refrigerants used in evaporators. This will not only benefit the environment but also enable the dimensions of refrigerant separators to be reduced. A start has been made for plate evaporators. In the coming years more developments in enhanced heat exchange may be expected.

References

B. W. Ingram (Custom Coils, UK) *The Selection and Application of Air Coolers for Refrigeration Use*, Meeting Institute of Refrigeration, 31 March 1977.

A. Tent (Helpman, The Netherlands), Capaciteits bepaling en capaciteits opgave van luchtkoelers, *Koeltechniek*, 72 (1979); 73 (1980).

C. A. Infante Ferreira (Helpman, The Netherlands), Toepassing van pompecirculatie in luchtkoelers, *Koeltechniek*, 79 (1986).

C. H. M. Machielsen (Universiteit Delft, The Netherlands), Theoretische gezichts punten van het ontdooien van luchtkoelers, *Koeltechniek*, 78 (1985).

C. Marvillet (Centre d'Etudes Nucleaire, Grenoble, France), Evaporateurs à ebullitions de paroi. Effect de l'etat de surface des tubes, *Revue General du Froid*, Octobre 1986.

J. van Male (Grasso, The Netherlands), De capaciteits opgave in de luchtkoeler catalogus, *Koeltechniek*, 61 (1968).

J. Berghmans, (Katholieke Universiteit Leuven, Belgium, Instituut Mechanica), *Verbetering van de Warmte overdracht in kondensors en verdampers*, Meeting BVK/ABF, 1973, 1974 and 28 Jan. 1987.

6

Vessel and Piping Design

6.1 Ammonia Circuits

One method of determining the size of ammonia piping is made simply on the basis of a maximum refrigerant velocity U m/s. First the refrigerant massflow m is determined for each section of pipework using the relationship

$$\dot{m} = Q_0/dh \text{ kg/s}$$

where Q_0 = refrigeration capacity of the installation or circuit concerned,
dh = difference in enthalpy between the refrigerant entering and leaving the evaporator(s).

Then with the value of refrigerant specific volume v for each section of the circuit, obtained from the $h/\log p$ chart or tables, we can calculate the various refrigerant flowrates V m^3/s using the equation

$$\dot{V} = \dot{m}v$$

The cross-sectional area A of each line can be determined from the equation

$$A = \dot{V}/U \text{ m}^2$$

In normal practice the following velocity values are used.

(a) Suction lines: 15–20 m/s.
(The high velocities apply to large refrigeration capacities and low evaporation temperatures, below −25°C)
(b) Vertical pipe runs between separator and pump: 0.5–1 m/s
(c) Pump suction line: 0.2 m/s
(d) Discharge line: 12–25 m/s
(e) Hot-gas defrost line: 20 m/s
(f) Liquid line: not to exceed 1 m/s

Ammonia suffers a relatively smaller pressure drop than do CFCs, but nevertheless it is essential to calculate the actual pressure drop, especially in very large installations with lengthy horizontal and vertical pipe runs. We need to do this in order to keep it to a reasonable limit, and to select the discharge and suction pressures at the compressor. Normally the pipe diameters are chosen so that the total pressure drop for both the suction and delivery lines does not exceed the equivalent of 1°C.

This calculation is necessary for forced circulation systems in order to select the pump with the correct pressure characteristics. The calculations for the return lines from the evaporators to the liquid separator or intermediate cooler are rather complicated because of the mixture of vapour and liquid in the lines – the so-called two-phase flow. The pressure drop calculations use formulae derived from Bernoulli's equation and some empirical formulae.

First let us refresh our knowledge of fluid mechanics. When a mass m with a velocity U_0 accelerates at a rate a m/s^2 to velocity U_1 in time t seconds, then distance S = velocity × time, or $S = Ut$ m, and acceleration $a = U/t$ m/s^2. Also

$$S = (U_0 + U_1)t/2$$

$$U_1 + U_0 = at_1 - at_0 = a(t_1 - t_0) = at \quad \text{or} \quad t = (U_1 - U_0)/a \text{ s}$$

$$S = (U_0 + U_1)/2 \times (U_1 - U_0)/a = (U_0 + U_1)(U_1 - U_0)/2a$$

$$S = (U_1^2 - U_0^2)/2a \quad \text{so} \quad 2aS = U_1^2 - U_0^2$$

$$\text{and if } U_0 = 0 \text{ then } 2aS = U^2 \quad \text{or} \quad a = U^2/2S$$

$$\text{Work} = \text{force} \times \text{distance} = F \times S$$

$$\text{and } F = \text{mass} \times \text{acceleration}$$
$$\text{therefore } FS = maS = mU^2S/2S = mU^2/2 \text{ (the kinetic energy)}$$

The pressure drop in a tube depends upon several factors: the kinetic energy $mU^2/2$, the ratio of length/diameter, the density of the fluid ρ, the dynamic viscosity of the fluid η and the fricton factor Γ resulting from the tube internal roughness.

Γ can be calculated from Blasius's equation for turbulent flow, $\Gamma = 0.3164Re^{-0.25}$ where Re is the Reynolds number given by the equation $Re = Ud\rho/\eta$.

Bernoulli combined all the above effects in the following equation for the total pressure drop in a tube:

$$dp = \Gamma\rho U^2 l/2d \text{ N/m}^2$$

A similar equation is used to estimate the pressure drop through pipeline size changes, bends, tees, valves, filters, thermostatic and solenoid valves and other accessories. The form of this equation is

$$dp = \Sigma \rho U^2/2 \text{ N/m}^2$$

There are tables published by manufacturers for the factor Σ for different fittings and accessories. As a rough guide, the following can be used in the absence of any manufactuers' figures:

Stop valves	$4 < \Sigma < 7$
Compressor discharge valves	$5 < \Sigma < 7$
Compressor suction valves	$12 < \Sigma < 15$
Bends 90°	

d/r	0.4	0.8	1	1.4	2
Σ	0.14	0.2	0.3	0.66	2

Several manufacters now publish the K_v values for their valves and controls. K_v is the quantity of refrigerant passing at 1 bar pressure difference:

$$dp = \rho(\dot{V}/K_v)^2 \text{ bar}$$

where $\dot{V}$ is the actual volume of fluid passing expressed in m^3/kg and ρ the density in kg/dm^3.

6.1.1 *Two-phase Flow*

In pumped circulation systems the amount of refrigerant circulated is anywhere between 2 and 4 times the amount actually evaporated. We say in such cases that $n = 2$, 3 or 4, where n = rate of circulation/rate of evaporation.

If v_f = specific volume of the liquid and v_g = specific volume of the vapour then we an determine the specific volume of the mixture, $v_{mixture}$, from the equation

$$v_{mixture} = (n - 1)v_f/n + v_g/n$$

Next we can determine the dynamic viscosity of the mixture $\eta_{mixture}$:

$$\eta_{mixture} = (n - 1)\eta_f/n + \eta_g/n$$

With values of $\rho_{mixture}$ and $\eta_{mixture}$ we can determine $Re_{mixture}$ using Blasius's equation and thence $\Gamma_{mixture}$ and finally dp for the mixture:

$$dp_{mixture} = \Gamma \rho_{mixture} U^2 l/2d \text{ N/m}^2 \text{ or Pa}$$

After the addition of the dp values in the various suction line sections, we must include the difference in pressure dp_h resulting from any difference in refrigerant liquid level h between the evaporator and separator using the formula $dp_h = \rho g h$ N/m², and to this add the pressure drop in any valves, bends or other accessories.

Note when calculating the lines between separator and evaporators that there is less refrigerant mass circulating than in the lines between separator and the circuit. The liquid that is pumped to the evaporators is free of flash gas. Also remember that dh'/kg of refrigerant circulating in the evaporator circuits is a different value from that used in the calculation for the compressor mass flowrate. dh' is in fact the difference between the enthalpy of saturated vapour and the enthalpy of saturated liquid at evaporation temperature. Thus

$$\dot{m}_{\text{evap}} = Q_{\text{evap}}/\mathrm{d}h' \text{ kg/s}$$

6.2 CFC Circuits

In principle we can calculate the refrigerant lines for CFC circuits in the same way as that described above. However, the pressure drop is significantly greater for these refrigerants and there are also other factors to consider when designing such circuits. CFC installations demand more care than do ammonia installations; not only during the installation, but also during design of the circuit.

Four important points must be considered:

1. Pressure drop
2. Oil return
3. Prevention of unwanted changes of state of the refrigerant
4. Compressor protection.

1. Any pressure drop between evaporator and compressor results in a lower suction pressure for the compressor than would be the case if the evaporation pressure were experienced at suction. Extra pressure drop results in higher power consumption, whether the suction pressure is too low or the discharge pressure too high. It is recommended that the pressure drop in the suction and discharge line does not normally exceed a figure equivalent to a 1°C change in the saturated temperature up to a maximum 2°C. In liquid lines only 0.5°C is accepted. This pressure drop is not detrimental to the power consumption but can be a contributary factor in the production of flash gas.
2. Oil is transported throughout the installation because most CFC/oil combinations are soluble to varying degrees. The compressor crankcase oil level must at all times be maintained, while any oil on heat-exchange surfaces must be kept to a minimum. Good system design ensures that the oil returns to the compressor positively and quickly. The method of achieving this is to arrange for the refrigerant velocity to be high enough in those parts of the circuit where the refrigerant is in the vapour state to

carry the oil through those parts. Where the refrigerant is in the liquid state the oil is carried along as a mixture or a solution.

3. The refrigerant must remain in a liquid state in the liquid line and in a gaseous state in the suction and discharge lines. Mixtures are to be avoided in all lines except where unavoidable, for example in expansion lines although it must always be kept to a minimum.
4. When the compressor stops and the velocity of the refrigerant falls to zero, the oil in vertical lines falls. In the case of a rising discharge line, oil can then enter the compressor; or in the case of the suction line, it can enter either the compressor or the evaporator depending on which item is at the lower level. To avoid this, oil traps are placed where the suction line leaves the evaporator and where the discharge line leaves the compressor. See figure 6.1 for details of such arrangements. A trap in the discharge line is not necessary where an oil separator is placed after the compressor as standard. Traps collect and hold the oil until the

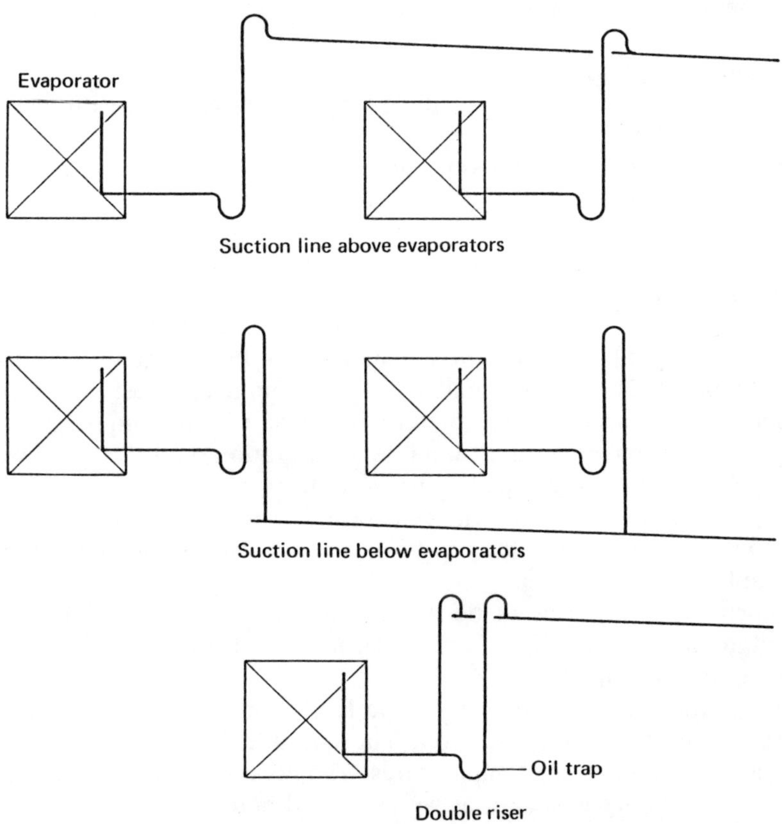

Figure 6.1 *Oil/liquid traps in CFC suction lines*

compressor restarts, and the flowing gas then carries it up the pipe. These traps should not be too big since large quantities of oil out of circulation can cause problems too. Any rising discharge line longer than 7.5 m must be fitted with an additional oil trap every 7.5 m. The same applies when the suction line rise is more than 2.5 m; in this case a trap must be fitted every 2.5 m.

All horizontal lines must be installed with a slope downwards in the direction of flow; the most critical line is that from evaporator to compressor.

The three main pipelines to be studied are:

1. Suction line
2. Discharge line
3. Liquid line.

Taking the suction and discharge lines together there are three important points to be observed:

(a) reasonable pressure drop
(b) oil-return to compressor including times of low capacity with resulting low velocities
(c) protection of the compressor against migration of large quantities of oil at one time.

The maximum allowable velocities in suction and discharge lines are 8 m/s and 10 m/s respectively.

Instead of calculating the pipe sizes by formulae charts may be used, such as those published by various manufacturers, see figures 6.2a and b (from Dupont). These charts enable us to determine the most suitable pipe diameter in mm or inches. The method involves starting from the required refrigeration capacity and a suitable evaporation temperature, which combine to provide a reasonable pressure drop.

Once the diameter d is known, we choose the nearest standard size pipe, and from this we can find the resulting pressure drop in the lower part of the graph.

A correction has to be made if the determined pipe size d_t has to be altered to a standard size d_s. The formula to calculate the corrected pressure drop is $dp_t\ (d_t/d_s)^5$.

Multiplying the calculated dp by the length of the line concerned, we determine the total pressure drop. The charts allow for readings to be taken at three condensing temperatures. The pressure drop caused by each valve and accessory is added, as we explained before.

In order to be sure that oil will flow up vertical pipelines, the so-called risers, a check must be made to see if the velocity in the line is high enough.

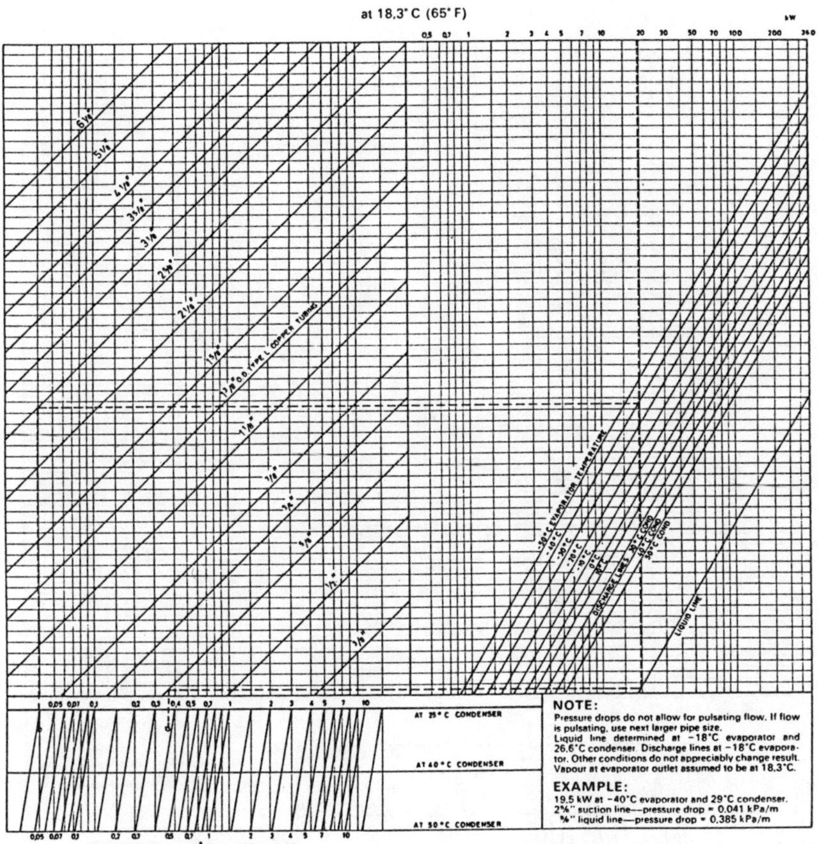

(a)

Figure 6.2 *(a) Pressure drop diagrams for R22 lines (Courtesy of Du Pont de Nemours)*

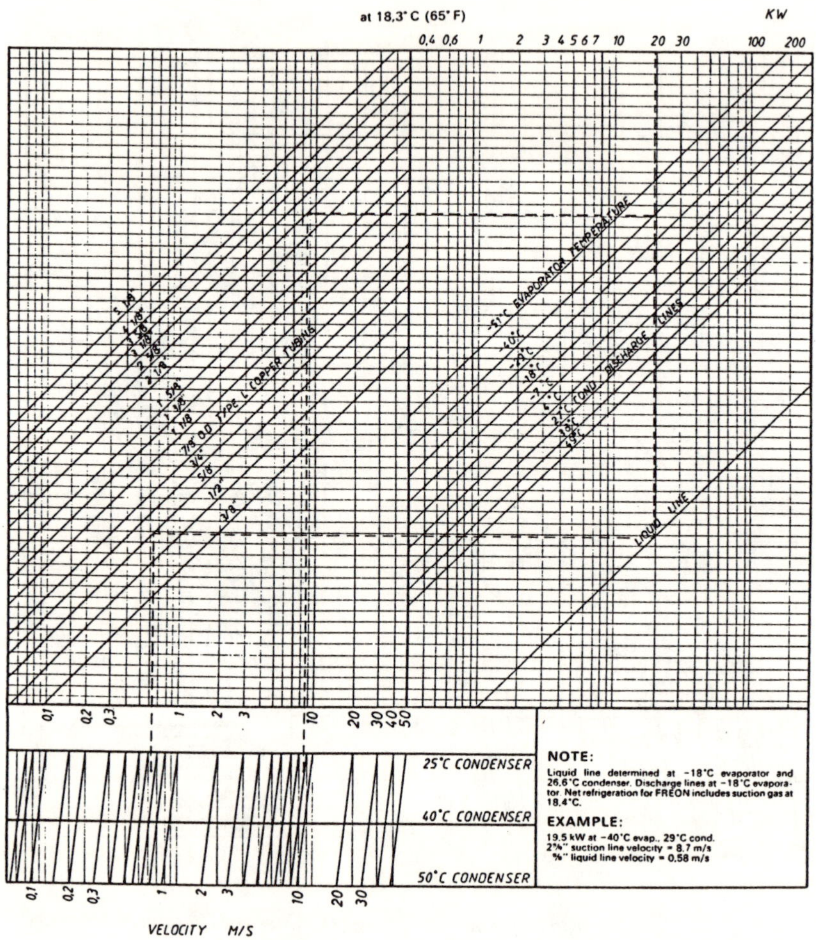

(b)

Figure 6.2 *(b) Velocity diagrams for R22 lines (Courtesy of Du Pont de Nemours)*

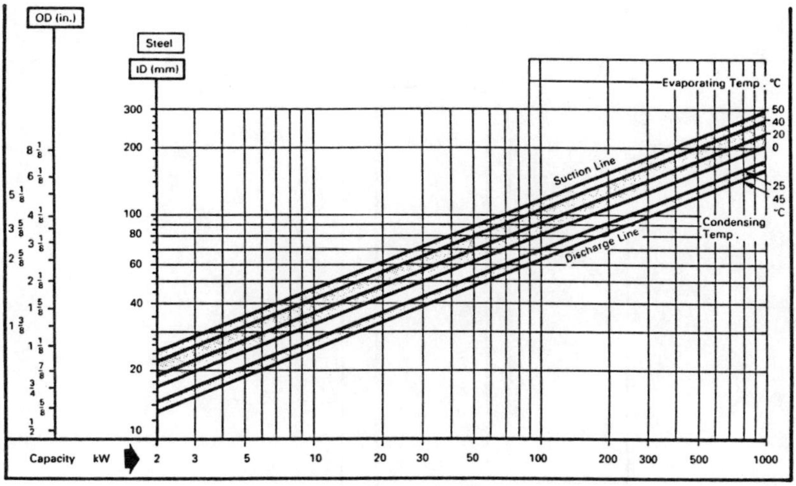

(c)

Figure 6.2 *(c) Maximum pipe size (diameter) diagrams to carry oil up R22 vertical risers (Courtesy of Du Pont de Nemours)*

This check can be made using manufacturers' charts. Figure 6.2c shows an example. If this check dictates the use of a smaller tube size, then this tube size has to be used to recalculate the pressure drop from the charts discussed above. If charts are not available then care must be taken to calculate whether the minimum velocity of 5 m/s in vertical lines and 2.5 m/s in horizontal lines is assured.

When there is a difference in level h m between two components in the installation, and the line connecting them contains liquid, then this pressure increase or decrease dp_h must be taken into account. The formula to calculate dp_h is

$$dp_h = \rho g h \text{ N/m}^2 \text{ or Pa}$$

where g is the acceleration due to gravity: 9.81 m/s^2. There will be a pressure increase when the refrigerant receiver is at a higher level than the evaporator.

Note that discharge lines are calculated in the same way as suction lines. Ensure that the correct lines on the charts are being used.

So far we have calculated for constant capacities. However, in most cases the capacity will vary. We must be sure that oil-return still takes place at the lowest capacity, which may mean that the gas velocity must be adjusted to the changed conditions.

The problem can be resolved simply by using a riser with a smaller section, so increasing the pressure drop for that part of the circuit at full load. Where there is a wide variation, long lines or a big level difference, a so-called 'double riser' is used. See figure 6.3 for details of this arrangement.

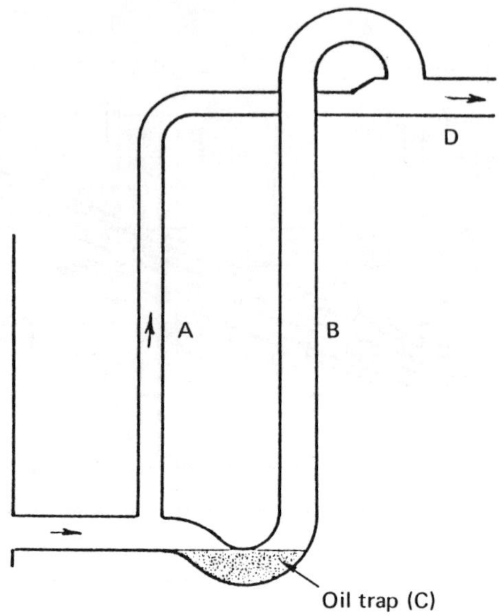

Figure 6.3 *Double riser system*

A is the smaller line sized for minimum load. Lines A and B working together on full load are selected in such a way that the total pressure is acceptable. C is an oil-trap, which collects oil at minimum load until the oil closes B. D is a bend to avoid oil return through B when only A is in action.

Double risers are needed where compressors are located above the evaporator level. They are sometimes used in rising discharge lines.

Note that there are occasions when risers are required in pumped systems in ammonia circuits, such as when evaporators are placed on a lower level than the liquid separator.

Liquid line calculations are made with the same charts, using the appropriate lines; condenser to receiver lines should not exceed 0.5 m/s and receiver to evaporator lines 1.8 m/s. However another check is necessary. Both friction in the liquid line and the static head pressure decrease can cause the formation of flash gas in the line. Subcooling of the refrigerant in the condenser or in a heat exchanger can prevent flashing of the liquid. The subcooling must be sufficient to ensure that the temperature of the liquid when it reaches the expansion device is still below the saturation temperature of the liquid at the lower pressure at the expansion device.

Additional practical advice follows:

- Avoid significant pressure drops in the suction lines to the refrigerant pumps. These can cause cavitation damage to the pumps.

- Keep bends, tees and stop-valves in this line to a minimum.
- Pressure drop in a sight glass is negligible.

6.3 Liquid Separators, Receivers, Refrigerant Pumps and Intercoolers

Although piston compressors are protected to some extent against liquid slugging by the built-in spring-loaded safety head assembly, extra precautions must be taken since the safety head only protects against very small quantities of liquid. For this reason, liquid separators are incorporated in most industrial installations and even in commercial installations of some importance. In the latter, simple liquid traps or liquid accumulators are sufficient. These are separators of medium size, placed between the evaporator outlet and the compressor inlet. The wide section of the separator creates a significant decrease in the gas velocity, so allowing drops of liquid refrigerant to separate out and collect in the bottom. See figure 6.4 which shows the internal arrangement of a typical separator. The ambient temperature will generally be enough to evaporate the collected liquid. Where this is inadequate a hot-gas coil or an electric heating element is incorporated.

Industrial installations with 'flooded' evaporators employing natural or pumped circulation require the installation of similar but much larger separators. These vessels are described below.

For large capacities the dry-gas outlet connection is placed in the middle of the vessel, with two wet-gas inlets, one at each end of the vessel. In this way the gas entering is divided into two flows, thus providing two separator sections. Two views of a typical horizontal liquid separator are shown in figure 6.5a.

For smaller capacities both the wet-vapour inlet coming from the evaporators and the dry-vapour outlet leading to the compressor are located at the end of the vessel. Sometimes this outlet connection is not welded directly to the shell of the vessel itself but to a small vertical drum built on the top of the vessel.

As we will see, the distance between the two connections must be calculated in accordance with the separation time and the gas velocity. The liquid refrigerant leaves the separator from the bottom of the vessel by the pump line connections shown in figure 6.5b. Horizontal runs in the line from separator to pump should be avoided.

Usually a small vertical drum or a sack-like liquid collector is fitted centrally under the vessel. The pump or pumps are connected to this receptacle, which also serves as an oil collector and oil-drain. The pump connection must be made on the side of the sack, not on the bottom. In this way turbulence above the liquid outlet is minimized and unwanted oil and sludges are not induced into the pump suction inlet.

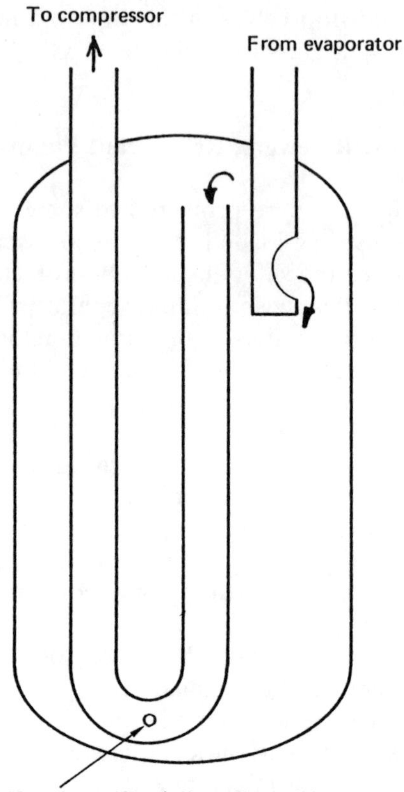

Figure 6.4 *Liquid trap*

The vessel is provided with a liquid inlet placed above the operating liquid level. The liquid level is normally controlled by a low-pressure float-operated expansion valve, or alternatively an electronic device operating a solenoid valve.

The liquid level in the vessel is indicated by means of a sight glass, a vertical tube of 50 mm diameter. The bottom of the sight glass is connected to a point in the vessel below the liquid level, and the other end to a point above the liquid. The upper connection to the vessel is restricted to a 3 mm orifice in order to minimize erroneous readings caused by boiling turbulence in the tube. Since the tube is not insulated, moisture in the air will be deposited as frost on the tube over the part where liquid fills the tube, thereby providing a reasonable indication of the liquid level. The tube must be made of stainless steel, as continuous fluctuations between the wet and dry states make it extremely prone to corrosion. Frequently, float valves and switches are fitted to this tube. These devices serve as level

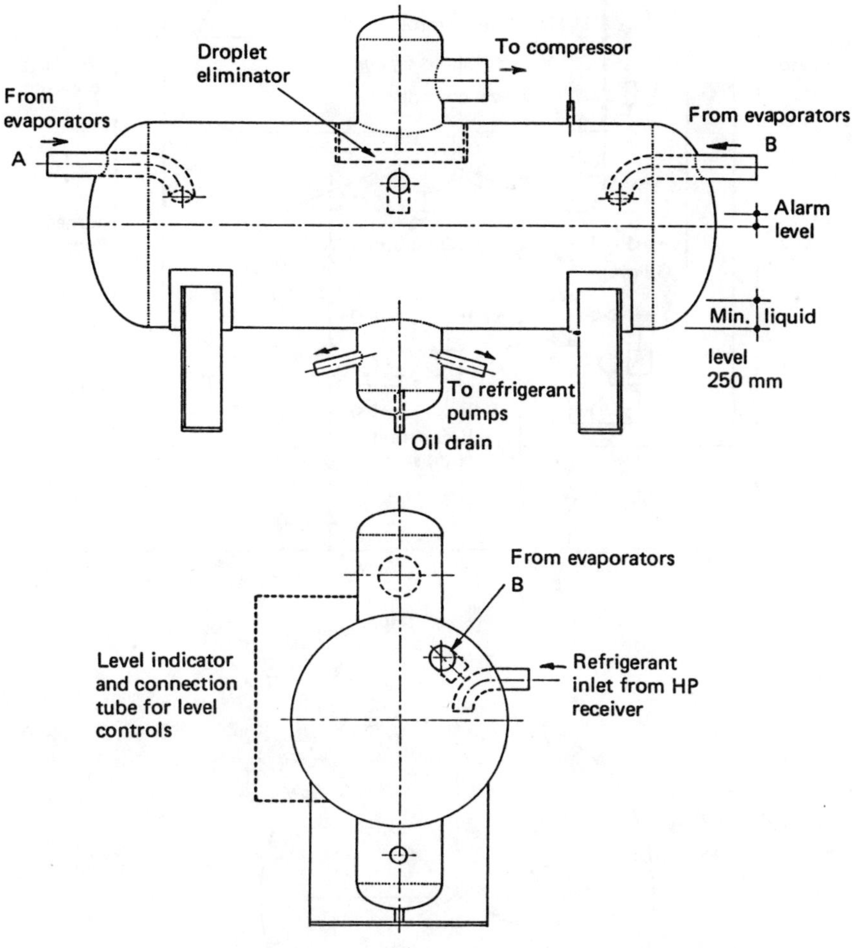

Figure 6.5 *(a) Liquid separator*

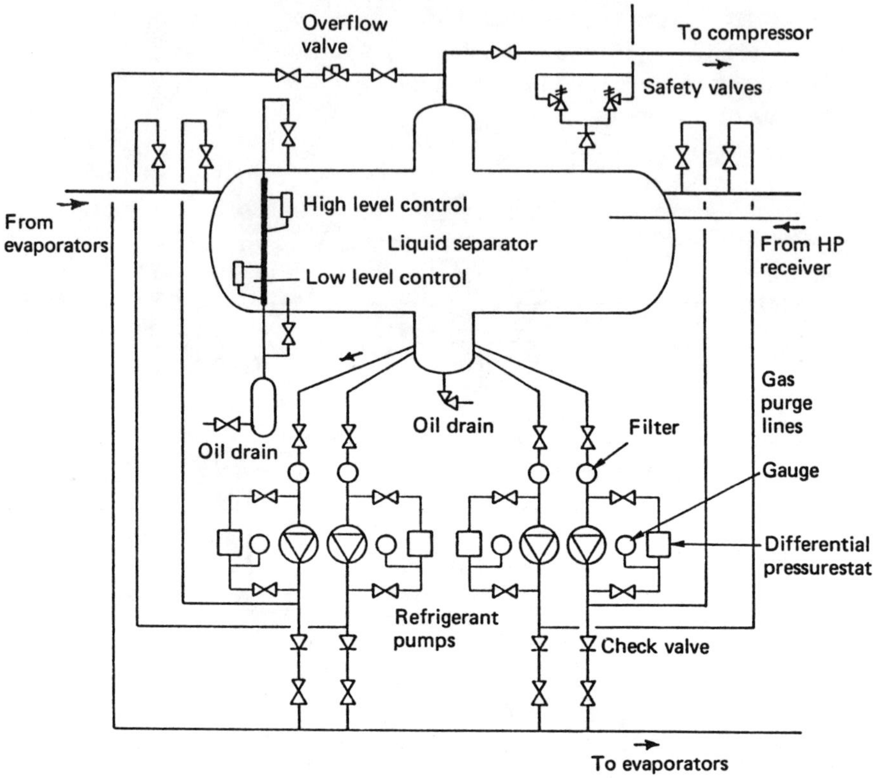

(b)

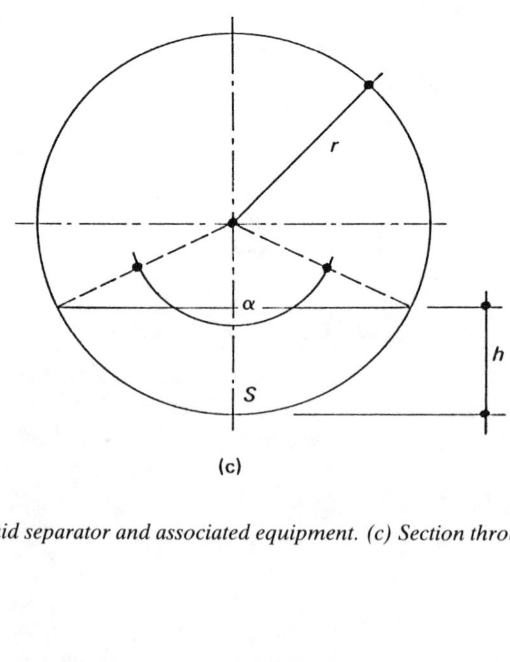

(c)

Figure 6.5 *(b) Liquid separator and associated equipment. (c) Section through liquid separator*

controllers or activators for low and high level alarms. Finally in the interests of safety, the separators are equipped with pressure relief valves.

Sometimes the separation process is reinforced by placing labyrinth-type eliminators or special drop-separators inside the vessels, close to the dry vapour outlet. In the case of high-pressure regulation, the float valve is placed at the bottom of the condenser. Whenever the liquid in the bottom of the condenser reaches a certain level, the float valve feeds the liquid to the separator. In this form of control there is usually no liquid receiver built under the condenser, but sometimes there is a small collector in which the float valve is assembled.

6.3.1 Pump Circulation

In large installations where the distance between the machine room and the evaporators is significant, pump circulation is normally used. Refrigerant liquid pumps are of the centrifugal or positive displacement type, specially designed for use with refrigerants. In order to avoid refrigerant leakage either a special shaft seal is used, or the pump and motor are built together in a hermetic unit similar in design to CFC hermetic compressors.

The advantages of forced circulation of the refrigerant are:

1. Better heat exchange efficiency because the oil is regularly washed from the inner surface of the evaporator tubes. The heat transfer coefficient α_i is also improved since the surface is always fully wetted.
2. The fact that oil is regularly taken from the evaporator coils makes it easy to purge it from the liquid separator. it can either be purged directly from the oil leg of the separator itself, or be collected in a special, small collecting vessel or oil dump tank. A larger vessel or oil still can be used to serve several evaporators or liquid receivers. By disconnecting this vessel from time to time from the rest of the installation by means of stop valves, recovery of the oil is possible without stopping the operation of that part of the system. The supply of liquid refrigerant to the individual evaporators is regulated by simple, individually hand-set needle valves.

6.3.2 Oil Separation in CFC Installations

As discussed above for R717 installations, it is very easy to purge the oil from the lowest part or from the oil down-leg of the appropriate vessel. However, since hydrocarbons always carry 2–5% of oil which cannot always be separated in a separator, measures must be taken to return this oil from time to time to the compressor. Therefore, on the pump-discharge line, a bleeding line is connected, which leads through a filter and an expansion valve to a small heat exchanger. On the outlet of the heat exchanger a solenoid valve is placed, which is activated by an oil level float valve on the compressor. The outlet of the heat exchanger is finally connected to the suction line of the circuit and the heat exchanger is equipped with a hot gas coil. When the oil level in the compressor crankcase is too low, the solenoid valve in the outlet opens, oil-rich refrigerant is injected and evaporated in the heat exchanger, and flows back through the suction line to the compressor.

In systems without pump circulation a special nozzle must be placed in the suction line to create the necessary low pressure at the point where the oil-rich gases enter the suction line.

6.3.3 Pump Selection and Cavitation Problems

Pump selection is made according to the circulating refrigerant flowrate and the required pump pressure head which is derived from the calculated total pressure drop of the system. If H = the 'pump height' then

$$H = \mathrm{d}p/\rho g \quad (\mathrm{m})$$

where $\mathrm{d}p$ is the pressure loss in the circuit, ρ is the density of the refrigerant, and g the acceleration due to gravity.

The problem of cavitation is taken into account by consideration of a special characteristic of refrigerant circulation pumps known as the 'net positive suction head', or NPSH. Cavitation occurs during pumping when gas or air is mixed with the liquid. When operating a forced refrigerant circulation system, we are pumping a liquid around which may be affected by any rise in temperature or drop in pressure. Cavitation problems might arise during pumping if liquid is transformed into gas as a result of a local pressure drop at certain spots inside the pump. At these spots the pressure becomes lower than the saturation pressure of the refrigerant, while at other spots where the pressure has risen again these gases will recondense, causing implosion of the gas bubbles. These implosions can be sufficiently strong to produce cavitation corrosion damage. Cavitation wears out pump parts prematurely and may ultimately cause breakdown of the installation.

Now look at figure 6.6. Note that during part of the operation the pressure inside the pump can be lower than that at the inlet of the pump. The maximum value of this pressure difference is the NPSH of the pump. Pump manufacturers usually publish datasheets giving the NPSH values for their pumps. Always ensure that the available pressure on the inlet of the pump is a good deal higher than the stated value of the NPSH, so that the internal pressure drop can be allowed for. The level difference between the liquid surface in the separator and the inlet of the pump must be at least equal to the sum of the pressure drops in the suction line of the pump and the pump NPSH value, plus a safety factor of 30% of the NPSH. In

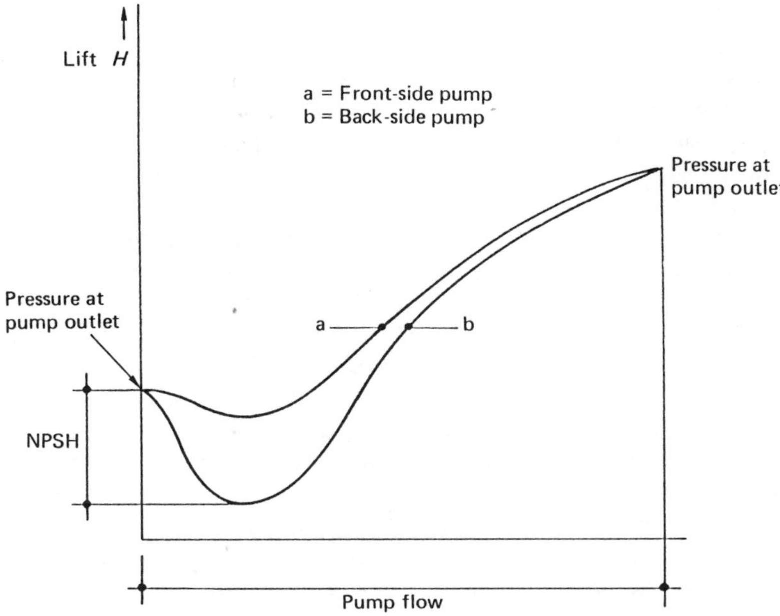

Figure 6.6 *Variation of internal pressure with flow for liquid pump*

practice this normally results in a 1.5 m level difference between the pump inlet and the refrigerant level in the separator. This is more than manufacturers advise and higher than the calculated value, because there is yet another reason for cavitation that must be taken into account – namely cavitation caused by pressure drop arising from restriction of liquid supply which results from operation of the expansion valve or sudden pressure changes. These effects may result in an extra pressure drop of between 0.1 and 0.3 bar.

In hermetic refrigerant pumps, another reason for cavitation may be the extra heating of the liquid refrigerant caused by friction in the pump bearings and heat generated in the motor windings. To ensure that this thermal energy is removed under all circumstances, the pumps must be equipped with a nozzle in a by-pass line between the suction side and the separator, in order to guarantee a minimum refrigerant flow at maximum pressure. Another nozzle is placed in the discharge line to keep the refrigerant flow to a certain maximum at a minimum pressure. This guarantees the minimum pressure necessary for a continuous flow of refrigerant and for the pressure to stabilize the hydraulic axial pressures. Finally all refrigerant pumps must be equipped with a differential pressure cut-out.

6.3.4 *Intercoolers*

Intermediate coolers for 'flooded' installations are mostly built in the same way as liquid separators. In fact they also act as liquid separators for the high stage of a two-stage installation. Mostly they are constructed in a vertical design; but for large capacities they are built horizontally. See figures 6.7 and 6.8 for details of both forms of design.

As we have seen earlier there are two main types of intercoolers, the open flash type and the closed type with a liquid cooling coil. This coil is placed in the lower part of the vessel. The introduction of the gas coming from the low stage of the compression system is always made under the liquid level by means of a perforated tube. In this way satisfactory heat exchange for desuperheating takes place. The total area of the holes is equal to twice the area of the tube. A small hole of 8 mm diameter is drilled in this tube at the top of the vapour section of the vessel in order to equalize the pressure should an abnormally high pressure occur at the liquid level. The same result can be obtained by adding a small tube of about 10 mm internal diameter between the gas inlet and the free gas section of the intercooler. If there were no equalizing arrangement, liquid

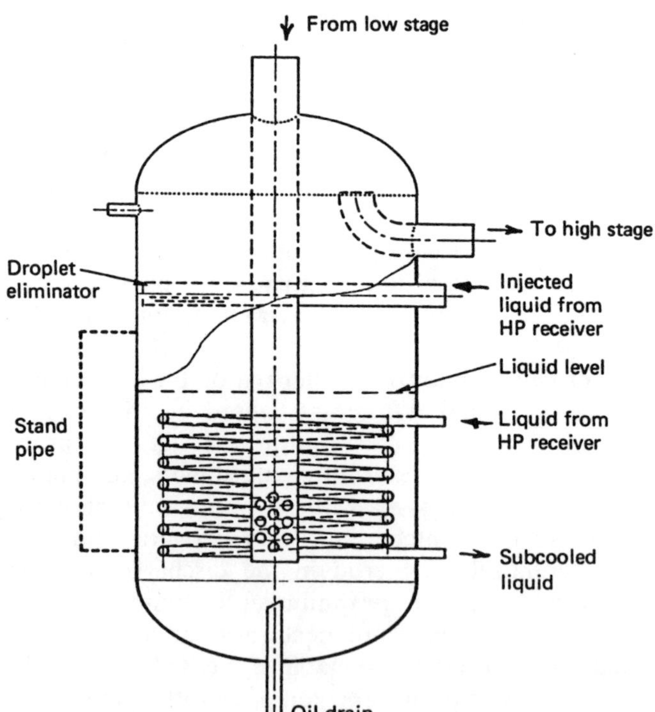

Figure 6.7 *Closed-type intercooler*

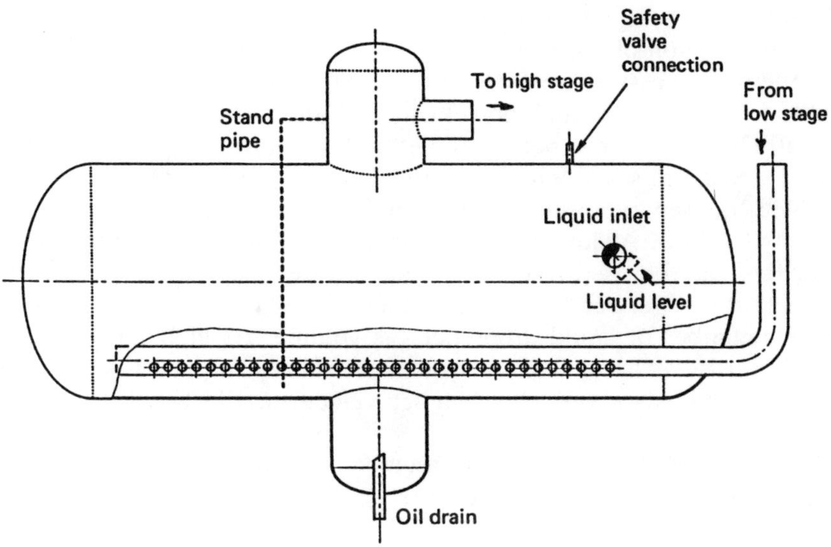

Figure 6.8 *Open-type intercooler*

might be pushed up the tube and so enter the low stage compressor to cause liquid slugging.

Intercoolers for CFC 'dry-evaporating' systems operate in a different way. In contrast to a flooded system, a 'dry' system must ensure that the gas and liquid maintain the velocity necessary for return of oil. A typical design is shown in figure 6.10. In this arrangement the cooling refrigerant flow is controlled by a thermostatic expansion valve as used with normal 'dry-evaporators'. The refrigerant is injected into the shell as the low stages gases are flowing through the coil, while the liquid refrigerant to subcool passes through.

6.3.5 *Calculation of the Liquid Separation Zone Dimensions*

Only the upper part of the separator vessel is available for the separation process. The lower part must be free to receive the extra amount of liquid refrigerant, which on some occasions does return to the separator, for example during the defrosting process.

The area A necessary for liquid separation is calculated as follows:

$$A = V/U \text{ m}^2$$

where V = suction volume flowrate m^3/s and U = gas velocity m/s.

For horizontal vessels, A is not the section of the vessel itself but only the section available for separation at the time of the highest liquid level that may occur during operation. Generally this section A is considered to be

45–50% of the complete section of the vessel. In vertical vessels the area A is the full section of the vessel.

If the time necessary for a drop of liquid to separate itself from the flow and to fall is 3 seconds and the velocity $U = 0.3$ m/s (for systems working at −10°C), the distance between inlet and outlet of the gases must be at least 3 s × 0.3 m/s = 0.9 m. When the distance between inlet and outlet is doubled, the separation time is also doubled, so we can then increase the velocity by 50%. At lower temperatures and/or higher capacities, higher velocities may be used. At lower temperatures the specific mass of the refrigerant is lower per capacity unit; there is less gas and so there will be a smaller number of liquid drops per gas volume. At high capacities, the vessel and tubes are relatively larger and post-evaporation is easier. In these conditions we can allow higher velocities, which will result in a smaller area A.

The table below can be used as a guide in selecting the intercooler gas velocity at different temperatures:

Evaporating temp. (°C)	*Gas velocity (m/s)*
−10	0.3
−20	0.4
−30	0.5
−40	0.7

However these values are not recommended when using vertical vessels. In vertical vessels, velocities about 20% lower than the values from the above table must be used.

The distance between liquid level and pump line connection must be at least $2d$ to $3d$, where d is the diameter of the pump line, if turbulence is to be avoided. It is also advisable to minimize obstructions and bends, as far as possible, in the connection line between vessel and pump.

6.3.6 Separator Liquid Capacity

Depending on the thickness of frost on the evaporator surface, the quantity of liquid held in the evaporator can vary between 20% and 90% of the total volume. From this we can see that 70% of the liquid content of the evaporator coils may return to the separator during normal operation. At other times, for example during defrosting, the condensed hot gas may flow to the separator, and under certain conditions when an evaporaor or a group of evaporators is out of service it is possible for most of the refrigerant charge to migrate to the separator.

There are two evaporator supply and return pipework arrangements: (a) bottom feed and (b) top feed.

With (a), the evaporator is fed with liquid refrigerant through the connection in the bottom part while the wet gases leave the evaporator through the connection in the top. In this arrangement, during defrosting or at other times when the evaporator is out of operation, the liquid will normally stay inside the coils. However, especially during a long stoppage, part or all of the refrigerant might flow back to the separator; for example, when the valves are leaking without it being known. In this case we must anticipate a large amount of refrigerant returning to the separator. Bottom feeding the evaporator also has the advantage that defrosting with hot gas will be more efficient since the hot gas is bubbled through the liquid.

With (b), the evaporator is fed with liquid through the top connection and the wet gas leaves the evaporator from the lower connection. Defrosting is less efficient, but on the other hand the cold refrigerant leaves the evaporator very quickly by gravity if the evaporators are at a higher level than the separator. Some designers see this as an advantage when rapid defrosting is required.

In each case then, we can be certain that all the liquid refrigerant could come back to the separator during defrosting or when the evaporators are out of service.

At any time there will be, of course, different evaporator conditions and not all will be at minimum filling. When many coolers are out of service, the refrigerant flow through the separation part of the separator is smaller, and a smaller section A could be sufficient. However, evaporator behaviour is unpredictable and if we want to make absolutely sure that there is a sufficiently large separation zone available at any time, it is better to calculate the volume in the following way.

The space which has to hold the contents of the evaporators shall be located between the operating level plus 50 mm modulating range and the alarm level. This method is highly recommended in the case of a high-pressure float valve system where there is no liquid receiver available to cope with liquid variations.

Do not forget that the normal content of an evaporator in a flooded system is about two-thirds of the coil volume. The actual content varies with the degree of frosting. The thicker the ice coating on the tubes, the more liquid will be held in the evaporator coils. For long suction lines, a factor of about 20% must be allowed for and added to the total capacity.

Above the alarm setting level there must be enough free separation space to provide the maximum allowable gas velocity. For practical reasons the maximum length of separators seldom exceeds 5 m and for economical reasons the maximum diameter is seldom more than 2.5 m. Dimensions in this range result in expensive construction costs, owing to the plate thickness and the price of vessel endplates. Normal diameters for separators with pump circulation are between 800 and 2000 mm.

The volume of intercoolers is simply based on a liquid level which just covers the subcooling coil or the horizontal perforated refrigerant inlet pipe, allowing for the level difference required by a regulating float valve of about 50 mm. To this volume, the necessary separation section *A* must be added.

In cases where the intercooler also serves as a liquid separator for a circuit working with the intermediate temperature as the evaporating temperature, the volume is calculated exactly as for low-pressure liquid separators.

High-pressure Liquid Receivers

The liquid receiver is placed immediately after the condenser, on a lower level than the liquid outlet of the condenser. It serves to collect the condensed refrigerant before sending it to the low-pressure side of the installation. There will never be any liquid refrigerant held in the bottom part of the condenser except for the operating charge, so keeping its surface completely free for the necessary heat exchange of the condensation process.

If liquid becomes blocked in a part of the condenser, the result is a decrease in effective surface and an increase in condensating pressure. In order to ensure sufficient draining by gravity forces, the condensate line between condenser and receiver must be sufficiently large and have few pressure-reducing accessories such as valves and bends.

Sometimes it is necessary to provide a pressure-equalizing line between the vapour space of the receiver and the top of the condenser to prevent vapour-locks.

The receiver is equipped with a sight glass for level inspection, an oil receptacle with oil drain on R717 installations, a safety valve and a refrigerant outlet. In CFC installations it may be equipped with a refrigerant filter/dryer in the outlet. An air purging valve may be connected to the top of this vessel.

The refrigerant charging connection for the installation is placed in the liquid line immediately after the liquid outlet. Sometimes a water-cooled coil is put in the lower part of the receiver to realize an extra degree of subcooling of the liquid refrigerant.

In systems working with high-pressure refrigerant float control, the receiver is just a simple vessel into which the float valve assembly is housed. The receiver also serves as a lock between the gas and the liquid side of the system.

Normally the function of the receiver is to hold a supply of refrigerant liquid in order to accommodate variations in the demands of the evaporator during operation. In some systems, particularly small installations, it is designed to hold the total refrigerant charge of the installation during stoppage or repair. These receivers are sometimes equipped with a refrigerant leakage detector or an alarm system, which will give a warning signal when the stock of liquid refrigerant falls below a safe level.

The system consists of an open-ended small capillary tube inserted in the receiver and almost touching the bottom. The other end of this tube is wound around the hot gas line and acts as a small evaporator. A thermosensitive bulb is connected to the tube leading from the receiver to this evaporator. The end of this evaporator tube is then connected to the low-pressure side of the system. When the liquid level falls below the opening of the inserted tube, gas will flow instead of liquid. The bulb reacts to the difference in temperature between liquid and gas, and an alarm is activated.

Calculation of the Receiver Capacity

R717 installations

The volume of the vessel depends on the design concept of the installation, and general rules are difficult to give. In each case a minimum layer of liquid must always be present (about 20% of the total volume of the vessel), on top of which a certain quantity of liquid must be stored to allow for fluctuations in the opening and closing levels of the level-regulating system in the separator. Theoretically this liquid capacity is sufficient, however there will always be situations when these fluctuations, normally dealt with by the separator, will have an influence on the receiver. When for some reason a lot of extra refrigerant is temporarily returned to the separator, its float valve system will stay closed for some time, so no liquid will leave the receiver. Extra space must be available to allow for surplus stock of the liquid, otherwise too much refrigerant will begin to collect at the bottom of the condenser. It would be easy to calculate the volume of receivers needed for the total refrigerant load of the installation, but this would result in enormous vessels.

The following treatment is therefore proposed. Assume that the amount of liquid refrigerant stored in the separator at standstill is M. The volume of the high-pressure receiver must be calculated in such a way that it will accommodate the quantity M, because we saw that when the installation is restarted, this amount of liquid will flow over to the receiver. This situation will prevail until the evaporators reach their normal liquid content. If the circulating mass of refrigerant in the installation is m kg/s and the refrigerant pumps are pumping nm kg/s of liquid around, $(n - 1)m$ kg/s will be used to fill up the evaporators and m kg/s will after evaporation pass through the compressor and condenser before finally reaching the receiver.

The filling time t of the evaporators will be:

$$M/(n-1)m \text{ seconds}$$

The quantity of liquid to be stored in the receiver will be:

$$Mm/(n-1)m \quad \text{or} \quad M/(n-1) \text{ kg}$$

With a minimum stand-by capacity of 20% liquid in the receiver and 20% free gas volume, the required receiver volume is:

$$1.67M/(n-1) \text{ kg}$$

When $n = 4$, this equation becomes:

$$0.555M \text{ kg or } 0.555M/\rho \text{ m}^3$$

Note that in this calculation the value of M is for all evaporators that are not operating. However, there may be evaporators which are constantly working, even over week-ends. Theoretically, the fillings of such evaporators should be subtracted from the total amount of liquid refrigerant of the system.

CFC installations
The method of control determines the capacity of the liquid receiver for CFC systems, resulting in two different procedures.

1. *Standard* For standard systems using the pump-down method of control, where the refrigerant charge is normally not very high, we specify the receiver volume that is able to receive the total charge of the installation plus an equal volume of free vapour space.

Dry evaporators normally operate with a charge of one-third of the coil capacity of the evaporator. Thus the volume of the vessel is given by $V = 2(\frac{1}{3}$ of evaporator coil capacity + the refrigerant in the pipelines and heat exchangers). In these kinds of installations the so-called pump-down system is frequently used as part of on–off control cycle. One advantage of this system is that all the charge of the installation is pumped into the liquid receiver to protect the compressor against liquid reaching the compressor during start-up.

2. *Exception* When condenser capacity is regulated by means of backing-up liquid in the condenser coils, the receiver must be capable of holding the extra quantity of refrigerant required for this technique.

Low-pressure Receiver System

A British contracting company offers a patented system that eliminates the use of the thermostatic expansion valves as well as refrigerant pump systems. It is called the Star Low Pressure receiver system.

The system design involves a low-pressure receiver placed in the suction line of a CFC plant. The receiver contains an arrangement of heat exchangers, traps and distillation devices which allow it to perform the following functions:

(a) It replaces the conventional high-pressure receiver.
(b) It acts as a suction linc superheater.
(c) It improves the efficiency of the refrigeration cycle by subcooling the liquid refrigerant coming from the condenser by heat exchange with the suction vapour returning to the compressor.
(d) It acts as a pump, allowing excess liquid refrigerant to be supplied to the evaporator which can then operate in a 'flooded' manner. It should be noted that the 'pumping' action of the low-pressure receiver has no direct effect on the thermodynamic efficiency of the refrigerating system because the amount of energy supplied to the excess refrigerant liquid coming back from the evaporator is returned to the evaporator in the form of additional subcooling of the liquid refrigerant supply to the evaporator.
(e) It holds the total refrigerant charge, thus preventing any risk of liquid being drawn into the compressor.
(f) It eliminates the need for evaporator superheat, thus obviating the extra evaporator surface required to give superheat.
(g) It eliminates the need for head pressure control.
(h) It makes the use of liquid-line solenoid valves unnecessary since the liquid will spill over into the low-pressure receiver.
(i) It allows rapid pressure equalization during the off-cycle, thus reducing load at start-up.
(j) It acts as an oil rectification device; however, special care must be taken to ensure satisfactory oil-return.

6.4 Air Purging

Air purging of a refrigeration installation is essential to provide economic and efficient operation of the plant.

The deleterious effects resulting from the presence of air in a system are:

1. The head pressure and the discharge temperature will increase, causing compressor wear and even break-down.
2. Capacity will decrease and power consumption will increase.
3. The oil and refrigerant composition resulting from the high temperatures will suffer, as explained earlier.

Purging is necessary to remove non-condensable gases, the presence of which produce the same effects as air.

Air and non-condensable gases may enter the installation by various means as listed below:

1. By ingress into installations operating at an evaporator pressure below atmospheric, through any small break in the seal.
2. During repair or routine maintenance.
3. When oil or refrigerant decomposes.
4. During the changing or adding of oil.

Figure 6.9 shows the effects of air on power consumption and plant capacity. The combination of increase in power consumption and decrease of capacity will obviously greatly increase the running costs.

The non-condensable gases can be removed manually or automatically:

(a) *Manually* – by the opening of a simple purging cock. This method involves labour and large refrigerant losses, and is particularly to be avoided with CFCs.
(b) *Automatically* – in this method equipment is used in which the mixture of refrigerant gas and air is cooled by refrigerant until most of the refrigerant is liquefied and separates out. The percentage of air increases in the remaining mixture and a float valve arrangement enables it to be blown to atmosphere.

There are several designs for this facility. In one, the cooling refrigerant is part of the main installation. Hence, it is only possible to work at the

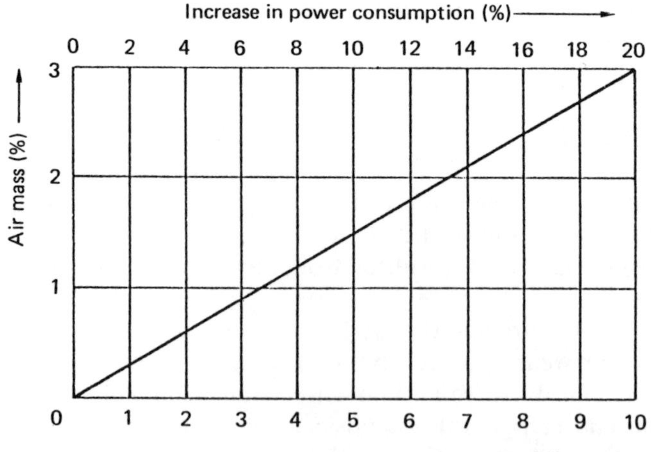

Figure 6.9 *Effect of pressure of air on the refrigeration cycle and the power consumption*

plant evaporation temperature and only when the main system is in operation.

Another is equipped with a small independent refrigeration system. Here the optimum evaporation temperature can be independently selected without wasting much refrigerant and with very high liquefaction efficiency. It will also work when the main installation is out of operation. This system is a more expensive installation, but the return or investment is soon realized.

A third system utilizes several purge points since air may collect in a number of places in the installation. This apparatus analyses the points where purging is most frequently required, thus enabling the purge times to be appropriately programmed for each point.

6.5 Heat Exchangers

We have already seen that CFCs have particularly poor efficiencies when the liquid enters the expansion device at high temperature, and how this effect could be reduced by subcooling. The improvement in efficiency by subcooling is much better for CFCs than for ammonia, because the saturated liquid line for the CFCs has the better shape.

In CFC systems, heat exchangers are installed inbetween the compressor and the evaporator to provide subcooling of the liquid as well as protection against liquid slugging, normally the function of separators or accumulators. A typical suction/liquid line heat exchanger is shown in figure 6.10. In fact, a heat exchanger is an intercooler without the secondary liquid inlet.

Other advantages of this type of heat exchanger when installed between evaporator and compressor are:

1. The evaporation process can be continued almost until the end of the last coil of the evaporator. No space is needed for superheating inside the coils, so the efficiency of the evaporator surface is improved.

   When superheating is the main objective, we calculate the enthalpy part (about 10%) of the evaporation plus superheat on the gas side of the heat exchanger, and this same quantity of enthalpy is now subtracted from the liquid passing through the heat exchanger. What we lose in efficiency on the evaporation side, we gain in subcooling. The result is a small gain in efficiency for R12 and R502, but a loss when using R22.

   It is important to feed warm superheated gas into the compressors in order to guarantee a sufficient flow of oil back to the compressor since the viscosity of the oil becomes too high at low gas temperatures. The thermal energy forces the CFC to leave the oil.

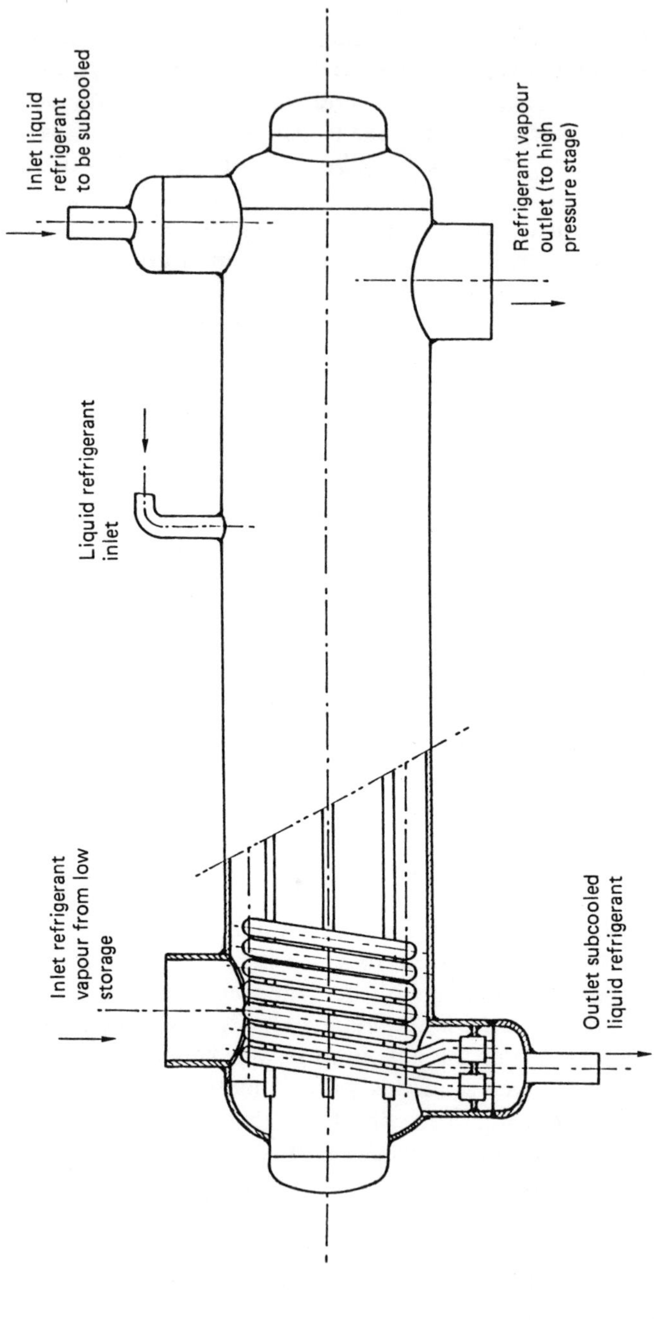

Figure 6.10 *Closed-type intercooler for (H)CFCs*

2. We can calculate the exchanger in such a way that the gases leaving it have a temperature above the dew point of the surrounding air and the insulation of the suction line will not be necessary. In this case we calculate the necessary enthalpy increase in order to reach the dew point temperature. The subcooling is then obtained with the same enthalpy difference value. Using a heat exchanger will in each case have the disadvantages of a higher superheat, higher specific volume of the gases and extra pressure drop in the suction line.

6.6 Oil Rectifiers

In practice an oil rectifier operates on the same principle as heat exchangers. The apparatus is used to achieve the automatic purging of oil without stopping the plant. This is especially important for continuously working installations. Firstly we bleed off the oil-rich liquid refrigerant which is then passed through a heat exchanger, and the evaporated refrigerant vapour and the warm oil are led back to the compressor (see the example at the end of this chapter).

In ammonia installations, bleeding from the low points of vessels is very simple. In this case, a small amount of refrigerant leaves with the oil.

In R12 and R22 installations, bleeding is done from the evaporator where the refrigerant is rich in oil.

In some R22 installations, the bleed-off point must be precisely located just below the liquid level in the vessel, a point which might prove difficult to determine in practice. In fact different layers of refrigerant contain different concentrations of oil, the upper layers having the highest oil concentration.

6.7 Some Practical Calculations

6.7.1 Capacity of a Vessel

The quantity of refrigerant liquid in a horizontal liquid separator is calculated as follows using figure 6.5c.

Draw a circle representing the separator section and find angle α by drawing the lowest liquid level h as shown in the figure. The area of cross-section S is given by

$$S = r^2[(\pi\alpha/180) - \sin\alpha]/2 \text{ m}^2$$

where α is the angle measured in degrees.
Thus the liquid capacity for a separator of length l is given by

$$V = lr^2[(\pi\alpha/180) - \sin\alpha]/2 \text{ m}^3$$

6.7.2 *The Overflow Valve*

The overflow valve serves to by-pass the amount of liquid that cannot be passed to the evaporators which are out of service for one reason or another. Those valves are automatic constant-pressure valves and are selected from the manufacturers' datasheets on the basis of a capacity expressed in kW. This capacity is *n* (total capacity of the installation – the capacity of the evaporator or group of evaporators that represents the minimum capacity in actual service).

6.7.3 *The Pressure Relief Valve*

The minimum required discharge capacity C related to air of any pressure relief device can be calculated from the formula

$$C = DLf$$

where C = minimum required discharge capacity related to air at each relief device (kg/s)
D = outside diameter of the vessel (m)
L = length of the vessel (m)
f = factor depending on the refrigerant.

R11	0.082
R12 and R22	0.130
R502	0.203
R717	0.041

With this massflow one can select a safety valve from the manufacturers' datasheets.

Actually, the calculations on safety valves and their discharge lines are now standardized. For full details consult ISO standard DIS5149 or CEN standard Pr EN 378.

6.7.4 *The Hot Gas Defrost Line*

If A = evaporator surface in m^2 and assuming a maximum hoar-frost thickness of 3 mm, we can calculate the volume of hoar-frost to be $0.003A$ m^3, which corresponds to a mass of $0.5 \times 0.003A$ kg of ice = m (kg). The quantity of thermal energy Q required to melt this amount of ice and turn it into water at 5°C is given by

$$m(c_{\text{ice}}\,\mathrm{d}T_1 + 1 + c_{\text{water}}\,\mathrm{d}T_2)$$

dT_1 depends on the surface temperature of the evaporator. A temperature difference dT_2 of 5°C ensures satisfactory drainage of water from the drain pan. This amount of thermal energy should be produced in about 30 minutes. Thus the rate at which the thermal energy must be supplied is $Q/1800$ kJ. Hot gas available for defrosting has a thermal energy dh'' kJ/kg which depends on the pressure and the discharge temperature. dh'' = final compression enthalpy – enthalpy at saturation temperature.

Now the amount of hot gas in kg/s required can bc calculated and converted into m^3/s. Allowing a velocity of 20 m/s in the hot gas line, the section of this line can be determined.

The amount of water produced during defrosting is known, so we can calculate the drainwater pipe system, allowing a water velocity of about 0.2 m/s.

6.7.5 Calculation of an Oil Rectifier in a R22 Pump Circulation Circuit

A part, about 4%, of the refrigerant-rich oil flowing from the refrigerant separator and the refrigerant pump is bled off. This is injected by means of an expansion valve into a heat exchanger connected to the suction line of the low-stage compressor. Here it evaporates while at the same time subcooling the high-pressure refrigerant liquid that passes through the heat exchanger on its way to the separator. The evaporated refrigerant vapour leaves the heat exchanger and enters the suction side of the compressor, bringing back the oil with it.

Assuming that the circulating refrigerant mass flowrate in the low stage of a R22 two-stage installation is 0.6 kg/s, the rate of refrigerant bleed-off will then be 0.04×0.6 kg/s $= 0.024$ kg/s. The evaporation temperature in the heat exchanger is -42°C. The latent thermal energy to convert 0.024 kg/s of refrigerant into vapour will be $0.024r = 0.024 \times 235 = 5.64$ kJ/s or 5.64 kW.

The intermediate temperature is in this case -4°C which gives a liquid temperature of $+6$°C on exit from the coil of the intercooler.

This liquid will be subcooled in the heat exchanger as part of the process of removing 5.640 W of thermal energy, which results in a specific enthalpy change of $5.64/0.6 = 9.4$ kJ/kg. The enthalpy of liquid at $+6$°C is 205 kJ/kg; after subcooling it will be $205 - 9.4 = 195.6$ kJ/kg. The temperature of the subcooled liquid will be -5°C. Now we can calculate the logarithmic dT across the heat exchanger. Given that $t_e = -42$°C, $t_{\text{liquid in}} = +6$°C and $t_{\text{liquid out}} = -5$°C, then dt log = 42°C. The value for K of the coil is 582 W/m^2 K. Thus the surface area A of the coil is given by $5640/42 \times 582 = 0.23$ m^2.

References

Ole Lassen (Nordborg, Danmark), Regeling der Flüssigkeitsumwälzung bei Anlagen mit Zwangumwälzung, *Die Kälte*, 9 (1983).

S. F. Pearson (Star Refrigeration Ltd, Glasgow), *Low Pressure Receiver, some applications*.

7

Controls

In a book about Refrigeration, it is essential to include a chapter on controls. However, there are good reasons to limit its extent. Firstly, if any particular aspect of refrigeration technology can be singled out because of its rapid pace of development, it is that of controls. Developments in electronics and data technology quickly make any detailed descriptions of controls and controlling systems outdated. Secondly, it is easy to obtain up-to-date documentation, leaflets, brochures and manuals from the manufacturers of such equipment, and the reader is recommended to do this. This chapter therefore merely surveys the essential features of control systems and indicates where improvements would be desirable. The examples and diagrams are included just to illustrate general principles and applications.

There is a wide range of components available to provide automatic regulation and control of refrigeration installations. The two main fields of application are:

(1) automatic function control,
(2) safety or security control.

Under (1) we find devices to control:

1. flow and level of liquids and gases
2. temperature
3. pressure
4. time and duration of certain operations
5. relative humidity
6. the capacity of the installation.

Under (2) we find devices to protect against:

1. extremes in liquid level
2. extremes of temperature
3. extremes of pressure
4. liquid slugging
5. oil-pressure failure
6. moisture in refrigerant
7. motor overload
8. freezing of water or brine
9. impurities in refrigerant.

7.1 Applications in Industrial Refrigeration Requiring Regulation and Control

1.1.1. Refrigerant flow to evaporators, separators and intercoolers:
 (a) the correct quantity
 (b) flow or non-flow.

1.1.2 Water flow to water-cooled condensers.

1.2. (a) The cold room temperature
 (b) The liquid temperature of shell and tube coolers or process temperature
 (c) The temperature of compressor oil-coolers.

1.3. (a) The evaporating pressure
 (b) The compressor suction pressure
 (c) The condensing or discharge pressure
 (d) The individual evaporator pressure when users requiring different temperatures are connected on the same suction line
 (e) The pressure in hot-gas defrost systems.

1.4. (a) The timing of defrost cycles. Activation of heaters or hot-gas valves, compressor motors, fans or solenoid valves
 (b) The evaporator fan start-up delay
 (c) The timing of oil pressurestat cut-out time-delay
 (d) The timing of loading of compressor at start-up
 (e) The timing of parallel compressor start-up
 (f) The timing of condenser fan or pump start-up
 (g) The timing of crankcase heater operation.

1.5. Relative humidity in air-conditioning installations of some meat stores. Activation of heaters.

1.6. (a) Activating of gas flows in compressor by-pass.
 (b) Activating of valve-lifting devices of piston compressors or regulation slides of screw compressors.
 (c) Stop and start of compressors and of the whole installation.

2.1. Protection against incorrect levels in:
(a) flooded shell and tube coolers,
(b) oil and liquid separators,
(c) water sumps of cooling towers and evaporative condensers.

2.2 Temperature control:
(a) defrosting system
(b) discharge temperature
(c) freezing temperature in shell and tube liquid coolers
(d) floor temperature in sub-zero stores
(e) oil temperature of compressors
(f) heating systems
(g) product.

2.3. Pressure control:
(a) high and low pressure of compressor
(b) differential pressure over air-conditioning air filters
(c) over-pressure in the refrigeration circuit
(d) over-pressure in the compressor
(e) low pressure of water supply to water-cooled condensers
(f) compressor oil pressure
(g) pressure over refrigerant pumps
(h) high enough pressure before expansion valve
(i) protection against too high suction pressure during pull-down, resulting in overload of the compressor motor.

2.4. (a) Damage to compressor components, resulting from liquid refrigerant or oil above the piston.
(b) Liquid separation and gas superheating before entry to the compressor.

2.5. Damage to compressor resulting from shortage of oil in the crankcase.

2.6. Refrigerant and oil deterioration and blocked expansion valves resulting from the presence of moisture.

2.7. Overheated motor windings and burn-out resulting from motor overload.

2.8. Freezing of water in drip pans of air-coolers, drain lines and siphons, on fan blades, on doors, on air-equalizing valves of the cold-store, contents of water-cooled or evaporative condensers and cooling tower sumps.

2.9. Prevention of metallic or other impurities from entering valves and compressors.

7.2 Devices and Components Available to Perform Regulation and Control

1.1.1. (a) Hand-regulated expansion valves, thermostatic expansion valves, electromagnetic valves
(b) Float valves.
1.1.2. Thermostatic or pressure operated water valves.
1.2. (a), (b) and (c) Thermostats.
1.3. (a) and (b) Pressurestats.
(c) and (d) Constant-pressure valves.
1.4. (a) Electromechanical timer or electronic equipment.
(b) to (g) Dashpot relays, electromechanical relays, electronic timers.
1.5. Humidistats.
1.6. (a) Constant-pressure regulators
(b) and (c) Step controllers with servomotors and electromagnetic valves, operated by pressurestats or electronic equipment
2.1. Float switches.
2.2. (a) to (f) Thermostats.
2.3. (a) Pressurestats
(b) Differential pressurestats
(c) Pressure relief valves (safety valves)
(d) Spring-controlled by-pass valve
(e), (f) and (g) Differential pressurestats
(h) and (i) Constant-pressure valves.
2.4. (a) Spring-loaded valve plate
(b) Liquid and oil separators and heat exchangers.
2.5. Differential pressurestats
Float valves.
2.6. Filters (strainers).
2.7. Overload protectors – bimetallic strips and thermistors.
2.8. Heaters.
2.9. Filters (strainers).

Note that there is a third set of static devices which allow observation and monitoring of the installation. These are:

1. sight glasses for oil, refrigerant and water level
2. manometers and gauges
3. thermometers
4. colour-changing sight glass for moisture control in refrigerant lines
5. signal lamps
6. thermographs

7. hydrographs
8. ampere meters.

Figure 7.1 illustrates a relatively complex refrigeration system and a wide range of controls used in the operation, control for and safeguarding of such a system.

It is not necessary to describe such simple and universal devices as float valves, timers, strainers, thermometers, safety valves and solenoid valves. However the following devices will be discussed:

(a) thermostatic expansion valves
(b) pressurestats and thermostats
(c) constant-pressure valves.

7.2.1 Thermostatic Expansion Valves

A typical thermostatic expansion valve (TEV) is shown in figure 7.2. This type of valve regulates the injection of liquid into the evaporator, depending on the refrigerant superheat in the suction line after the evaporator at the point where the TEV bulb is located. A thermostatic expansion valve is temperature controlled and opens on rising bulb temperature.

The pressure over the diaphragm (1) is determined by the bulb temperature, the thermostatic element being filled with a charge that in principle reacts according to the same physical laws as refrigerants in the saturated condition. In other words, for a particular temperature there is a corresponding pressure. This pressure is transferred by the diaphragm to the push rod (2) in the form of a downward-opening force. The push rod is also subjected to an upward-closing force from the regulating spring (3). When the bulb temperature increases it produces a greater force downwards than the force of the regulating spring; the thermostatic expansion valve will then open wider. Valve operation is therefore predominantly determined by the bulb temperature and setting of the regulating spring.

Most valves of large capacity, or used in conjunction with an evaporator designed with a large refrigerant pressure drop, have an equalizing connection between a point under the diaphragm and a point on the suction line immediately after the bulb location. With this arrangement, the pressure under the diaphragm will correspond to the refrigerant vapour pressure at the point where the bulb is located, rather than at entry to the evaporator.

A thermostatic expansion valve with a maximum operating pressure (MOP) charge will only begin to open at an evaporating temperature below the set point of the MOP. The valve is closed until the compressor

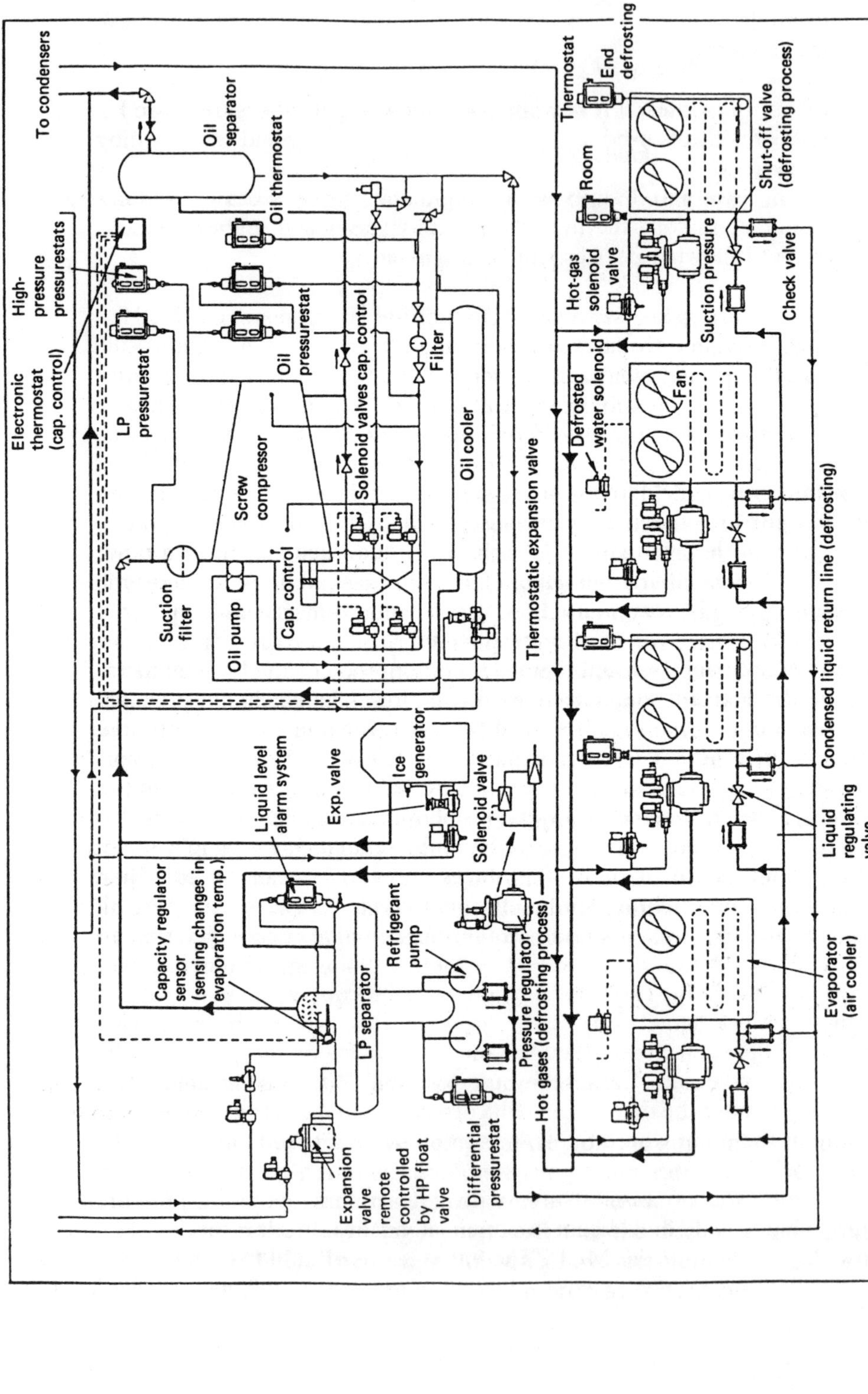

Figure 7.1 Comprehensive safety and control arrangement for a large refrigeration application (Courtesy of Danfoss)

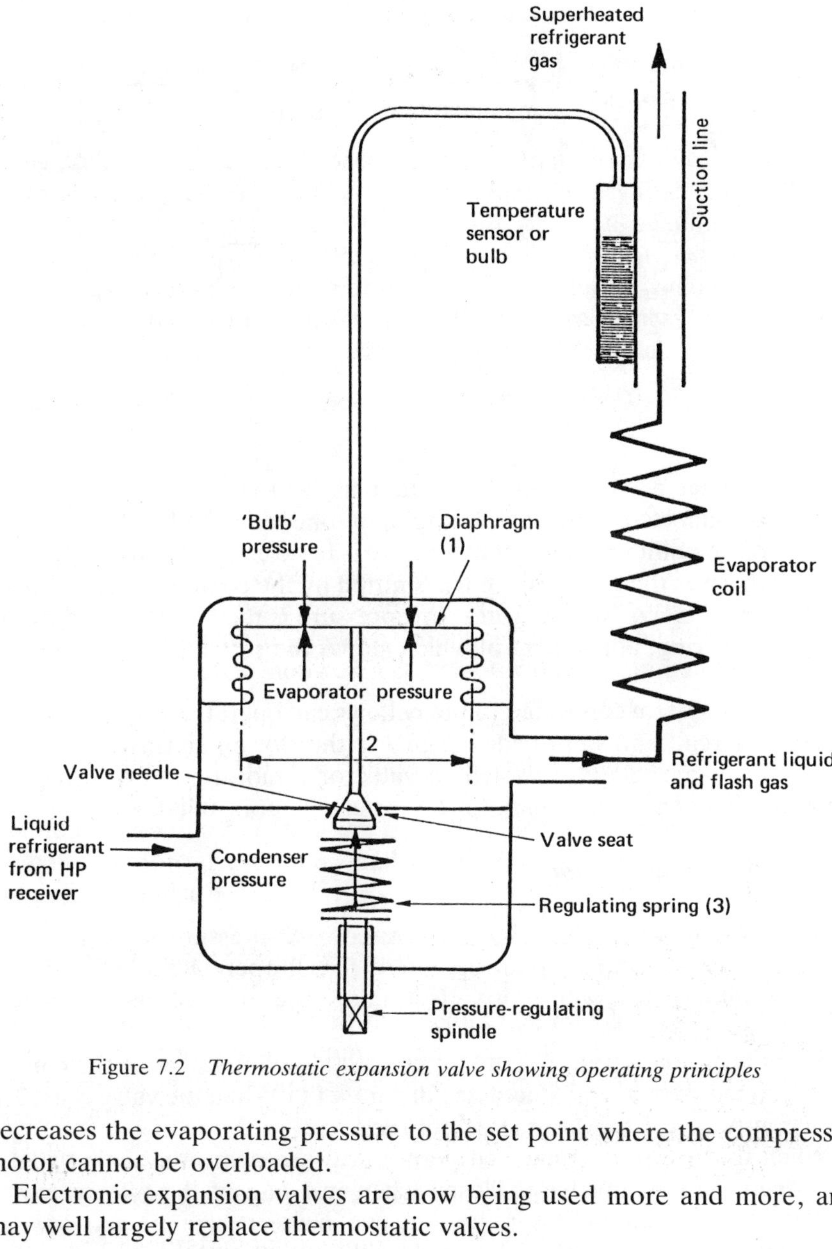

Figure 7.2 *Thermostatic expansion valve showing operating principles*

decreases the evaporating pressure to the set point where the compressor motor cannot be overloaded.

Electronic expansion valves are now being used more and more, and may well largely replace thermostatic valves.

7.2.2 Pressure and Temperature Controls

Both of these controls are based on the same operating principle that figures 7.3a and b show in simplified form. In the case of the pressurestat, a capillary tube connects the bellows to the measuring point. In the case of

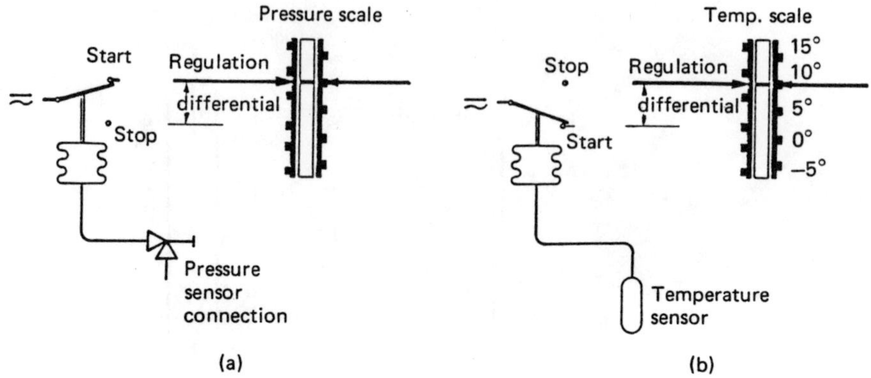

Figure 7.3 *Operating principles of (a) pressurestat and (b) thermostat*

the thermostat a bulb, identical to that used in thermostatic expansion valves, is connected by means of a capillary tube to the bellows. The bulb is placed at the point where the temperature is to be controlled or monitored.

In both cases the gas pressure transmitted by the connecting tube causes the bellows to flex. Acting against this pressure is the force exerted by the regulating spring, adjustment of which allows the set point to be raised or lowered.

The lever system connected to the bellows can operate the contacts in an electrical circuit and so provide a signal, either for an alarm system or an electromechanic relay to a solenoid valve or a motor starter. Using two opposing bellows, a differential pressure can be controlled.

7.2.3 Pressure Regulators

Figure 7.4 shows a Danfoss servo-operated back-pressure regulator. This kind of valve is normally pilot operated. It will operate according to the type of pilot valve, which in effect takes the role of the regulating mechanism.

There are three arrangements for connecting pilot valves to the control port, two in series S′ and S″ and one in parallel P. When the valve is used as a suction pressure regulator, P is not used.

When the two semi-connected pilot valves are open, the main valve is also open. The mainvalve is closed when just one of the pilot valves is closed. The constant-pressure pilot valve protects against too low an evaporating temperature. It closes when the pressure drops so much that the spring presses the diaphragm against the valve seat. So we can adjust the spring pressure in a way that enables us to maintain a certain suction pressure in the evaporator, regardless of the suction pressure on the compressor. Using a thermostatic pilot valve, the evaporating temperature

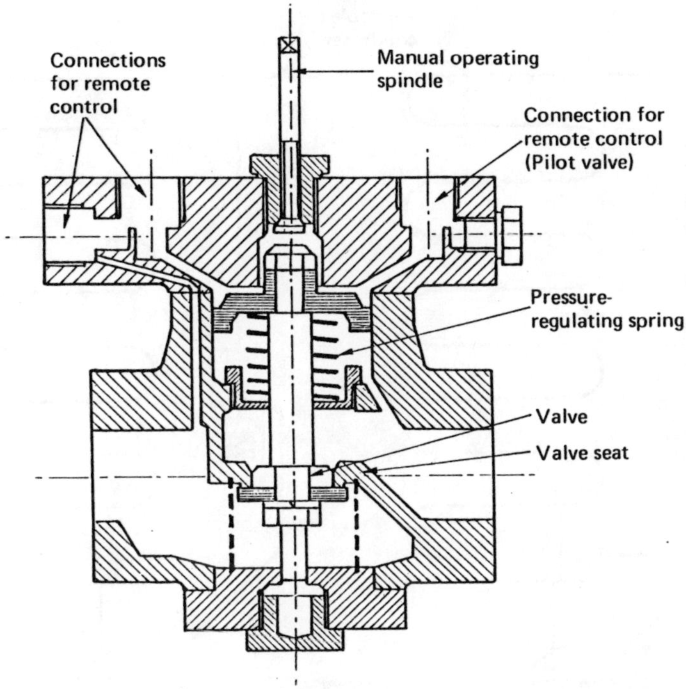

Figure 7.4 *Section through a pilot-operated pressure and temperature regulating valve*

can be regulated so as to maintain the temperature of the cooled fluid or the coldstore air.

These valves can also be used for maintaining the discharge pressure in the hot-gas line at the necessary level for the defrost period, or as an on–off regulation for liquid supply to an evaporator or separator with a float valve system, or to control a gas flow on an on–off basis. In both cases the pilot valve is a solenoid valve connected to the hot-gas line.

The valve is then hot-gas operated. This arrangement is used for large refrigerating capacities when a normal solenoid valve is too small. They can also be used as differential pressure regulators when a differential pressure pilot valve is used.

7.3 Condensing Pressure Control by Flooding the Condenser with Liquid Refrigerant

In the earlier treatment of condensers, reference was made to the need in winter to flood part of the condenser with liquid refrigerant. This can be achieved using back-pressure regulators. Figure 7.5 shows a simple arrangement for this.

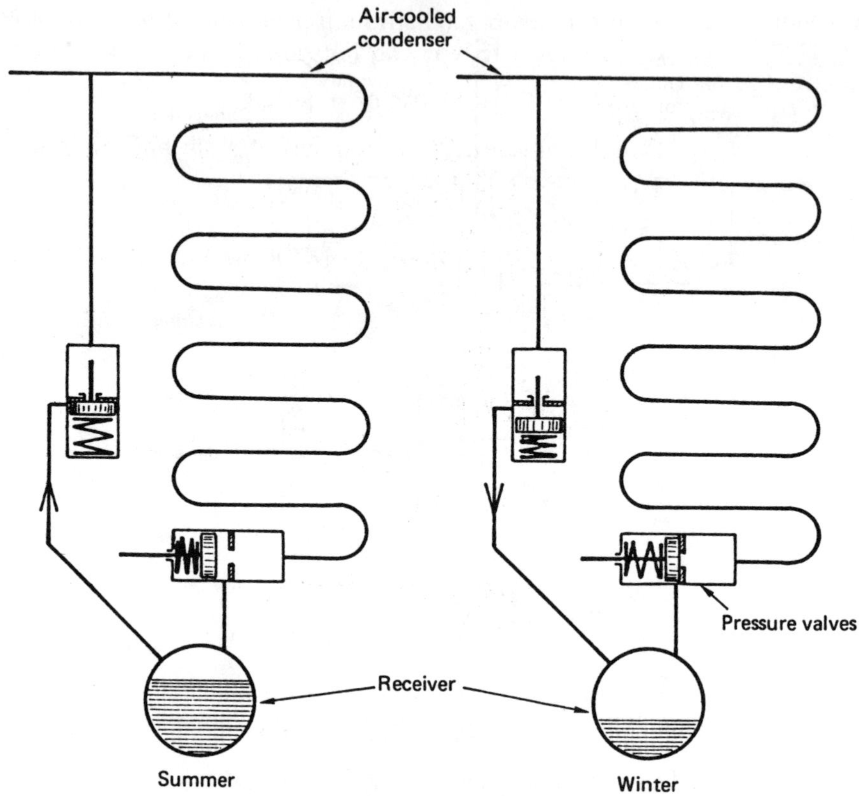

Figure 7.5 *Principles of condensing pressure control*

When the condensing pressure falls below the setting of the regulator A, the degree of opening is reduced correspondingly. This results in the condenser being partially filled with liquid and the required condensing pressure is thereby maintained. Since the actual task during winter operation is to keep the receiver pressure at a suitably high value, valve B is mounted in a by-pass line as shown in the figure. The degree of opening of B, and thereby the receiver pressure, is determined primarily by the discharge pressure and secondly by the setting of A. Because B can be adjusted directly, adjustable control of the receiver pressure is possible.

Summer operation at high air temperatures normally results in the condensing pressure exceeding the setting point. Under such conditions, A is therefore completely open and there is no liquid charge in the condenser. In this case the receiver pressure exceeds the setting point and B is therefore closed.

This close control of the receiver pressure is necessary to ensure an adequate supply of liquid from the receiver to the expansion valve with sufficient pressure. More details about these controls are available in manufacturers' catalogues.

At the time of writing the use of microprocessors, either by PLCs or computers, is in continuous development. The resulting changes brought about by these will be considerable in the field of refrigeration control.

Bibliography

Advances in refrigeration and heat pump technology achieved by application of micro-electronics and control systems by micro-electronic devices, International Institute of Refrigeration, 177 Boulevard Malherbes, F75017 Paris.

Continuous monitoring of industrial refrigeration system efficiency, Institute of Refrigeration, UK, 84(1987/88).

Danfoss, Nordborg, Denmark.

Microprocessor control of refrigerator compressors, The Institute of Refrigeration, UK, 79(1982/83).

The application of microprocessors to industrial refrigeration, The Institute of Refrigeration, UK, 82(1985/86).

John van de Vechte, *Feedback Control Systems*, 2nd edn, Prentice-Hall.

8

Food Products and their Preservation by Refrigeration

8.1 What are Food Products?

Why and how do we use refrigerating techniques to conserve food? Refrigeration is not a goal in itself but simply a means of prolonging the usable life of food. Of course, it does achieve other goals, but to a great extent, dealing with refrigeration is dealing with improvements in the quality of human life and optimizing conditions for human comfort: it serves computer installations, laboratories and hospitals, as well as conserving medicines and food. It is essential for a refrigeration engineer to have some background knowledge about life in general and the composition of food products in particular.

When life began on this planet, the amino acids were the first form to appear. Of these, some 24 different kinds are important since out of these combinations many forms of proteins were created. One important group, the so-called enzymes, play a very important role as the 'workers' that start and regulate all kinds of actions and processes. Their 'work' depends on how they are 'programmed', in other words how the 24 different amino acids are arranged in their particular molecular structure. Amino acids also create nucleotides which are the components from which nucleic acids are built. One of the nucleic acids is the famous deoxyribonucleic acid molecule DNA. It was discovered in 1869 by the Swiss scientist Niescher and modelled in 1953 by the American James Watson and his British colleague Francis Crick. This DNA molecule is in fact the 'central processing unit' of living matter. The DNA molecule has the appearance of a double helix, a double spiral staircase as can be seen in figure 8.1. The longitudinal 'stringers' of the staircase consist of sugar and phosphate groups, while the cross links are composed of different types of nitrogenous bases, namely adenine, thymine, cytosine and guanine. Their

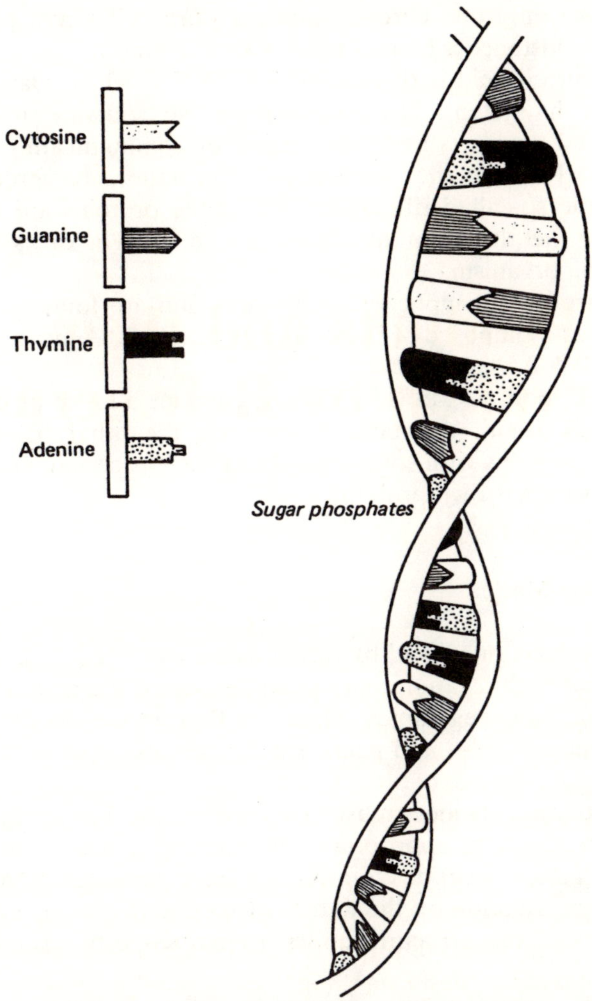

Figure 8.1 *Model of the double helix DNA molecule*

molecular structure determines the genetic characteristics and the metabolism of the living matter of which they form a part.

The key to heredity lies in the pattern of this molecule. The sequences of the bases can be seen as an instruction code 'software', or program for living matter, whether it be a bacterium, a yeast cell, a plant or a human being. The DNA commands the biochemical equipment of the cells in such a manner as to control the way a living thing functions. The DNA is inside the so-called 'genes', which are found in great numbers on the 'chromosomes'. Each species has a different number of chromosomes and it is this number which determines whether the living matter is a bacterium (some of which have only one chromosome), a yeast cell (having 17 chromosomes) or for instance a human being (with 23 pairs).

Another nucleic acid is ribonucleic acid, RNA, which plays the role of 'interface' in the system. The mRNA reads the 'software' instructions of the DNA, and then acts as a 'messenger' by communicating the instructions to the tRNA in the 'workshops'. The latter are spread over the cytoplasm, the so-called ribosomes, where the proteins are synthesized. Finally there are the vitamins which can be thought of as the 'control system' of the organism.

Proteins are very complicated substances and are found in very many forms. They are composed of 50–55% carbon, 15–17.6% nitrogen, 18–28% oxygen and 0.4–2.5% sulphur.

Meat-based food, as we will see later, contains a lot of proteins, which form the bulk of it. Vegatable matter, on the other hand, generally contains few proteins with some exceptions such as certain bean varieties. Plants produce their own proteins.

8.2 Vegetable Matter

Of all the products preserved by refrigerating methods, vegetable matter, especially fruit such as apples and pears, is one of the most complicated. When storing fruit, vegetables, plants or flowers we are dealing with a living substance. Apples and pears are the best example of this phenomenon.

The refrigeration engineer must have considerable knowledge of biology if he wants to design the appropriate refrigerating system for the storage of such living matter. For this reason we have provided here biological background information on the nature of vegetable matter. Let us have a look at the vegetable structure under a microscope by examining figure 8.2.

We can see a cell structure, formed by cell walls mainly consisting of water and cellulose, a glucid $(C_6H_{10}O_5)_n$, separated by a lamella consisting of fat (lipids) and protein (enzymes), and 'glued' together by grains of carbohydrate (pectins). Between these walls and even on the outer layers we see the cytoplasm, a viscous liquid in colloidal form and mainly consisting of protoplasm which contains some 80–90% water with traces of fat (lipids) and mitochondria, small grains of about 1 micron. These are in fact the 'power-plants' in which the respiration process takes place. Furthermore, grains of carbohydrates, minerals, chlorophyll, carotenes and plasms can be seen in which the photosynthesis takes place.

In the middle of this protoplasm we see a nucleus which acts as the 'control room'. Here are the chromosomes, the DNA, the RNA and the vitamins, in addition to the so-called vacuoles – spaces in the middle of the

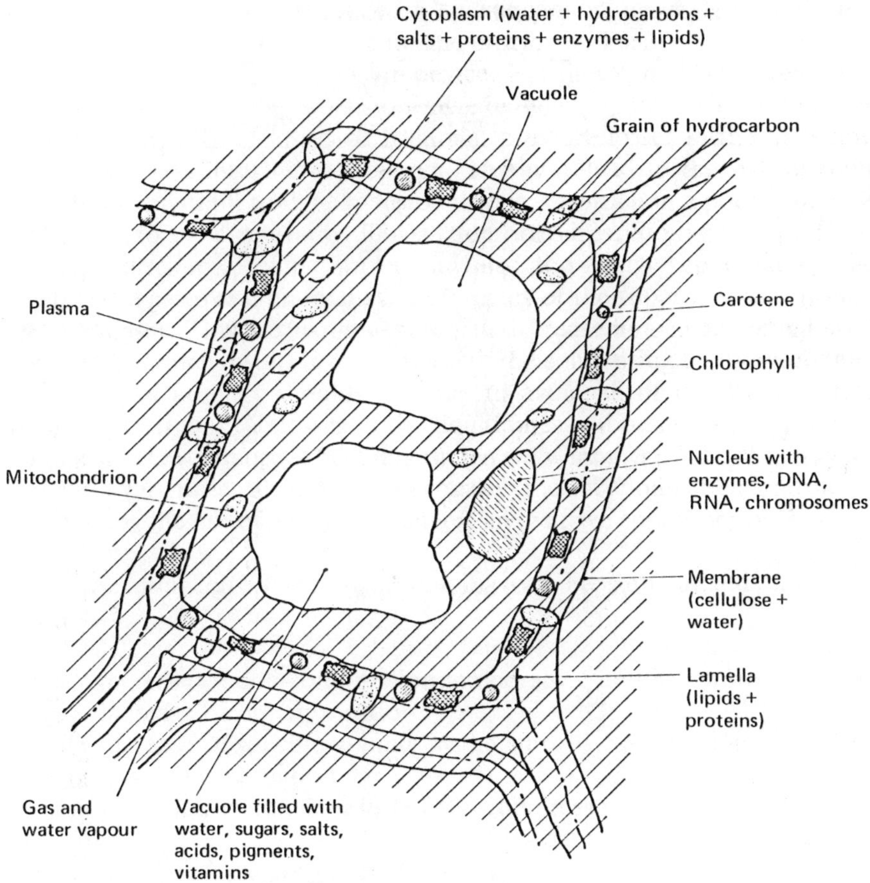

Figure 8.2 *Vegetable cell structure*

protoplasm, filled with cell-juice, containing flavours, sugars, acids, minerals, pigments, more vitamins, but mostly water.

In general terms we can say that vegetable matter consists mainly of water, carbohydrates ($C_nH_{2n}O_n$) and small quantities of other matter. Fuller information is provided in tables 8.1 and 8.2.

8.3 How Plants Live

Plants are unique systems in that they do not need to destroy other living matter in order to be created or to continue living. Using solar energy, water, minerals – in some cases our refuse – and carbon dioxide they produce growth with the help of a catalyst called chlorophyll. This process

Table 8.1 Composition of general food products

	Mass (kg/kg)				
Product	H_2O	*Protein*	*Fat*	*Hydrocarbon*	*Remainder*
Potatoes	0.79	0.02	—	0.19	0.01
Peas	0.82	0.05	—	0.10	0.03
Carrots	0.90	0.01	—	0.06	0.03
Beans	0.90	0.03	—	0.05	0.02
Tomatoes	0.95	0.01	—	0.03	0.01
Onions	0.86	0.01	—	0.10	0.03
Strawberries	0.91	0.01	—	0.05	0.033
Apples	0.87	0.007	—	0.10	0.03
Bananas	0.75	0.01	—	0.22	0.02
Berries	0.85	0.015	—	0.05	0.055
Grapefruit	0.91	0.005	—	0.07	0.015
Melons	0.94	0.005	—	0.03	0.025
Pears	0.85	0.005	—	0.10	0.045
Prunes	0.85	0.005	—	0.10	0.045
Oranges	0.86	0.005	—	0.10	0.035
Brown bread	0.40	0.079	0.015	0.43	0.076
White bread	0.40	0.093	0.040	0.44	0.027
Cake	0.15	0.06	0.240	0.54	0.010
Milk	0.88	0.033	0.035	0.046	0.006
Butter	0.16	0.006	0.83	—	0.004
Beef lean	0.62	0.18	0.05	—	0.15
Beef fat	0.45	0.10	0.20	—	0.25
Veal lean	0.58	0.17	0.10	—	0.15
Veal fat	0.52	0.13	0.15	—	0.20
Pork lean	0.62	0.18	0.10	—	0.10
Pork fat	0.47	0.13	0.25	—	0.15
Hamburger lean	0.70	0.20	0.09	—	0.01
Hamburger fat	0.47	0.12	0.35	0.04	0.02
Chicken	0.667	0.17	0.143	0.003	0.009
Duck	0.566	0.158	0.242	—	0.333
Turkey	0.746	0.238	0.008	0.006	0.012
Egg yolk	0.487	0.162	0.300	0.003	0.010
Egg	0.759	0.129	0.105	0.033	0.009
Herring	0.620	0.175	0.185	—	0.020
Fish	0.790	0.180	0.010	—	0.020
Shrimps	0.730	0.165	0.020	0.020	0.065

is called photosynthesis or assimilation and proceeds as follows: $6CO_2 + 6H_2O$ with some minerals + solar energy with chlorophyll as catalyst gives $C_nH_{2n}O_n$.

Carbohydrates, of which there are many kinds, can be starch or sugar. As we saw before, plants can create their own proteins. As living matter, plants, fruit, and flowers all respire. Respiration is a process in which heat is generated when sugars react with oxygen to produce water and carbon dioxide according to the equation:

$$C_6H_{12}O_6 + 6O_2 \rightarrow 6H_2O + 6CO_2 + \text{thermal energy}$$

Another process is the ripening of fruit, in which carbohydrates are changed to sugars, such as fructose and glucose, and aromatic matter.

Remembering that the carbohydrate is the 'glue' between the cells, and knowing that sugar is a weaker 'glue' than pectin, we can appreciate that fruit softens as it matures. This gives us a method to test the degree of ripening of fruit by measuring its firmness. This can be done using a penetrometer, a device that contains a steel pin which penetrates the fruit. Using a simple or an electronic meter, we can read the pressure needed to penetrate the fruit; this gives an accurate assessment of the degree of ripeness. We need to know this to determine the right moment for picking and storing. Another method is the use of a colour chart by which one can judge the degree of ripeness from the colour of the fruit.

Figure 8.3 shows how the rate of respiration of a typical fruit varies during its life cycle. Note how very similar is the pattern to a human life cycle. In humans, after birth there is a temporary decrease of energy and CO_2 production; during puberty an increase, with a maximum around middle age: for some fruits, such as apples and pears, there is a similar maximum, also called the climacteric.

After this point during the maturation of a fruit, decay starts, and CO_2 and energy production decrease until all activity ceases at death. Most fruits must be harvested just before the climacteric in order to avoid storage diseases. If we leave the fruit on the plant beyond this point, while

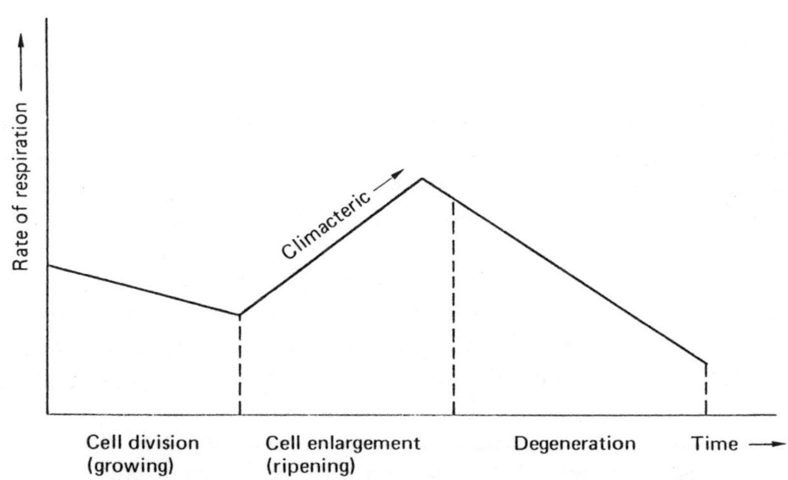

Figure 8.3 *Rate of respiration of fruit during its life cycle*

it increases in weight, more fruit is lost through disease in storage. In addition to CO_2, fruit also produces ethylene (C_2H_4) of which immature fruit contains 0.01 ppm. A ton of immature apples produces 0.01 millilitre of ethylene; a ton of mature apples produces 5000 times that quantity. Recent research showed that at low temperatures more ethylene is produced, which is unfortunately counter-productive during low-temperature storage.

Comprehensive information regarding the production of CO_2 and the rate of thermal energy produced, the so-called respiratory heat, is available for a wide range of fruit products. Table 8.2 shows how these rates of production depend on the kind of fruit and the storage temperature. This is not the case, however, for water and ethylene production.

Table 8.2 Thermal energy and CO_2 production rates for some fruits

Product	*Temperature (°C)*	*Production of CO_2 g/kg h*	*Thermal energy (W/kg)*
Apples	0	0.003–0.004	0.003–0.010
	4.4	0.005–0.008	0.014–0.022
	15.6	0.020–0.030	0.056–0.084
Green bananas	12.2	0.015	0.042
Oranges	1.7	0.012	0.0056
	26.7	0.015	0.042
Strawberries	0	0.016	0.0448
	15.6	0.070	0.196
Gooseberries	15.6	0.080	0.22

Research is continuous in the field of fruit storage. Work on ethylene production is being carried out at universities in Britain and Leuven, Belgium. The Sprenger Institute in The Netherlands is investigating water transfer from fruit to the atmosphere. See figure 8.4 for information in this area.

In a cold store a specified air temperature and relative humidity is maintained. In the boundary layer of air at the surface of the product, where the respiratory heat leaves the product, there are different conditions from the rest of the cold store: namely higher air temperature and higher water vapour pressure. In addition, the actual water vapour content of this air is higher than in the main body of air of the cold store.

Mass transfer of water vapour takes place in accordance with the equation $m = \beta A dp$ where β is the permeability of the product at the surface. Compare the similar equation for heat transfer, $Q = \alpha A dT$.

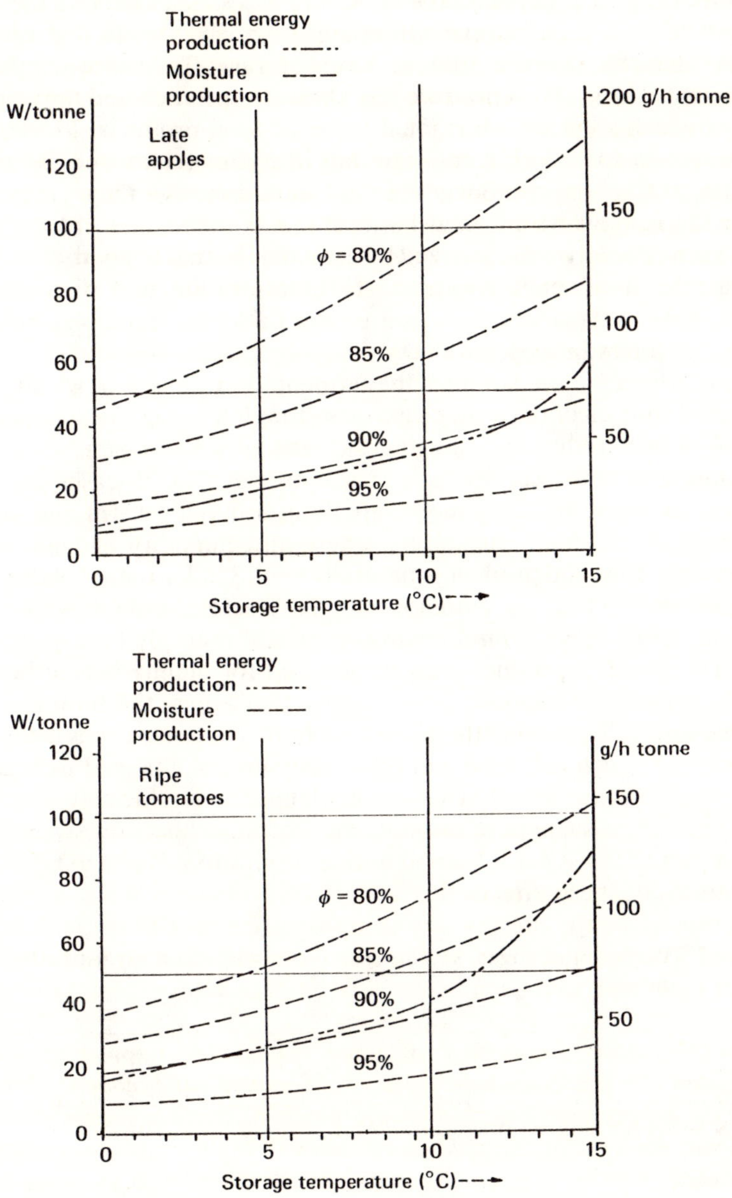

Figure 8.4 *Heat and moisture production rates for apples and tomatoes*

8.4 Storage Diseases

It is important to understand the intricate processes underlying the growth of plants during their lifespan. In doing so we are able to appreciate the dangers that threaten the product during storage. To start with, there are the normal chemical reactions we saw above, respiration and fermentation, both of which spoil the nutritional value of food products. Moisture loss causes weight loss, with a resulting fall in commercial value. Mould and bacteria are to be reckoned with as well, because they create other chemical reactions with similar harmful consequences.

All such reactions are catalyzed by enzymes; both the 'good' processes as well as the 'bad' ones which create deterioration and discolouration. Undesirable enzymatic processes are found a lot in vegetables but less in fish and still less in meat products.

Yeast, for example, catalyzes the fermentation process in which sugar is converted into alcohol. Other enzymes stimulate reactions which change the pH of the product by changing the ratio of acids:sugars.

In addition to these enzymatic reactions there are diseases created by bacteria. Bacteria multiply more rapidly in a humid, warm and non-acid environment. In fruit coldstores, where the humidity is high and the temperature low, the product is naturally acid. Under these conditions the presence of bacteria is a minor danger whereas mould is the major problem. Both like a humid environment and multiply best on damaged fruit. Therefore, a product must be selected for quality before being put into the coldstore. Mould creates poisons that attack the protoplasm and dissolve cell walls, causing the fruits to soften. They also act as catalysts for processes in which cellulose and other starches are changed to sugars.

Saccharose and natural acids are depleted by mould, and since at the same time more sugars are created, the fruit also becomes sweeter, even though part of the sugar is burned during respiration. Refer to table 8.3 for illustrations of these effects.

Table 8.3 Variation of sugar, saccharose, and sugar:acid ratio with cold-store duration

	At start	*After 2 months*	*After 3.5 months*	*After 6.5 months*
Original sugar to be converted	6.91	6.8	6.65	5.36
Saccharose	1.23	0.92	0.68	0.12
Total amount of sugar	0.79	0.66	0.55	0.35
Ratio sugar:acid	10.3	11.7	13.3	15.7

Another storage disease is Low Temperature Disease (LTD), also called 'cold injury'; it colours the fruit flesh brown by disturbing the metabolism of the plant. The reactions of some enzymes producing harmful substances are no longer neutralized by the reaction of those enzymes which normally destroy harmful substances; as a result, pears, for example, cannot continue to ripen. When the temperature becomes too low, subcooling and finally freezing of the product occurs as the protecting enzymes are no longer performing their function at this low temperature. Another disease colours the core of the fruit brown; this is caused by a high toxic concentration of CO_2 and insufficient O_2. This condition, called brown core, occurs usually at the beginning of storage.

Finally, there is a disease called scald that affects the skin of fruit and is caused by a very high percentage of ethylene gas. Today the ethylene in the atmosphere is suspected of being responsible for many problems of fruit storage. As fruit ripens, it produces more ethylene.

Ethylene is a hormone in the gaseous state; a hormone is a substance which controls the growth of the plant. Hormones translate the information stored in the DNA into instructions for the biochemical equipment of the cells.

At low O_2 rates, thc production of ethylene seems to be decreased; however the amount of the precursor of aminocyclopropane, ACC, which is the predictor of ethylene, is increased. ACC forms EFE, an enzyme complex which produces ethylene. The formula of ACC is

$$\begin{array}{ccc} H_2C \diagdown & & \diagup COO \\ & C & \\ H_2C \diagup & & \diagdown NH_3^+ \end{array} .$$

Removal from the coldstore creates an explosion of ethylene production, which results in the shelf-life of the fruit being decreased.

8.5 Meat Products

After an animal's death the composition of its meat changes; the physical, chemical and organoleptic characteristics change naturally. Enzymatic activities commence, bacteria attack and the ageing process starts, finally leading to putrefaction if nothing is done to prevent the action. Low temperature is very effective in limiting those reactions which result in deterioration of the meat's condition.

Immediately after slaughtering, meat has a temperature around 37°C; it is still supple, contractable and rich in glucids. It remains in this state on average between 1 and 3 hours, and if eaten in this state is found to be extremely tough.

After this, the muscles set hard and because of glucid degradation the acidity increases. This state, known as *rigor mortis*, lasts between 12 and 48 hours. During this period meat is unsuitable for normal consumption but can be used for some industrial foods such as dry sausages.

Next, important biochemical changes occur under the influence of certain enzymes. The period over which these changes take place varies from 2 days to 3 weeks, depending on species, age, fat content, storage temperature and humidity. Meat is then considered mature or aged. It offers the ideal combination of tenderness and flavour since the muscles are relaxed and deterioration due to contamination has not yet had any significant effect.

Alternatively, internal and external bacterial attack coupled with poor storage conditions can lead to rapid deterioration. The colour fades, taste is lost and a fusty smell is emitted; in the extreme it smells of ammonia and putrefaction sets in.

Ideally, meat is stored at a temperature between 0 and 2°C (7°C maximum). Under these conditions, ageing is slowed down and the growth of bacteria greatly reduced. This so-called fresh meat is usually achieved by refrigeration.

Freezing and deep-freezing between −18 and −50°C stop the enzyme activity and the ageing process. These temperatures allow meats to be preserved for periods of between 6 and 18 months. However, we should not forget that deterioration continues after thawing.

The initial quality and condition of meat and meat products are of considerable importance in determining the storage life. Under the same storage conditions, good quality meat loses less weight by shrinkage than does poor quality meat. Moreover, conditions during slaughtering and bleeding of the animal and the dressing and cooling of the carcasses largely determine the number of spoilage organisms in the meat at the time that storage begins. To ensure a good storage life: the animal must be in good condition; the slaughtering, bleeding and dressing operations must be carried out quickly and hygienically; and the cooling down of the hot carcass to a temperature approximately of the storage temperature should be rapid and commence immediately after dressing. The physical conditions to be maintained during this initial chilling depend to some extent on the type of trade for which the meat is intended.

In the case of meat intended for the domestic market or for short-distance chilled transport, the aim should be to restrict the shrinkage due to chilling as far as possible. This is usually achieved by chilling beef carcasses with cold air as low as −5°C, at a high velocity of 3 m/s and a relative humidity as high as 95%. However, a relatively high rate of evaporation is required in the later stages of chilling of beef destined for transport lasting several weeks. This assists in controlling microbial growth on the meat by producing a thin dry surface layer. For longer storage, meat should be kept at a temperature as close as possible to its freezing point at about −1.5°C.

The structure of meat is different from that of vegetable matter, but in principle it is somewhat similar in that both forms consist of cell structures. Compare figures 8.2 and 8.5. In addition to water, the main constituent is protein, in contrast to vegetable matter where it is carbohydrate.

During the storage of meat, the breakdown of protein into amino acids, nucleic acids, NH_3 and N_2 takes place. This process, called hydrolysis, cannot be stopped at low temperatures, because the enzymes continue to

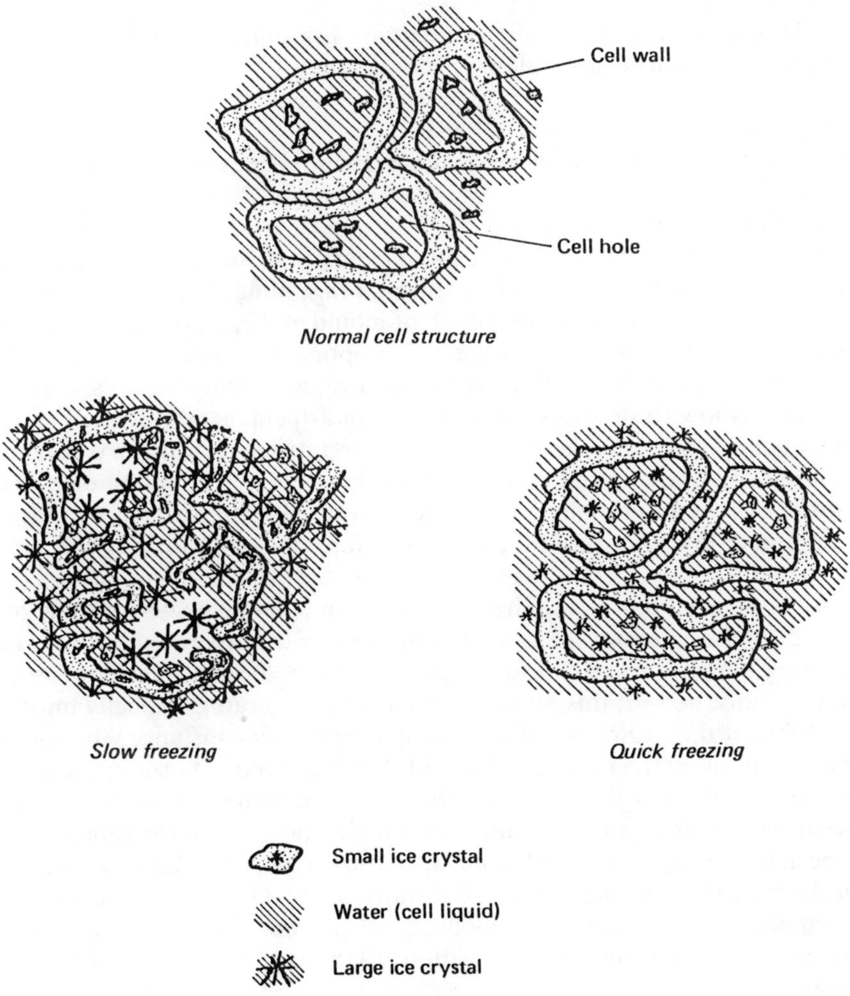

Figure 8.5 *Effects of freezing rate upon meat cell structure*

work; only after deep freezing can this activity be arrested. Bacteria have a greater effect on the storage life of meat than on that of vegetable matter.

Another constituent of meat is fat, which is also to be found in very small amounts in vegetable matter. In fact, only a few plants such as peanuts and maize contain much vegetable fat. Fats are combinations of glycerol, fatty acids, H, C, O and combinations of CH. The metabolism of the body burns fats, so producing a large amount of energy. In fact, fat serves as an energy and nutritive reserve. In storage it must be protected against oxidation which would cause it to become rancid. There is a difference in behaviour between meat fats and fish fats. Fish fats are mainly non-saturated acids which oxidize faster than meat fats.

Like vegetable food products, meat and fish must be protected against weight loss due to evaporation of body water.

8.6 How Refrigeration Techniques can Help to Conserve Food Products

Now that we have some background information to understand the behaviour of food products and the dangers that threaten them, we can consider measures for extending their storage time. We have seen that most of the reactions and the growth of mould and bacteria are stopped or slowed down by low temperatures. The optimum temperature varies for different kinds of food. With fruit, for instance, temperatures must be sufficiently low to slow down respiration and ripening, yet high enough to burn enough sugar to keep the acid:sugar ratio in satisfactory balance. Slowing down respiration means slowing down the consumption of its own nutritive stock of carbohydrates. The fruit needs this stock to be able to continue to live after harvesting, but we want this stock for our consumption.

As most reactions create thermal energy, metabolic heat – sensible and latent – must be removed, as well as the heat of the environment. In order to bring the product as quickly as possible to the best storage conditions, chilling must be fast; this means that a large refrigerating capacity must be available at the beginning of the storage period. For instance, the sooner the conditions in the boundary layer of the air around the product are close to the conditions of the coldstore, the earlier the minimum weight loss will occur. However lowering the temperature of a hot carcass too rapidly may, especially for beef, veal and lamb, result in a severe localized contraction of the muscle fibres, known as 'cold shortening'. This causes considerable toughness in the meat after cooling. To prevent this, the temperature should not fall below +10°C within 10 hours of slaughter for these products.

Another condition to be controlled is the relative humidity. For fruit we need a high relative humidity in the region of 97–98%, for vegetables such as lettuce, close to 100% in order to minimize the weight loss. In contrast, for meat products we need low relative humidity, around 85%, in order to decrease the growth of bacteria. At the same time, a thin layer of dried-up skin will protect the rest of the meat from moisture loss.

In modern fruit and vegetable coldstores it is possible to prolong storage life from a few months to almost a year, and at the same time improve the shelf life of the product after it leaves the coldstore. The constituents of the store atmosphere which need to be controlled are:

(a) fresh air intake
(b) CO_2
(c) O_2
(d) ethylene.

The actual quantity of each constituent is not only different for vegetable and meat products, but also for varieties of apple and different kinds of fish. It depends on characteristics such as:

- specific heat
- heat transmission coefficient
- water/salts composition
- density
- viscosity
- variety
- desired storage time.

To present a good fresh-tasting product to the consumer, the conditions created in the coldstore must be maintained continuously until the product reaches the consumer. This is called the 'cold-chain', but the chain is only as strong as its weakest link. Well designed equipment must be used at every stage: the coldstore, the transport medium, the wholesaler, the supermarket and the household.

Storage temperatures can be selected in a range from above freezing to below freezing, depending on the storage time required. In the upper range we speak of refrigerating; in the lower range of deep-freezing. Both systems will now be studied in more detail.

8.7 Refrigeration or Cooling

Weight loss is an important factor when using refrigeration for conservation. As low relative humidity retards bacterial growth and high relative humidity reduces weight loss, a compromise is necessary when specifying the humidity of the air. In all cases the cooling process causes a weight loss.

To some extent, refrigerating equipment is as much a drying as a cooling process because moisture from the environment, and thus from the product, condenses and freezes on the evaporator tubes. This water is drained away from the coldstore after being defrosted from the evaporators.

To summarize, we must prevent:

- excessive weight loss by a large temperature difference and a water vapour pressure difference between product and environment
- Excessive growth of micro-organisms.

Rapid chilling or freezing is obtained by:

- a thin product layer, or small dimensions
- good contact, in general, between the cold medium and the product
- good heat transfer conditions, little insulation around the product and high velocities of the medium whether it be air, brine, water or refrigerant
- big temperature difference $\mathrm{d}T$ between product and environment
- sufficient refrigerating capacity.

However, a large $\mathrm{d}T$ is not economical, nor are high air velocities from the energy consumption point of view.

Large temperature differences are only permissible for short periods, such as when bringing down the temperature; after that, during the storage period, $\mathrm{d}T$ is reduced.

Let us see why weight loss occurs and what we can do to limit it. When we refer to the sections of chapter 6 dealing with mass and heat transfer, we see that the quantity of transferred water from a wet surface A is

$$m = \beta A \mathrm{d}p = \beta A (P_\mathrm{p} - P_\mathrm{c})$$

and transferred heat

$$Q = \alpha A \mathrm{d}T = \alpha A (T_\mathrm{p} - T_\mathrm{c})$$

where c denotes coldstore and p product. When the latent heat of evaporation is L_v it means that $m = Q/L_\mathrm{v}$.

The air of the coldstore carries moisture to the evaporator or air cooler surface at the rate

$$m = \beta_\mathrm{e} A_\mathrm{e} (P_\mathrm{c} - P_\mathrm{e})$$

where e denotes evaporator. So

$$Q/r = \alpha A (T_\mathrm{c} - T_\mathrm{p})/L_\mathrm{v} = \beta_\mathrm{e} A_\mathrm{e} (P_\mathrm{c} - P_\mathrm{e})$$

Remember that $\beta = \alpha/c_\mathrm{p}$, so

$$(A/A_\mathrm{e})(\alpha/\alpha_\mathrm{e}) = (L/c_\mathrm{p})[(P_\mathrm{c} - P_\mathrm{e})/(T_\mathrm{c} - T_\mathrm{p})]$$

If $A\alpha/A_e\alpha_e = 0$ then there is no vapour transfer. In that case A_e has to be infinite; this proves that the bigger the air cooler surface, the smaller the water vapour or weight loss. On the other hand, the bigger $A\alpha/A_e\alpha_e$, the higher is the weight loss.

A small surface A and small heat transfer coefficient α means stacking the product close together and packaging it. A partly filled coldstore gives more weight loss than a fully packed one. With natural convection, cooling the outside surface A_e is very large; subsequently there will be little weight loss. Finally we can conclude that a low value of T_c results in less weight loss.

When chilling freshly stored goods, the evaporation on the surface of the warm product is higher than that during the following periods, so during chilling, we must work with low air temperature and a big dT.

It is not yet possible to give exact values for vapour transfer, though some experiments have been made and certain values are already available. There are, however, as yet too many unknown variables, such as the value of the dry skin formed by strong vapour transfer at the beginning of the cooling period. Figure 8.4 shows some results of experiments carried out by the Sprenger Institute, Wageningen in The Netherlands. As indicated, much depends on the characteristics of the fruit, which differ every year and for each region of origin.

Modern theories about the quick chilling of meat are based on the formation of a layer of dry skin, especially in pork meat.

The Danish Meat Research Institute recommends for pigs of 70 kg after slaughtering:

Chilling time: 60–70 minutes, spread over 30–40 minutes during the first stage and 30–40 minutes during the second stage, and an air temperature during both stages of −24°C to −26°C, an evaporating temperature of −38°C, and about 3000 kg/h circulating air per pig.
Refrigerating capacity: 2.5–3 kW/pig/hour. This will limit the weight loss (or shrinkage) to between 0.50% and 0.70%.
Evaporator surface: about 23 m^2/pig/hour.
Air velocity: 2.5–3.5 m/s.
Surface temperature of the pig after shock chilling: −4°C to −8°C.
Core temperature of the meat: +30°C to +28°C.
Average temperature: +6°C to +4°C.
After removal from the tunnel, the meat shall remain 12 h in an equalizing room at an air temperature of +4°C.
Air volume: 0.7 m^3/min/pig
Evaporator surface: 0.45 m^2/pig.
Evaporation temperature: −1°C.
Total weight loss after chilling and equalizing: 0.52–0.75%.
The rails on which the pigs are hung to have a height of 2.5 m and be 0.6 m away from the walls.

Loading: 2 or 3 pigs/m.
Distance between the rails: 0.7–1 m.

Another method, with lower investment costs, results in higher shrinkage of about 1.8% after removal from the tunnel. This method involves cooling down the pigs in a tunnel at a temperature not lower than −10°C, a relative humidity of between 90 and 95% and an air-change rate of 500 per hour.
Capacity: 2.3 kW/pig/hour.
Evaporation temperature: −15°C.
Cooling time: 2 h.
Final equalizing time: between 10 and 12 h in an air temperature of +3°C.
Air circulation rate: about 80/h in the first period and 40/h in the second period.

Shock chilling also can be used for beef but, as it is very expensive, intermediate solutions are frequently used. For instance, chilling in a coldroom at a rate of 1K/h. Beef can be chilled in an airstream of −2°C to −3°C at 90% RH in 4 hours with an air change rate of 200 per h. Followed by 8 hours between 0 and +1°C with an air change rate of between 80 and 90 per h and the last 12 h at 40 air changes per h and an RH of 85%. The meat will have a core temperature of about +7°C.
Capacity: about 5 kW/beef/hour.
Loading: 3 beef quarter carcasses per m rail or 2 to 3 calf carcasses.
Distance between rails at 3.7 m height: 1.2–1.5 m.
Loading: 2 or 3 halves of beef carcass may be hung per m.

As we can see, the main consideration when using refrigeration is how to limit weight loss, which is the only negative effect of cooling.

The measures that should be taken to limit weight loss are:

1. Good stacking; putting the products close together, but leaving enough space for cold air to circulate. Goods should not be placed against walls and ceilings.

   Apples and pears are now stored in big box pallets, 1.25 × 1.25 × 0.75 m, each containing 300 kg of fruit. They are stacked 5–7 box pallets high. Between the rows of pallets 10 cm of free space is necessary, between the pallets and the walls 25 cm, and between the highest pallet and the air cooler 75 cm.
2. Appropriate temperature and relative humidity selected for each product; temperatures as low as possible without freezing.
3. A different coldstore for each kind of product.
4. Rapid cooling down to storage conditions should be applied. For example, apples should be brought to a temperature of +3°C in 3–4 days, pears to +5°C in 4 days and then within a week to −1°C. A large dT, 10K, is allowed only during this period.

The heat load is calculated so that, following the pull-down period, the refrigerating system has to work only 4 hours a day during the storage period to maintain the temperature.

The pull-down heat load is based on the assumption that during loading of the store only 20% of the storage capacity will be handled each day, and that this quantity will be cooled down in 22 hours with a dT on the evaporators of about 9K.

Note that such fruit coldstores are insulated with 12 cm thick polyurethane and the floors are insulated as well.

5. Once the products have reached the ideal temperature, then the dT must be limited to 4K.
6. High air velocities are allowed only during the pull-down period, with reduced fan capacity during storage.
7. Temperature fluctuations in coldstores and during the storage period should be avoided.
8. Short operation time per 24 h of the compressor limits weight loss for vegetables and fruit stored at a high RH of 98%. Meat coldstores, however, operate at a low RH of 85% and then it is advisable to have longer running times per 24 h while the cooling is creating the thin dry protective layer. Sometimes extra cooling is enforced by switching in heating elements on the evaporator and controlling them by means of a hygrostat.

   The meat surfaces must not touch. Forced fresh air is required at a rate of three times store volume per day.

8.7.1 Determination of the Cooling Time

The simplest equation that can be used in practice to give reasonable results is:

$$t = \ln[(t_1 - t_0)/(t_2 - t_0)]/p \quad \text{where} \quad p = \alpha A/mc$$

where p was defined by Plank as the specific cooling velocity

α = heat transfer coefficient
A = product surface
m = product mass
c = specific heat of the product
t_1 = temperature before chilling
t_2 = final temperature
t_0 = temperature of the refrigerant or intermediate medium
t = cooling time.

When t_1 and t are known, the temperature t' at any time can be found by using $t' = t_0 + (t_1 - t_0)e^{-pt}$ where e = 2.718.

The cooling time curve is an exponential curve. The larger the value of p is, the steeper is the beginning of the curve, which then flattens out later. This is shown by the curves for pork in figure 8.6.

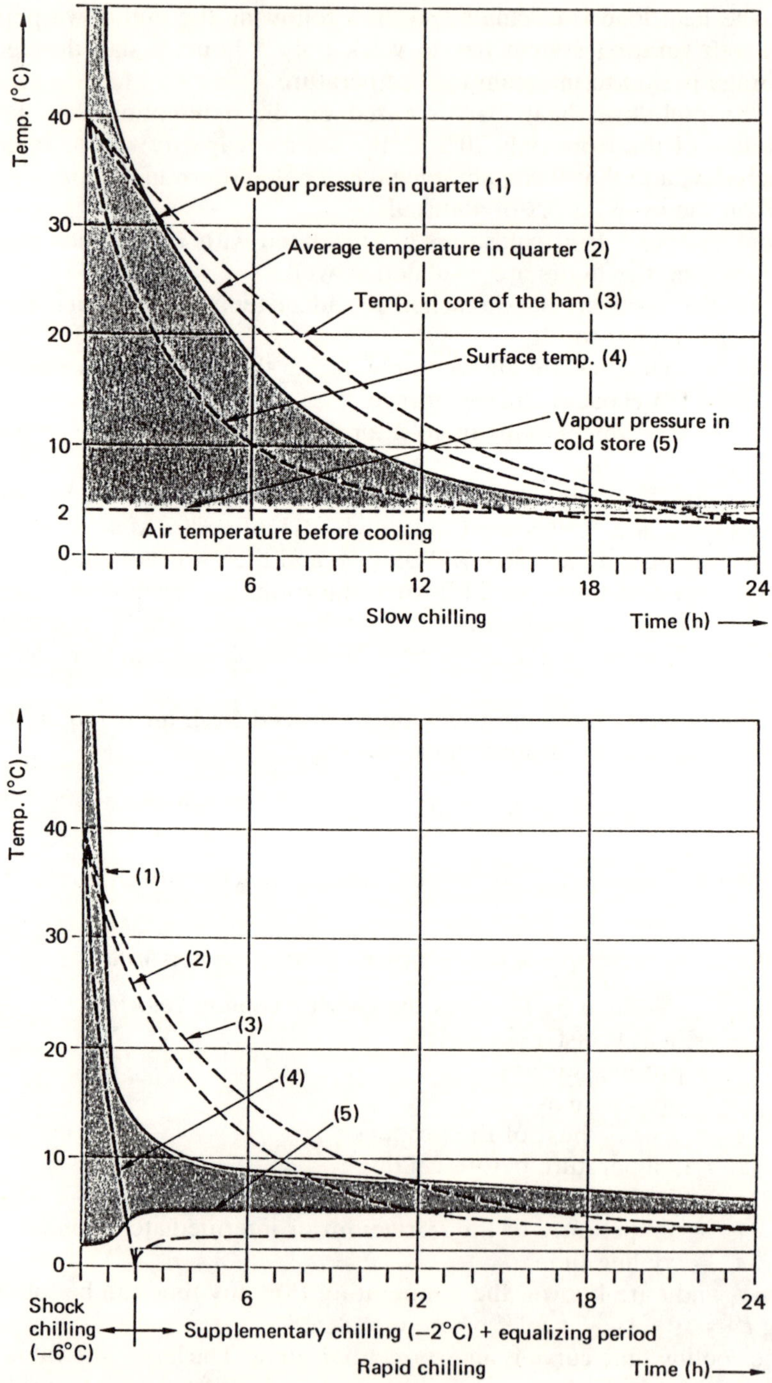

Figure 8.6 *Cooling curves for pork*

To obtain the final product temperature, put $t_2 = t_0$ and the time t is found to be infinite.

The average temperature difference between product and refrigerant is

$$dt_m = (t_1 - t_0)/\ln[(t_1 - t_0)/(t_2 - t_0)]$$

The specific cooling velocity p can also be defined as:

$$p = (t_1 - t_0)/t dt'm$$

8.7.2 Half Cooling Time

If we were to bring the final temperature t_2 of a product to the ambient temperature t_0, then

$$t = \ln[(t_1 - t_0)/(t_0 - t_0)]/p = \ln[(t_1 - t_0)/(0)]/p$$

and t is then equal to ∞.

Close to the end of the real cooling time t, the difference between t_2 and t_0 is very small. If it were one degree, the expression can be written:

$$t = \ln(t_1 - t_0)/p$$

Thevenot defined the half cooling time as the time necessary to reduce the value of $(t_1 - t_0)$ by half. In fact we find that the same period of time t' is required to reduce the temperature difference by half, as is shown on figure 8.7.

For example, if $t_1 = +30°C$, the air is at 0°C and the half cooling time is 4 h, then in 4 h the product reaches +15°C. In the next 4 h it cools from +15°C to +7.5°C, in the next 4 hrs to +3.75°C, and so on. Finally, we can express the half cooling time t_1 as $\ln(2)/p$ or $0.69/p$.

This method was published during the IXth International Refrigeration Congress in 1955 by Thevenot. Merlin, Richard and Sainsbury have determined by experiments the half cooling time of many products, such as for beef in a tunnel, veal, etc.

Bäckström has described a practical method to calculate the cooling time. He starts with the idea that the average temperature of a product is that temperature measured from the surface at a depth equal to one-quarter of its total thickness.

To be cooled down to t_2 the thermal energy must penetrate from this point towards the surface, which depends on the heat transmission coefficient k of the product. Plank's equation $p = \alpha A/mc$ then becomes $p = kA/mc$. Bäckström stated that

$$1/U = 1/\alpha + 3\delta/16\lambda$$

where δ = product thickness in m
λ = heat transmission coefficient by conduction of the product.

α can be calculated as follows. The value of α from a surface of a product

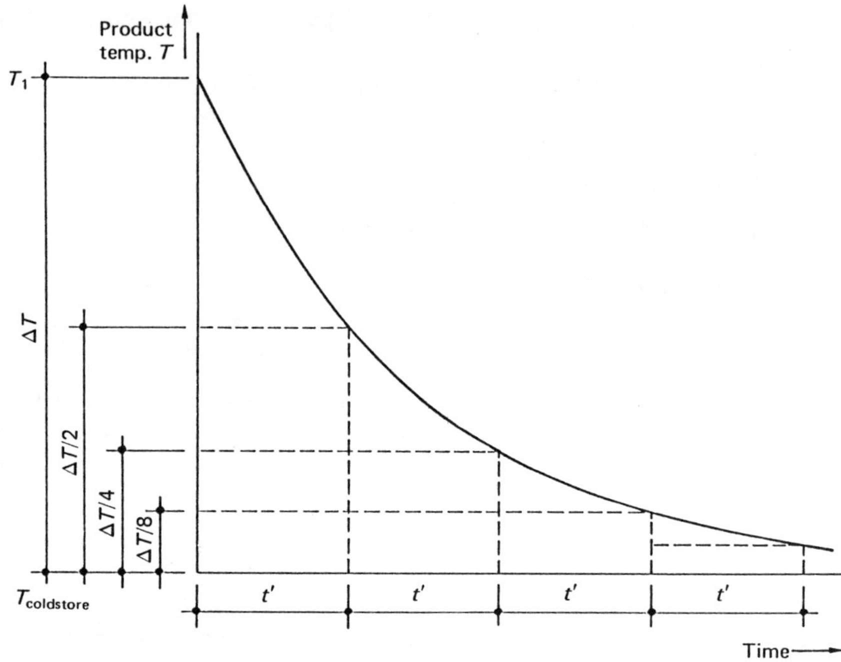

Figure 8.7 *Cooling curve illustrating the half cooling time* t'

containing water and taking into account evaporation of surface water to the coldstore air is given by

$$\alpha = (5 + 3.4c)(1 + 10\mathrm{d}p)/(t_1 - t_0) + 3.5 \text{ kcal/m}^2 \text{ h K}$$

Note that this equation is given in the old metric system. dp is the water vapour pressure difference in cm of Hg between the surface with a temperature of $t_0 + (t_1 - t_0)/5$ and the ambient air.

A typical value of A/m for meat is 0.02 m^2/kg. The average specific heat $c_{\text{average}} = (h_{t1} - h_{t2})/(t_1 - t_2)$. Here Bäckström introduced t_k, the ideal temperature of the product core:

$$t_2 - t_0 = (t_k - t_0)/\{1 + [\alpha\delta/4\lambda]\}$$

With $t = \ln[(t_1 - t_0)/(t_2 - t_0)]/p$ we can calculate the cooling time or the half cooling time $t_1 = 0.69/p$. The surface temperature is given by $t_0 + (t_k - t_0)/\{1 + [\alpha\delta/4\lambda]\}$

8.8 Deep-freezing of Food

As we saw before, food consists of between 50 and 90% water; animal or vegetable organisms contain a low percentage of salts and other solid matters. This means that food freezing is in fact the freezing of low concentration brines with freezing points just below 0°C. When between 80 and 90% of the water content is frozen, this is enough to hamper the growth of micro-organisms. The development of enzymes in vegetables is stopped by blanching before the freezing process, otherwise vegetables lose their green colouring and become brown. Enzyme development in meat is stopped by freezing. For this reason it should be allowed to age before freezing. Fish, in contrast, should be frozen immediately after killing, which means while on board the vessel or factory-ship.

Other methods to stop enzymes from multiplying are smoking and sterilization at high temperature levels. As the temperature of a solution goes down, ice formation increases, while the concentration of solids in the remaining liquid increases until a maximum is reached, as is shown in figure 8.8. This maximum is different for every material and is called the eutectic or cryohydratic point. At this point the temperature will not fall any further owing to heat extraction as the remaining material solidifies. Once all crystals are formed, the temperature will begin to fall again.

However, in practice, foodstuffs are never frozen to such a degree that the eutectic point is reached, never mind exceeded. This would require a temperature of −60°C or lower. It is accepted that a product may be considered frozen when betwen 80 and 90% of its water content is solidified.

To what temperature should a product be frozen? For microbiological reasons the product temperature after deep freezing must be at least −18°C but, for chemical and enzymatic reasons, it must be stored at −24°C or lower.

The majority of harmful bacteria are not destroyed completely by the freezing process but their multiplying process is reduced or stopped. Some bacteria are damaged. Others are even destroyed at −18°C owing to the freezing of the free water; the ice crystals damage the fine cell-structure and change the viscosity of the cytoplasm in the cell, cytoplasmic gases like O_2 and CO_2 escape, the pH changes and there is a concentration of toxic electrolytes in the cells. Denaturation of protein and flocculation occur and, last but not least, the bacteria suffer 'cold-shock'. The metabolism is disturbed, even stopped.

The killing effect on bacteria is even greater in the range −4°C to −10°C than in the range −15°C to −30°C. When the freezing process is slow, the effect of the 'cold-shock' is lost. The bacteria have time to adapt themselves to the new conditions.

There are mainly three kinds of harmful bacteria having the ability to thrive at different temperatures:

- Phychofile
- Mesofile
- Thermofile.

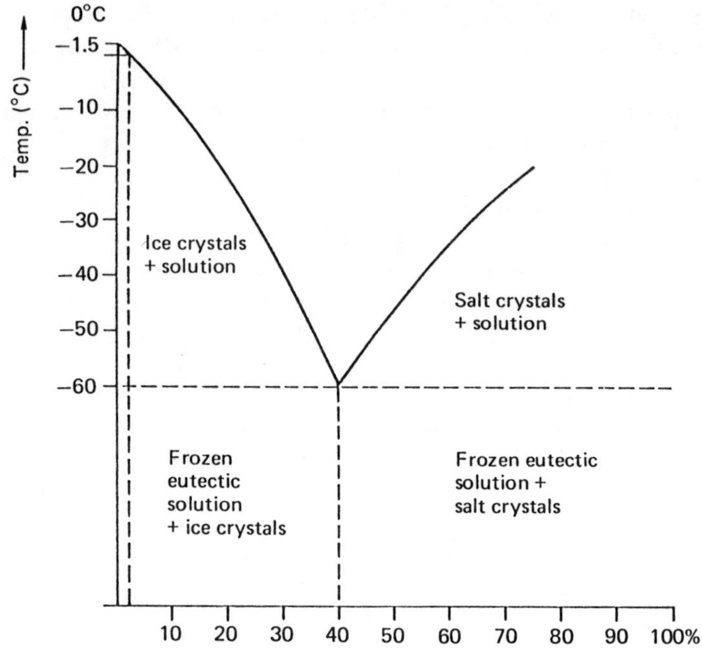

Phase diagram of a typical salt solution

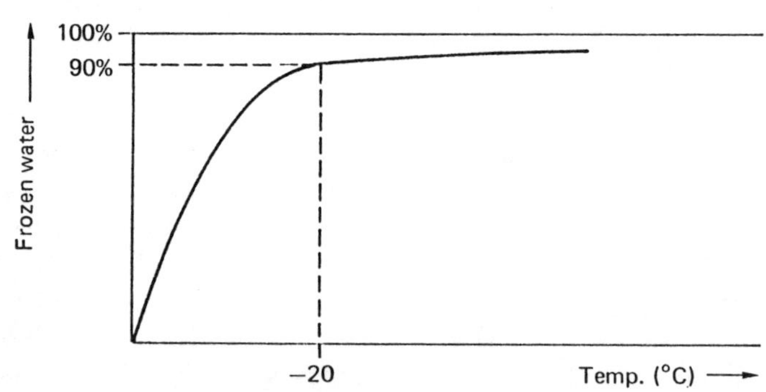

Figure 8.8 *Freezing process of a brine solution*

The first develops best between −18°C and +25°C, the second between +6°C and +52°C and the third between +37°C and +70°C. This accounts for the above-mentioned temperature of at least −18°C.

8.8.1 Freezing Speed

An important element in the evaluation of the quality of freezing methods is the average freezing speed. Freezing speed is the speed at which the ice-front moves forward through a product. Supposing the freezing time is t and the thickness of the layer to be frozen is δ, then the freezing speed will be

$$U_f = \delta/t \text{ cm/h}$$

The further the ice-front moves forward in a product, the higher the freezing speed will be, because the thermal conduction coefficient λ of the frozen layer is higher than that of the water-bearing layer. It should be noted that λ is not constant but is dependent on the temperature. Furthermore, the freezing point is not constant, but depends on the degree of concentration. This is why we speak of an average freezing time between the outer layers and the core of the product. $U_f = \delta_0/t$, where δ_0 is the shortest distance from core to surface.

The freezing speed must be high and consequently the freezing time t short if we want to maintain the quality of the product. When freezing slowly the product will be damaged because as the concentration of minerals in the protoplasm between the cells is higher than inside the cells and the vacuoles, the freezing point there is lower than inside the cells. So when freezing starts there, the liquid is subcooled and water from inside the cells has a tendency to migrate by osmosis through the cell walls to the space inbetween the cells. The ice crystals formed there will grow and become very large; the bigger they are the more water they attract and they can become so big that they perforate and damage the cell wall. Ice has 9% more volume than water, so it can be imagined what happens in the small space between the cells.

In the meantime, the concentration of minerals and other elements in the remaining water in the cells will increase the pH value and the salt concentration. So the cell structure of the product is totally changed and can never be restored. In addition, this dehydration causes changes in the enzymatic activity as in the case of freezer damage to fresh fruit.

Fischer's law says that every change in condition leaves indelible marks.

It is impossible to reverse the freezing process completely by defrosting and, because structure changes must be prevented if possible, quick freezing is necessary; crystals will form simultaneously throughout the product, inbetween as well as inside the cells, but they will stay very small.

Some products such as vegetables are blanched at a temperature near boiling point before deep-freezing. This will arrest the enzymatic activity and has the side effect of drying the cell walls which no longer allows osmosis; thus leakage by osmosis of moisture from the cells to the space between the cells is prevented. At a high freezing speed the ice crystals remain small and will not burst the cell membranes. Consequently the juices and the taste are not lost during defrosting and the structure of the product remains unaltered. The speed required for achieving a good quality final product varies with different foods and is not always equally high. Soft fruit, for example, requires a very high freezing speed, fish and poultry a high speed, but meat requires a much lower one. Some products may even be damaged if the freezing speed is too high.

Freezing speed is considered to be extremely high when U_f is between 300 and 600 cm/h, high when U_f is between 5 and 20 cm/h, medium when U_f is between 1 and 5 cm/h, and slow when U_f is between 0.1 and 1 cm/h.

With vegetables frozen at the rate of 18 cm/h, as much as 90% of the ice crystals remain within the cells. For fish we need a speed of between 0.6 and 5 cm/h; for meat a speed of between 1 and 4 cm/h is satisfactory. At freezing speeds below 1 cm/h, the cell membranes of most products will be destroyed. Usually, large pieces of meat such as beef quarters or pig halves cannot actually be frozen fast enough and under these conditions the best speed is $U_f = 0.1$ cm/h.

In practice the freezing speed varies from a minimum of 0.2 cm/h, when bulk freezing in a coldroom, to 600 cm/h when freezing by contact with liquid gases.

With special refrigerating equipment, the freezing speeds obtained are usually between 0.5 and 5 cm/h. For obvious reasons, the design freezing time t is very important to the cooling plant engineer.

Plank has proposed formulae to calculate the freezing times for various dimensions of products to be frozen. All formulae are based on the four factors which have the biggest influence on the freezing time t:

(a) the solidification heat l calculated on the percentage of water to be frozen,
(b) the thermal conduction coefficient λ which is determined by the speed at which the heat will penetrate the layer already frozen,
(c) the thermal transmission coefficient α between the product and the refrigerant, and of course
(d) the thickness of the layer to be frozen, h_o.

8.8.2 Freezing Time Calculation

Let us look at the freezing process of part of the cylindrically shaped food item shown in figure 8.9. Consider the part with a thickness dr and a length of L.

The thermal energy to be extracted is given by:

$$dQ = l2\pi rL dr \tag{8.1}$$

where l is the latent heat of solidification of the product in W/m^3.

This thermal energy has to pass through the already frozen mass of thickness $(r_o - r)$. Regarding the coefficients of heat transfer α (W/m^2 K) by convection and λ by conduction in W/(m K):

$$dQ = 2\pi L dT dt/\{(1/\alpha r_0) + (1/\lambda)(\ln[r_0/r])\} \tag{8.2}$$

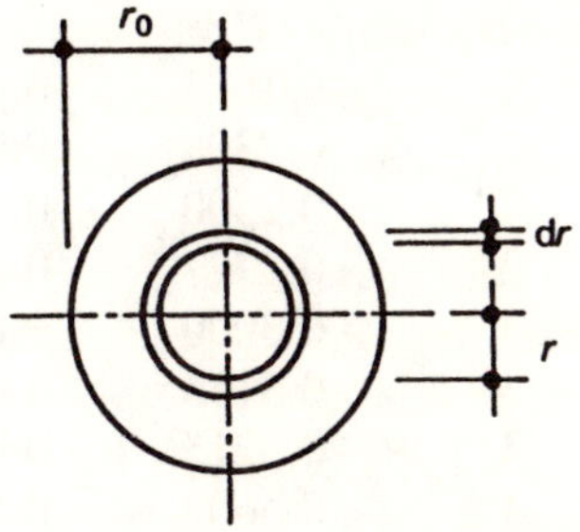

Figure 8.9 *Geometry of a cylindrically shaped food product*

From (8.1) and (8.2) the freezing time becomes:

$$dt = (l/dT)[(1/\alpha r_0) + (1/\lambda)\ln(r_0/r)]r\,dr \tag{8.3}$$

$$\int_0^{t_0} dt = (l/dT)\left[(1/\alpha r_0)\int_{r_0}^{0} r\,dr + 1/\lambda\int_{r_0}^{0} r\ln(r_0/r)\,dr\right]$$

After integration and rearrangement:

$$t_0 = (l/4\lambda dT)r_0[r_0 + (2\lambda/\alpha)]$$

When we put a diameter D_0 in place of the radius, we get a cylinder:

$$t_0 = (l/16\lambda dT)D_0[D_0 + (4\lambda/\alpha)]$$

Example

Fish contains 82% H_2O of which we freeze 90%:
$l = 0.82 \times 334 \times 10^3 \times 1000 \times 0.9 = 246\ 492 \times 10^3$ J/m^3
$\lambda = 1.45$ W/(m K)
$(4\lambda/\alpha) = 0.03$ when freezing in a brine tank.

$$t_0 = (246\ 492 \times 10^3)D_0(D_0 + 0.03)/(16 \times 1.45 dT)$$

D_0 is the least thickness on the biggest section of the fish. If $dT = 30$ and $D_0 = 0.01$ m, then

$$t_0 = (246\ 492 \times 10^3)0.01(0.01 + 0.03)/(16 \times 1.45 \times 30) = 13\ 501 \text{ s or } 22.5 \text{ min}$$

For blocks cooled on all sides and entering the freezer at their freezing point, Plank proposed the equation:

$$t_0 = (l/dT)[(Ph_0/\alpha) + (Rh_0/\lambda)]$$

where h_0 = smallest dimension.
He also provided a table for P and R, depending on the dimension ratios B_1 and B_2, where B_1 = length/h_0 and B_2 = weight/h_0.

B_1	B_2	P	R
1	1	0.1677	0.0417
2	1.5	0.2308	0.0656
3	2	0.2500	0.0719
3	2	0.2727	0.0776
3	3	0.3000	0.0849
4	3	0.3156	0.0887
4.5	3	0.3215	0.0902
6	4.5	0.3602	0.0490

For freezing between two refrigerated plates, $p = +1/2$ and $R = 1/8$

Example

For fish: h_0 = 8 cm
length = 48 cm
width = 36 cm
If air velocity U = 5 m/s:

α = (10 + 3.5U)1.163 = 32 W/(m^2 K)
λ = 1.45 W/(m K)
B_1 = 48/8 = 6
B_2 = 36/8 = 4.5
air temperature = −35°C
P = 0.362
R = 0.0490

then

$$t_0 = 246\,492 \times 10^3[(0.3602(0.08/32.4) + 0.0490(0.08/1.45]) = 7747 \text{ s}$$
$$= 2.15 \text{ h}$$

So far we have calculated only for products which are entering the freezer at their freezing point. Normally time is needed for pre-cooling to the freezing point, as well as time for post-cooling.

Plank stated for blocks of meat that the pre-cooling time is given by

$$t_p = t_0 0.0053 T_i$$

where T_i = initial temperature of the product
and for fruit and vegetables $t_p = 0.01\ T_i$.

Rjutoff gives the following equation for the postcooling time t_a:

$$t_a = \frac{\rho n c}{\lambda \pi^2} \ln\left(\frac{T - T_0}{T_e - T_0} - 0.21\right) h_0\left(h_0 + \frac{4\lambda}{\alpha}\right)$$

where n = correction factor = 1.03 for quick freezing, 1.16 for slow freezing
ρ = density of the product
c = specific heat of the product
λ = heat transfer coefficient by conduction of the product
α = heat transfer coefficient by convection
T = product temperature (at freezing point)
T_0 = air, plate or brine temperature
T_e = final temperature of the product.

Another source gives an approximate equation:

$$t_p + t_a = t_0 \times 0.039(T - T_0) \log\left(\frac{T_1 - T_0}{T_e - T_0}\right) + 0.114\left(\frac{\rho c}{\lambda}\right)\left(\frac{h_0^2}{4}\right)$$

All these equations apply to cooling on two sides of the product; if all sides are in contact t should be reduced by 1/3.

The theories on freezing time prediction discussed in this section are the basic theories that have existed for several years. Today, many sophisticated equations are available, mostly involving computer software. Some appropriate references are given at the end of this chapter. However, despite this the equations given here are still valuable background information, necessary for a good understanding of the parameters influencing freezing times.

Influence of Packing Material

Where added packing is involved, Plank's equations must be adapted. Adding the effects of the heat transfer by conduction of the packing material, the equation above involving P and R becomes:

$$t_u = [(lh_0)/(T_p - T_0)]\{P[(1/\alpha) + \Sigma(\delta f/\lambda f)] + R(h_0/\lambda)\}$$

Here l = $A_v(l'_{water}A_w + c_{water}dT)$ in units of kJ/m^3
dT = $T_i - T_{end}$
A_v = percentage of packing-volume filled with the product
A_w = percentage of water in the product
c = specific heat of the product in W/m^3.

Some values for δ/λ when freezing between two plates are:

Waxed carton:	$\delta = 0.625$, $\delta/\lambda = 0.0096$
Waxed carton with 4 layers of cellophane:	$\delta = 0.568$, $\delta/\lambda = 0.0109$
Aluminium foil:	$\delta = 0.568$, $\delta/\lambda = 0.0075$
	$\delta = 0.599$, $\delta/\lambda = 0.0095$

$\delta = 0.509$, $\delta/\lambda = 0.007$

Double-waxed greaseproof paper: $\delta = 0.212$, $\delta/\lambda = 0.0035$

Using a plate freezer, we must add δ/λ for a small air-film, which depends on the pressure on the plates as follows:

$$(\delta/\lambda) = 0.00018/(p + 0.004)$$

in which p is the pressure in bar on the product.

When we are dealing with a blast freezer tunnel, $(\delta/\lambda)_{air} = 0.045$ must be added.

Chistodulo and Rjutoff determined the following values for δ/λ:

Grease-proof paper:	0.0016
Waxed, half grease-proof paper:	0.0042
Waxed carton + grease-proof paper:	0.0060

Watzinger derived an equation for the freezing time of fish fillets between plates as follows:

$$t_0 = (1 + 0.008T_1)\{(l + q_n)/(-t_0 + 1.5)\}[(h_0/8) + (1/2\alpha)]h_0.$$

where q_n = thermal energy which must be removed for postchilling. $1/\alpha$ is determined mainly by the air-film between package and plate = 0.0018. The average heat transfer coefficient α for hoar-frost on the plates is 0.025 W/(m^2 K).

In the recommendations for the treatment of deep frozen products of the International Institute for Refrigeration (1972) we find:

$$t = l(\mathrm{d}h/\mathrm{d}T)(h_0/N)[(h_0/4\lambda) + (1/\alpha)]$$

where dh = enthalpy difference before and after freezing
dT = temperature difference between cooling medium and freezing point of the product
h_0 = thickness of the product in m
N = factor that depends on the shape of the product
= 2 for plates
= 4 for cylinders
= 6 for a sphere.

Scheffer gives a simple equation for meat:

$$t = (lph_0/\mathrm{d}T)[(0.32/\alpha) + (0.1h_0/\lambda)]$$

where t is the time to reach −5°C in the core of the product.

Examination of these equations leads us to the following. The freezing time will be shorter when the heat transfer coefficient between product and medium α is high, and α is high at higher medium velocities. Freezing time is shorter for a small thickness of product, either h_0 or r_0. A big dT gives a shorter freezing time and the freezing time of unpacked products is shorter too. Air velocities must be kept within economical limits. Velocities above 5 m/s result in high fan energy consumption and much extra thermal energy removal, whereas only slight improvement in the quality of the frozen product is obtained. Finally we must emphasize the importance of allowing the chilling medium, air or liquid, to touch all sides of the product. It is essential that products should not be placed too close together.

8.8.3 Freezing Equipment

From the theory we conclude that the entire freezing process has to be fast, and the freezing speed as high as possible. The following elements are necessary if a high freezing speed is to be achieved:

(a) low temperatures
(b) good thermal transmission
(c) at least one small product dimension so that the thermal energy can leave the product by a short route.

In the case of (c) the element is fully dependent on the product, especially its shape. However we can do something about this if we do not increase the total mass thickness of the product by piling it up or by using wrong packing methods for the freezing process. The thermal transmission (b) can be increased by providing the best possible contact between the product and the cooling medium. The best thermal transmission is achieved by contact with cold liquid; as we shall see later, many practical freezing systems are based on contact between a gas, air, and the product. In order to improve the transmission, cold air must be blown over the product at a suitable velocity. Contact with liquid creates problems of hygiene and contamination. So an alternative freezing method may be used in which direct contact is made with the cold metallic solid surface of the evaporator.

Which equipment can we use for freezing purposes? In broad terms there are five groups:

1. Blast freezing tunnels with carriage or conveyor systems, with either automatic or manual transport of the carriages as shown on figure 8.10.

2. Blast freezing tunnels with perforated conveyor belts or rails, the first one for bulk goods, the second one for packaged products. See figures 8.11 and 8.12.

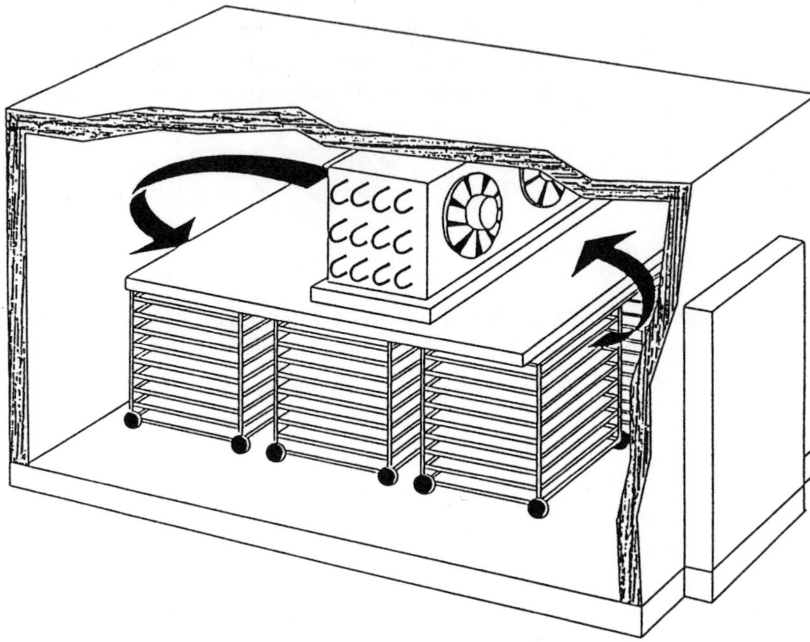

Figure 8.10 *Trolley-system blast freezing tunnel*

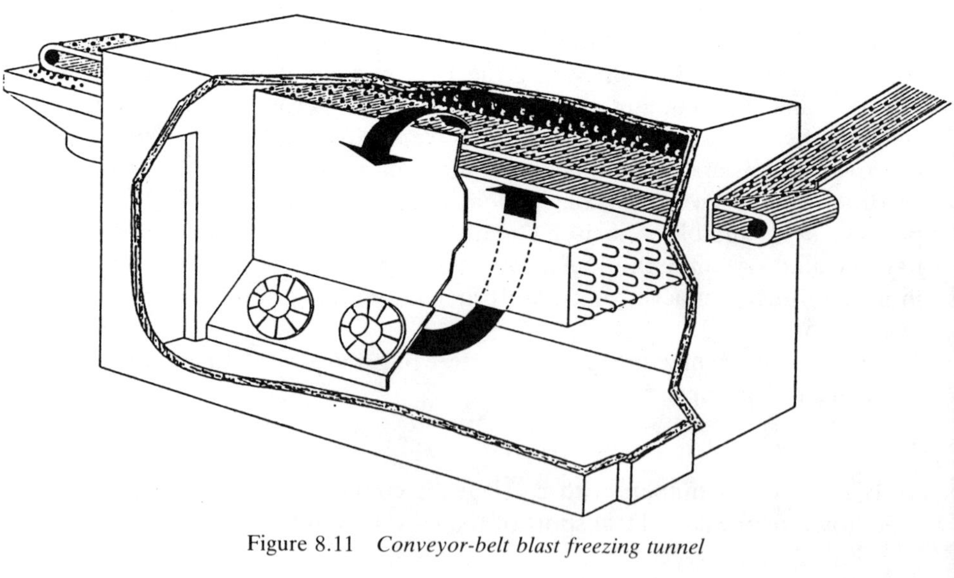

Figure 8.11 *Conveyor-belt blast freezing tunnel*

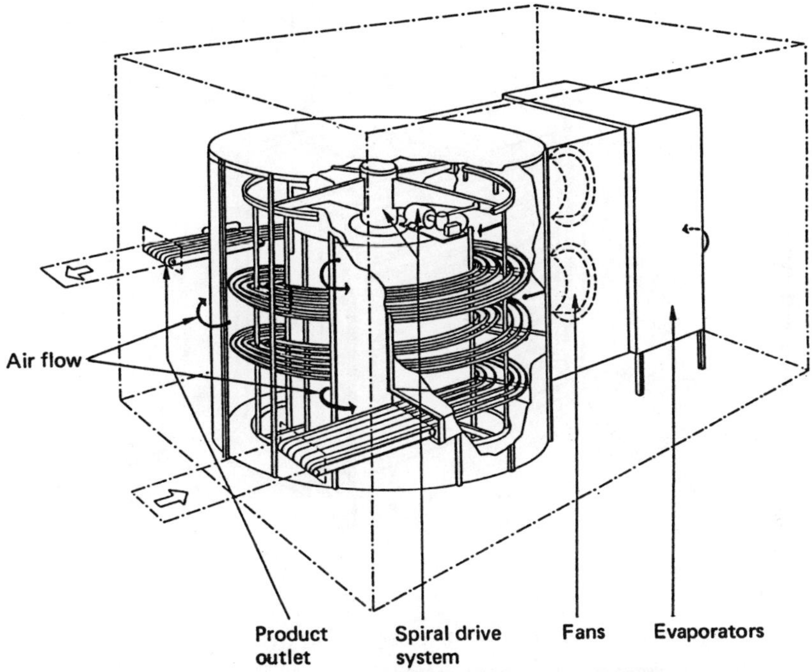

Figure 8.12 *Spiral conveyor blast freezer*

3. Plate freezers with or without an automatic loading system either for vertical or horizontal loading. See figures 8.13, 8.14 and 8.15.

In addition to the equipment above other freezing machines have been devised, especially for certain products such as:

(a) Drum freezers cooled by cold brine on the inside, producing frozen food flakes on its surface from liquid food as shown on figure 8.16.
(b) Blast freezing tunnels in which the belt conveyors are replaced by a fluidized bed; these are used in the processing of small food particles such as peas (see figure 8.17).
(c) Ice-cream freezers, consisting of either a cylinder with a refrigerated jacket, or a brine tank for immersion of the metal ice-cream moulds. Direct immersion in cold brine is no longer common, although the system is still being used for freezing tuna fish and similar foods.
(d) Stainless steel conveyors cooled by spraying cold brine against the underside of the conveyors.
(e) Freeze-drying equipment. After freezing the water, the product is placed under a vacuum and by warming up the product the ice is sublimated and a porous product obtained with 70–90% less weight.

The majority of frozen food production is handled by the equipment 1, 2 or 3 listed above. For outputs below 2000 kg/h,

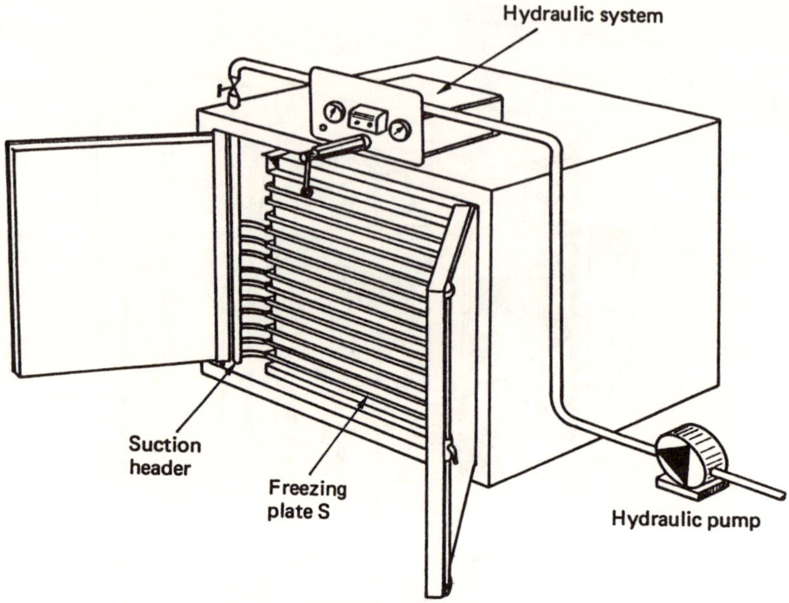

Figure 8.13 *Horizontal plate freezer*

Figure 8.14 *Automatic horizontal plate freezer*

Figure 8.15 *Automatic vertical plate freezer*

Figure 8.16 *Drum freezer*

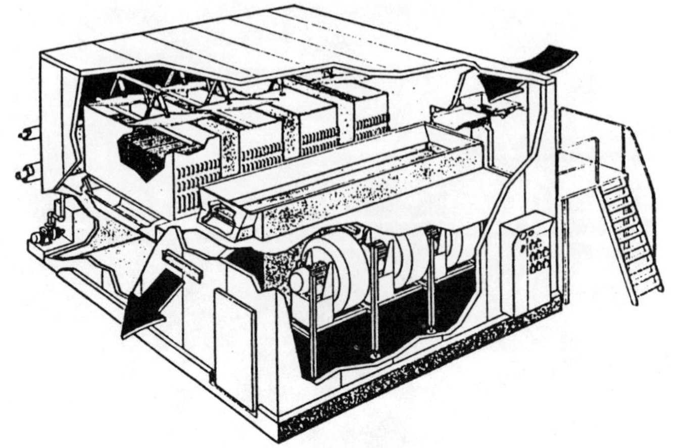

Figure 8.17 *Fluidized bed freezer*

continuous working, only a tunnel with carriages provides a good return on investment.

Although automatic conveyor belt tunnels (figure 8.18) and plate freezers have the advantage of shorter freezing times and give a better quality product, the cost of investment will be too high if production is below the above-mentioned figure, because the running costs and therefore the cost of freezing are too high.

A freezing tunnel equipped with carriage conveyor must be designed in such a way that the products are exposed to an air velocity of approximately 2 or 3 m/s. The calculation may also be based on an air velocity of 1 m/s in an empty tunnel. Air temperature must be at least −35°C, and in some cases even as low as −45°C, which means that the evaporation temperature in the plant should be between −42 and −52°C. Freezing time is dependent on the nature of the product and varies from 1 to 4 hours for small size products and up to 10 and even 18 hours for large pieces of meat such as ham or beef quarters. Nowadays, meat is often frozen on metal or plastic trays or in boxes, and quite often it is pre-packed in plastic film. Stacking must be done in such a way that cold air can circulate between the trays or boxes at a suitable speed.

The thickness of packing may double the freezing time not only because of the packing material used but also, more importantly, as a result of the insulating power of air enclosed in the packing. It is preferable to use boxes 10 cm high rather than 20 cm: the empty weight is proportionally the same, so by using adequate stacking methods and provided that the cooling plant has been designed

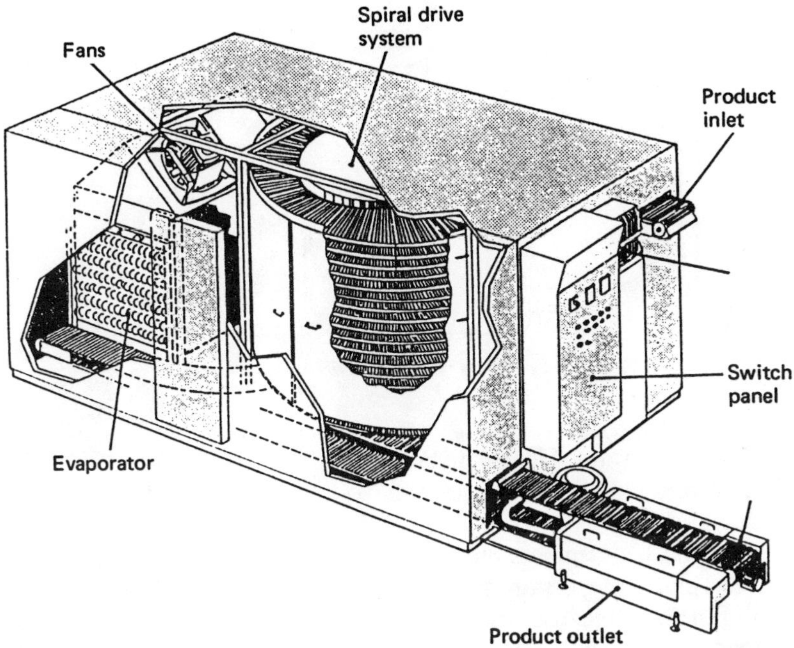

Figure 8.18 *Spiral belt freezing tunnel*

correctly, it is quite possible to freeze meat to a core temperature of −10°C in about 15 hours or −18°C in 24 hours.

When air coolers are mounted above a false ceiling, it is necessary to ensure a uniform air flow over the products by means of air deflectors which must be spaced logarithmically over the freezing room for best results. See figure 8.19. Transverse air blowing is preferable because long air passages not only require too much ventilation capacity on account of the resistance of ambient air, but also because of the large temperature difference between the first and the last product covered by a long distance current of air.

Freezing tunnels have the additional advantage of being suitable for many different products.

4. For large-scale production, especially small-sized products in bulk, the individual quick freezing, or IQF, system is used. In this system food particles, such as diced carrots and peas, are conveyed through the tunnel on a perforated conveyor belt for a few minutes, while cold air at −35°C is blown at high speed through the tunnel. The particles are lifted by the upward air current and hover over the belt, thus being totally in contact with cold air; the evaporation temperature required is

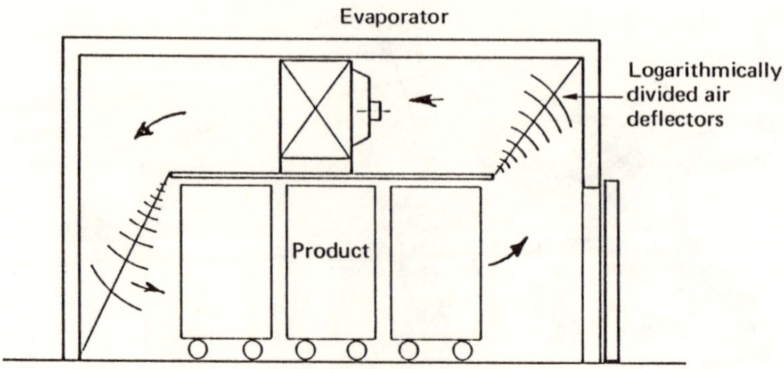

Figure 8.19 *Blast freezing tunnel with air deflectors*

between −40 and −43°C. The belt velocity can be adjusted according to the requirements of the product, so that the tunnel may be called universal, at least for a range of products. This is not so, however, with the specific IQF tunnels mentioned under (b) above.

5. Finally, plate freezers are known to give very efficient freezing times for small portions because these are in close contact with the refrigerant as it evaporates inside the plates against which the product rests. On average, it may safely be said that a thickness of 25 mm requires 1 hour of freezing time. As is true with all of the above mentioned systems, the correct freezing time has to be calculated or established experimentally. One disadvantage with plate freezers is that all the products must be equally dimensioned and the thickness is limited to between approximately 75 and 150 mm, related to the distances between the plates. When freezing thicker blocks of product, these distances can only be achieved between vertical plates using a so-called vertical plate freezer. These are used for freezing big blocks of fish or fish fillets.

   Cold consumption is about 450 kJ/kg or less, depending on the water content of the product for all the equipment mentioned above. However, since plate freezers require no air circulation, the power consumption is lower with this system than with the others. Plate freezers, since they are operating without air as an intercooling medium, can work with a higher evaporation temperature of approximately −35°C.

   The systems mentioned under (a), (c) and (d) do not have fans either, which makes a difference in power consumption, but they generally work with a secondary refrigerant, which results in extra power consumption because of the lower evaporating temperatures and energy for pump circulation of the brine.

   All of these are best operated as 'flooded systems', which makes them most suitable for operation with pumped ammonia rather than with an HCFC.

In the comparative evaluation of freezing costs per kg, the refrigerating capacity required and the necessary evaporation temperature are more important than the cost of investment, which is made up of the cost of capital and amortization. In this connection, for a given production run it may prove interesting to examine the cost of freezing entailed by the above systems, and then compare the figures with the cost of the liquid gas system. With nitrogen freezing we may ignore the cost of the compression energy, but this is replaced by the heavy cost of nitrogen consumption of approximately 1 kg nitrogen per kg product.

With CFC freezing, compression energy remains and a CFC consumption of 3% must be reckoned with, which makes this system impossible to use in the near future because of ozone depletion, global warming problems and the rules limiting the use of CFCs.

Carbon dioxide requires less than 1 kg per kg product. The enthalpy is higher but 1 kg of CO_2 is generally more expensive than 1 kg N_2. When using N_2 there is a loss of gas when refrigerating during storage, whereas with CO_2 a small cooling plant is necessary for this purpose.

A comparison between conventional systems and the last-mentioned using liquid nitrogen is illustrated by the diagram shown in figure 8.20. Freezing with a liquid CFC is somewhat less expensive than with liquid nitrogen because the gas used can be recovered by means of a conventional NH_3 cooling plant, in which the evaporator acts as a condenser for the evaporated CFC. Liquid carbon dioxide is usually the most expensive system, although for very large installations it may prove comparable, and in some cases cheaper. The cost of the liquid gas used is the main determinant in the evaluation of the above systems. They yield somewhat lower freezing temperatures: while CFCs and CO_2 are not much lower than the conventional systems, N_2 goes as low as somewhere between −80 and −196°C, although the temperature can be controlled at high values. It must be said that such extremely low temperatures are not always beneficial to the quality of the product. Too high a freezing speed can result in permanent changes to the intercellular structure, a phenomen called ‘intercrystalline collapse’. Those systems are mainly used either for luxury food products having a high retail value, or as an extra plant for meeting peak production requirements.

As an argument in favour of fluid gas systems it is often said that desiccation is kept down to a lower level. In practice this is true to some extent, but the difference is not so big as is sometimes claimed. As against 2 or 3% with cold air systems, desiccation is between 0.5 and 1.5% with liquid gas systems. Of course, the desiccation factor may be disregarded when freezing packed products. The difference in the effects of desiccation on packaged and loose products is shown in detail in figure 8.20.

Freeze-drying systems, which occupy a special place in this selection of freezing equipment, are only economically viable for very special applications such as expensive and important pharmaceutical products.

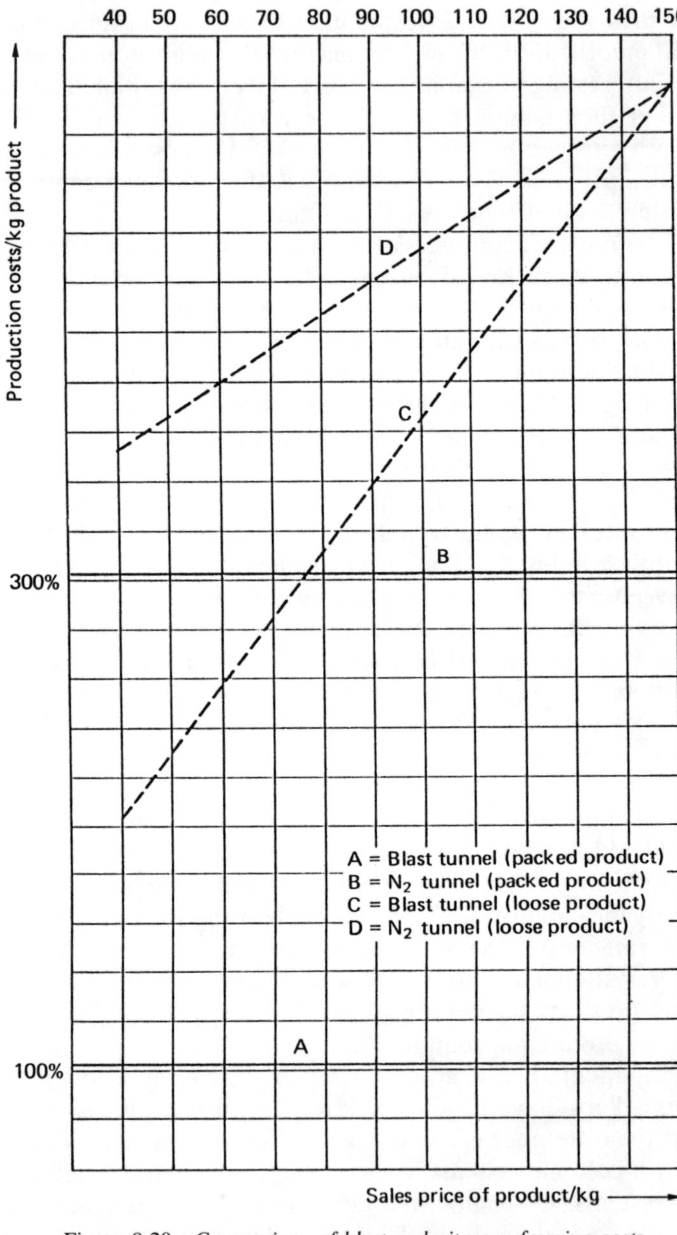

Figure 8.20 *Comparison of blast and nitrogen freezing costs*

8.9 Defrosting of Frozen Food

In many cases the frozen products, if these are raw, must be defrosted fully or partially before being prepared for final consumption. Efficient defrosting is as important as efficient freezing. Unfortunately, for reasons of economy, this principle has not been put into practice up to now.

Defrosting is usually done by water spraying or immersion, or in defrosting rooms; but it is impossible to reverse the freezing process completely. As Fischer's rule says: 'every change of condition leaves indelible marks'.

Defrosting in a cooling room equipped with a cooling and heating system leaves many prejudicial effects, in other words this method of defrosting is detrimental to the product. Too much heating will cause surface dehydration, whereas, if the surface remains too humid because of condensation, micro-organisms will develop on it. As the ice crystals melt, open pores may remain which offer more contact with the ambient air and thus encourage desiccation. Surface moisture evaporation also means a loss of vitamins; the loss of weight can in some cases amount to 3% and even up to 10%. Therefore, rapid defrosting is advisable, especially for smaller pieces, although this is not feasible with large pieces such as beef quarters etc. As explained earlier, the freezing process is never ideal for those products, and slow defrosting is beneficial because it gives the meat cells more time to recover some of their natural moisture content.

Generally speaking there are two important factors involved in defrosting. The first is based on the thermal conductivity of the material and is called the thermal diffusion of the product. Thermal diffusion is defined as

$$\text{thermal conductivity/density} \times \text{specific heat}$$

The second factor is based on the electric properties of the product, such as the dielectric constant and the specific resistance.

The following systems rely on the influence of the first factor mentioned above:

(a) defrosting by conditioned air
(b) water defrosting
(c) vacuum steam defrosting.

The second factor finds application in the following processes:

(d) high-frequency heating
(e) microwaves
(f) electrical resistance heating.

A closer examination of these different systems reveals the following particulars:

(a) When defrosting in special rooms, the rooms are first cooled down to approximately 2 or 3°C above storage room temperature, cooling is then stopped and after filling the defrosting rooms the heating cycle is started. The air circulation is as much as 10–15 times room volume/h and the air temperature is between 15 and 25°C.

Defrosting can take from 24 hours to several days. A saturated air flow is used in some cases with the air-defrosting system. Saturation of the air is achieved as the air passes through an air-washer; air flow velocities of approximately 5 m/s are appropriate. Since the air is saturated with water vapour, desiccation is reduced to a minimum and the velocity over the product surface can be kept high.

When large quantities are involved, a reversible air current system is used to achieve a uniform spreading of the defrosting effect over the products. The disadvantage with this method, however, is once again the long defrosting period of several days and the amount of space required.

(b) In some instances, defrosting is done in drag tanks using water at 20°C and circulating at a speed of between 0.3 and 1.5 m/s. As pollution and bacterial growth in the defrosting water may be caused by drip losses carrying organisms, this method is not always applicable.

(c) The vacuum heating system is another defrosting method in which products are put under a vacuum and water vapour is forced to condense on the surface. In fact, the water vapour is steam, generated at a low pressure and therefore at a low temperature. Temperatures over 21°C are not suitable since they would overheat the surface of the product.

We will now examine the second group of systems. The main difference from the first group is that it results in a shorter defrosting time. Moreover, this group is suitable for continuous operation and can therefore be used on production runs. Consequently, it is safe to say that these defrosting systems are certain to find greater use in the future. However, in the case of systems (d) and (e), which have been developed fairly recently, the investment costs as well as the operating expenses are still quite high. Until recently they were mainly used for 'tempering' purposes, which involves raising the temperature to just below the freezing point, so facilitating further treatment of the frozen products, while the cost of energy is kept down to an acceptable level.

The dielectric method (d) operates with a high-frequency 25 kV electrical supply with a frequency between 1 and 100 MHz.

The microwave defrosting method (e) will certainly be preferred in the future to method (d), in which defrosting time is longer and product heating is unequal. For this reason it is important to examine the microwave defrosting method more closely. The system mentioned in (f) is based on heating by electrical resistance between two electrodes and requires pre-heating by conventional means for most products in order to reduce the electric resistance of the product to an acceptable level. For this reason it is less efficient.

Defrosting by Means of Microwaves (e)

For a better idea of what happens when using this process, let us first recall what is meant by 'temperature'. In fact, the temperature of a material is determined by the degree of vibration of its molecules. An increase of the vibration frequency causes more intermolecular friction, by which energy is produced in the form of thermal energy. Let us suppose that a molecule is a small magnet having a north and a south pole. When placing this magnet in an alternating magnetic field, the magnet will follow the alternations at the same frequency, although it will be slowed down to some extent by friction. This phenomenon is called *dephasing* in electrical ac theory. Friction uses up some of the field energy, and so the question of efficiency arises, as only 50% of the amount of energy required is efficiently used. Microwaves can be described as high-frequency vibrations used for accelerating the frequency of vibration of the molecules of a material between two excited poles or magnetrons.

Let us first examine the frequency of these vibrations. When looking at the spectrum shown in figure 8.21, we see that microwaves have a low-frequency range close to that of normal telecommunications. *We can see for ourselves that there is no question of radiation at all.*

It is obvious that the authorities in most countries have promulgated laws to prevent any disturbance to their telecommunication system. In Great Britain, for instance, microwaves are permitted at a frequency of 915 MHz, whereas in the Continental Europe the lowest frequency is 2450 MHz.

The degree of penetration of the waves into the product depends among other things on the wavelength. A frequency of 915 MHz results in a wavelength of 33 cm, 2450 MHz in 12 cm, 5800 MHz in 5 cm and 22125 MHz in 1.35 cm. From this we can see that the ability to handle a given product thickness is determined by the frequency permitted by the authorities.

With the different frequencies used, the waves penetrate through glass, air, plastic and chinaware and are reflected by metal and absorbed by water, albumens, proteins, glucose and food in general.

The friction losses between the molecules depend on the dielectric strength, which is different for all foods that require defrosting. This dielectric strength is a form of electrical viscosity which causes the dephasing and energy loss. The dielectric constant uses the symbol ϵ. For air, $\epsilon = 1$. As we know, $P = UI \sin Q$ or when the angle is small

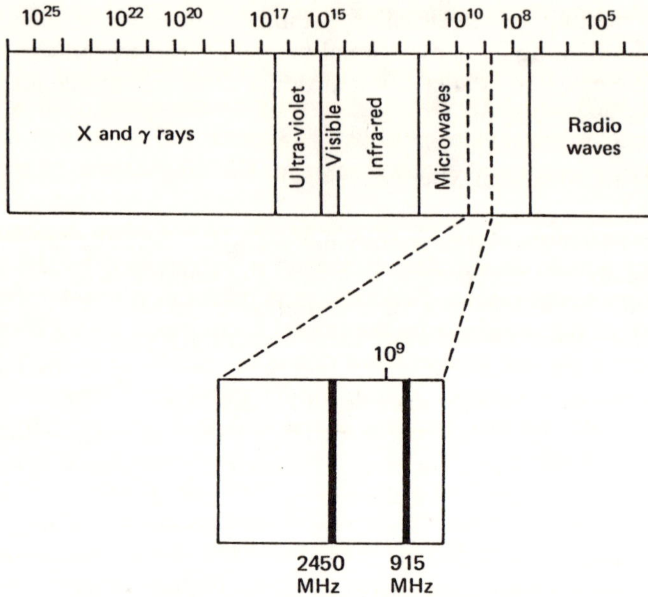

Figure 8.21 *Electromagnetic spectrum*

$P = UI\tan\delta$. The angle δ is a measure for the energy loss, as also is ϵ. When ϵ'' is the so-called imaginary vector and ϵ' the active vector, then $\tan\delta = \epsilon''/\epsilon'$.

The dielectric conductivity is denoted by σ, and is a function of the frequency f and of ϵ'':

$$\sigma = 2f\epsilon''$$

Analogously to $P = I^2R$, we have the equation $P = U^2\sigma$ or $U^2 2f\epsilon''$. We also have empirical values of $\epsilon' r$, and $\epsilon' r = \epsilon'/\epsilon_0$ where ϵ_0 is ϵ for air or $10^{-6}/36\pi$ Farads/m.

Now the equation becomes $P = 55.61 \times 10^{-4} f\epsilon' r \tan\delta\ U^2$ (W/cm^3). The increase in temperature is

$$dT = P/c_p = 8 \times 10^{-12} U^2 \epsilon' r \tan\delta\ f/c_p \text{ (K/min)}$$

Also

$$\text{Penetration} = \lambda_0/2\pi\sqrt{\epsilon' r}\tan\delta$$

where λ_0 = wavelength in air. As the usual permitted $\lambda_0 = 12.2$ cm:

$$\text{Penetration} = 1.94/(\sqrt{\epsilon' r}\tan\delta)$$

The main parts of a microwave oven are the magnetrons, which can be compared to the positive and negative plates of a capacitor with a non-conducting intermediate material or a dielectric between them in the magnetic field.

At present, the magnetrons have a limited lifetime of somewhere between 4000 and 6000 hours. The only maintenance cost with a microwave tunnel is the replacement of its magnetrons; the energy requirements are electric current and cooling water for the magnetrons of approximately 6 litres/kW.

The penetrating capacity of the waves is not only dependent on the frequency which, as we already know, is determined by the wavelength, but also on temperature. Penetration is better at a low temperature; this being one of the reasons why the tunnel is equipped with a cooling system, which keeps the air temperature down to −25°C. It is quite remarkable that defrosting is achieved at freezing temperature. This has the additional advantage that surface heating is reduced and local overheating is prevented as a result.

The microwave system is different from all others in that the rise in temperature is not achieved by a common heat-transfer process, but by direct action on the molecules of the materials. Moreover, defrosting times are extremely short; there is no question of days or even hours, but of minutes and seconds.

Quick defrosting prevents loss of cell juice, in other words it prevents nutrition and quality losses. Loss of cell juice stimulates the development of microbiological cultures. A high pH value also reduces loss of cell juice.

The freezing process has more effect on the pH of a product than the defrosting process, but fortunately the pH value can be increased by artificial additives such as insulin and organic salts to achieve the desired pH value of between 6.7 and 7.

A typical industrial microwave tunnel is shown in figure 8.22.

8.10 Packing Methods

We will now look at the main effects of the different packing methods on the cold chain. It is a fact that on the one hand packing is a hindrance and on the other hand it is a help. It is certainly a hindrance whenever quick heat transmission is required, namely with quick-freezing or quick-defrosting processes.

The wrong type of packing can:

(a) Be an obstacle to correct stacking of products in freezing tunnels.
(b) Obstruct heat transfer and thereby decrease freezing speed. This can be caused by the insulating properties of the packing material itself as well as by an excess of enclosed air. The latter impediment is due to

Figure 8.22 *Microwave thawing tunnel*

inadequate shape or to dimensions of packing which are unsuitable for the product.

(c) Make certain defrosting methods totally ineffective, for instance metal packing for microwave defrosting and heating, or thermoplastic packing materials for products treated in convection ovens.

Packing can be a help for cold storage. It can prevent desiccation, freezer burn, adherence by freezing and oxidation. It is helpful in creating the right atmospheric condition or even maintaining a controlled atmosphere. In some cases surface asphyxiation, through lack of CO_2, must be prevented by removing the ethylene or the respiration heat, whereas at the same time exudation from the product must be avoided. As such, packing has a direct influence on the development of micro-organisms, while on the other hand it provides indirect control on these by mechanical protection. Therefore, packing must be done in such a way that exchange of water vapour or gas between the product and the cold atmosphere is either prevented, slowed down or only allowed to a limited extent, depending on the nature of the product stored and the storing technique used.

An important characteristic of packing is its mass transmission coefficient, in other words the permeability coefficient to water vapour, oxygen, carbon dioxide and nitrogen. In theory, it is possible to calculate the amount of water vapour or gas flow transmitted if the above-mentioned coefficients are known.

$V = \mu_v dp/\delta$, in which V stands for the flow transmitted, which is of course directly proportional to the permeability coefficient μ_v, the surface A and the vapour pressure difference dp across the packing, whereas it is in inverse proportion to the thickness δ of the packing material. Practically however, the construction of the packing, the method of closure and the efficiency of the packing machine all have a significant effect on the gas flow through the packing. Finally, the mechanical strength of the packing is very important too, if damaged and weak spots, which are the favourite breeding grounds for micro-organisms and decay, are to be prevented. From this, it is quite clear that adequate packing is of the utmost importance in the cold chain. Though packing is only one link in this chain, it definitely deserves as much attention as the others.

References

M. Manche, France (lectures on food technology).

Freezing time calculations bibliography

D. J. Cleland, A. C. Cleland and R. L. Earle, *International journal of Refrigeration* 10 (1987) 22–31, 156–164; *Journal Food Science*, 44 (1979) 964–970.

Q. T. Pham, *Food Technology*; 21 (1986) 209–220.

V. O. Salvadori and R. H. Mascheroni, *Journal Food Science* (1990).

C. Ilicalli and N. Salgam. *J. Food Science Proc. Eng.*, (1987) 299–314.

9

Special Food Preservation Methods and Other Applications

9.1 Controlled Atmosphere Storage

In the past the only conditions that were controlled in a coldstore for fruit or vegetables were temperature and relative humidity. These conditions are obtained by using efficient refrigerating plant and good insulation of the room. The only other action taken was refreshing the air from time to time by extracting air through a ventilator. This action was particularly necessary for citrus fruit, which produces a large quantity of volatile gases.

After some time, engineers became aware of the necessity to pay more attention to the composition of the air in the coldstore if the life of the products was to be extended. As we saw, fruit and vegetables are composed of living matter that respires, and it can well be imagined that their quality is considerably impaired after being sealed in a coldstore for some time. A simple comparison would be the condition of humans after being placed in a closed room without adequate ventilation.

As we saw before, the presence and especially the quantity of gases like CO_2, O_2 and ethylene have an influence on the behaviour of products, and techniques were developed to control the presence and quantity of those gases; hence the term controlled atmosphere storage (CA). By putting the product in optimum conditions according to its type, variety, origin and climatological background, product life can be extended by 30% and losses from diseases are reduced by 30–80%.

Although CA storage can be used for many kinds of food products, such as fruit, vegetables and even meat, until recently in Europe it was mostly used for apples and pears, although tests are now being carried out to extend its use to meat products. High CO_2 rates, about 10%, prolong the storage life of meat. Too much CO_2, however, discolours the meat brown.

In Europe the most popular apple until the mid 1980s was the Golden Delicious; later, other varieties such as Jonah Gold, Elstar, Gloster, Cox and Boskoop, became more popular, together with pears such as Lombards Calville. Golden Delicious apples can be stored successfully in about 3% O_2 and 3–4% CO_2. The new varieties, however, demand a lower oxygen rate, down to between 1.8% and 2.2%. This is called Ultra Low Oxygen (ULO) Storage. A low oxygen percentage decreases respiration, so less thermal energy is produced and less nutritive points are modified. See figure 9.1 for illustration of this point. A certain amount of CO_2 in the atmosphere protects the fruit against disease; however, if the CO_2 content is too high, this introduces diseases. The recommended CO_2 percentage for new apple varieties is also very low, 2% and even lower. CA installations have had to be redesigned to accommodate these lower values for O_2 and CO_2, necessitating better gas-tight insulation techniques and computerized control systems.

Table 9.1 provides data on experimental CA conditions for a number of apple and pear varieties.

Various types of apparatus are used to control the atmosphere; so-called 'scrubbers', 'burners' and also, more recently, membrane gas separators and molecular sieve systems.

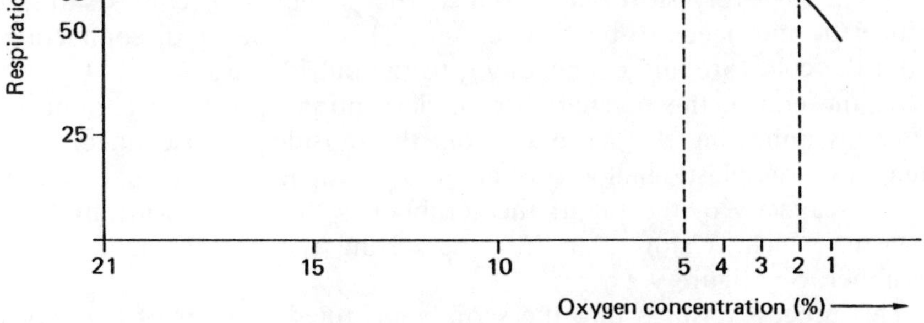

Figure 9.1 *Respiration rate of apples in controlled oxygen storage*

Table 9.1 Optimum storage conditions for apples and pears

Fruit variety	*C (%)*	*O_2 (%)*	*CO_2 (%)*
Cox's Orange	3–3.5	2–2.2	0.5–1.5
Boskoop	3–3.5	2–2.2	0.5–1.5
Golden Delicious	0.5–1	2–2.2	1.0–3.0
Jonah Gold	0.5–1	2–2.2	1.0–3.0
Conference	0.5	2–2.2	0.5–1.0

9.1.1 Scrubbers

A scrubber is a gas absorber. A quality scrubber must be able to:

(a) bring the concentration of CO_2 down to suitable values in a short time after loading the storage room or after opening the room for repairs
(b) maintain the value within tight limits during storage without affecting the O_2 concentration
(c) regulate the atmosphere during ripening periods
(d) absorb other volatile gases when storing vegetables or citrus fruit.

In addition it must be reliable, cheap, simple and have little need for maintenance.

The old system, utilizing 25 kg bags of high calcium lime, $Ca(OH_2)$, is still used sometimes. For each tonne of Cox's apples five bags are required, while three bags are needed for 10 tonne of Boskoop, Lombards, Winston, Melrose or Golden Delicious. A bag lasts between 6 and 10 weeks.

From all systems used or tried to date, the active carbon scrubber seems to meet most of these requirements. In this system a ventilator sucks the air out of the coldstore via a plastic tube system, and pushes it over an active carbon filter which absorbs the CO_2. After some time, which may be different for every store and product, the 'scrubber' becomes saturated with CO_2 and needs to be regenerated. This is done by disconnecting it from the coldstore and connecting it to the outside air.

At the end of the regeneration or desorption process, it is important after disconnecting the scrubber from the outside air to connect it to a 'lung' (a large plastic balloon) of about 5 m^3 capacity or more, depending on the capacity of the plant; the scrubber is then filled with air low in oxygen, which is stored in the lung during the normal action of the scrubber. See figure 9.2

This process ensures that the scrubber is filled with air of low oxygen content after the regeneration operation. In the old scrubber types, this system was not available and air with 21% O_2 was blown into the coldstore

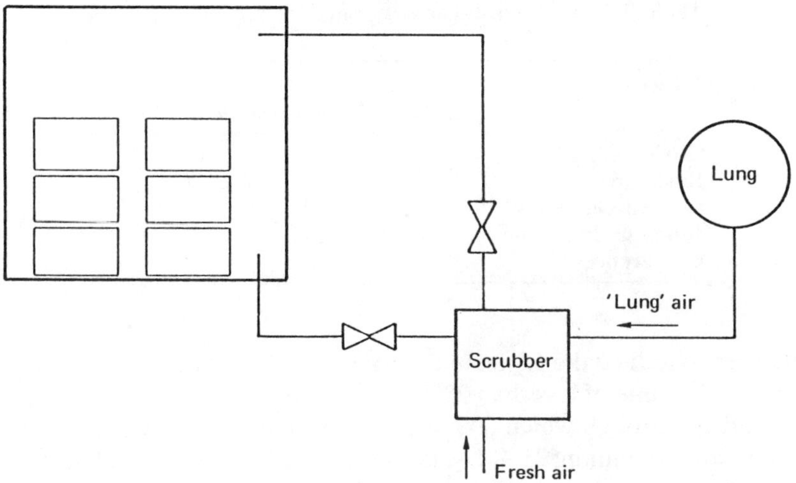

Figure 9.2 *Scrubber system with 'lung'*

just after the regeneration process; this of course had a deleterious effect on the composition of the atmosphere.

For the same reason, the equalizing valves of coldstores for CA storage are also equipped with a small lung, the volume of which is calculated in such a way that it can cope with the volume changes of the air in the coldstore. More details of this are given in chapter 10.

The capacity of the scrubber is calculated on the basis of values of CO_2 respiration for different fruits and vegetables. The capacity must be adequate to bring the room to the desired CO_2 percentage within 3 days. Table 9.2 gives recommended rates of CO_2 scrub capacity.

When the products are brought into the store, respiration values are 20% higher than the values given in table 9.2.

9.1.2 Burners

Two methods are used to reduce the O_2 percentage. The first is the so-called natural way, in which the fruit consumes O_2 as part of its natural respiration process. Using this method in a gas-tight room reduces the O_2 percentage to about 2% in three weeks. The second method uses forced burning of O_2 by means of special burners. There are three types of burners mainly in use today:

(1) the catalytic burner
(2) the NH_3 cracker
(3) the molecular sieve system.

Table 9.2 CO_2 scrub capacity rates for apples and pears

Variety	*Rate (kg CO_2/100 tonne day)*
Cox's Orange	35
Boskoop	29
Golden Delicious	10
Jonah Gold	14
Conference	10

In order to reduce the O_2 from the original 21% to the desired 3%, or to today's ULO value of less than 2%, the room is connected to the burner by plastic tubes, through which air, or rather nitrogen, is blown. The original coldstore air, containing 21% O_2, leaves the coldstore via an exhaust valve.

During the storage period when fruit uses oxygen for respiration, the O_2 percentage decreases and it is possible that it may fall below the concentration we would ideally maintain. Therefore it is necessary from time to time to inject fresh air containing 21% O_2 by means of a fan. This fan must be constructed so that no air can enter the room when it is not operating.

In modern practice, the burner first reduces the O_2 concentration down to 5% and then the fruit itself is allowed to bring the value down to 2% by natural respiration.

The operation of the catalytic burner system is shown in figure 9.3. Propane or methane, natural gas, is burned, using outside air, over a catalyst at high temperature without a flame. The air leaving the burner will consist mostly of nitrogen (86.5%), the remainder being O_2 (1.5%) and CO_2 (12%); this is removed by passing it over the scrubber before it enters the coldstore.

The fuel must contain 95% methane, with less than 5% propane, butane etc. There must be a maximum of between 1.5 and 2% propylene, butadiene etc. The sulphur content should not exceed a value of between 0.25 and 0.3 g/m^3.

An alternative fuel specification is 95% propane, with between 1.5 and 2% non-saturated hydrocarbons and 0.25 gr/m^3 sulphur, the remainder being made up of butane, n- and iso-butane.

When using propane, it is better to use a tank rather than bottles. In a bottle, the heaviest elements collect at the bottom and the lightest at the top. The resulting gas composition makes it difficult to regulate the burning process.

The operation of the more complicated ammonia cracker system burner is shown in figure 9.4. The ammonia is first cracked as follows:

$$2NH_3 \rightarrow N_2 + 3H_2.$$

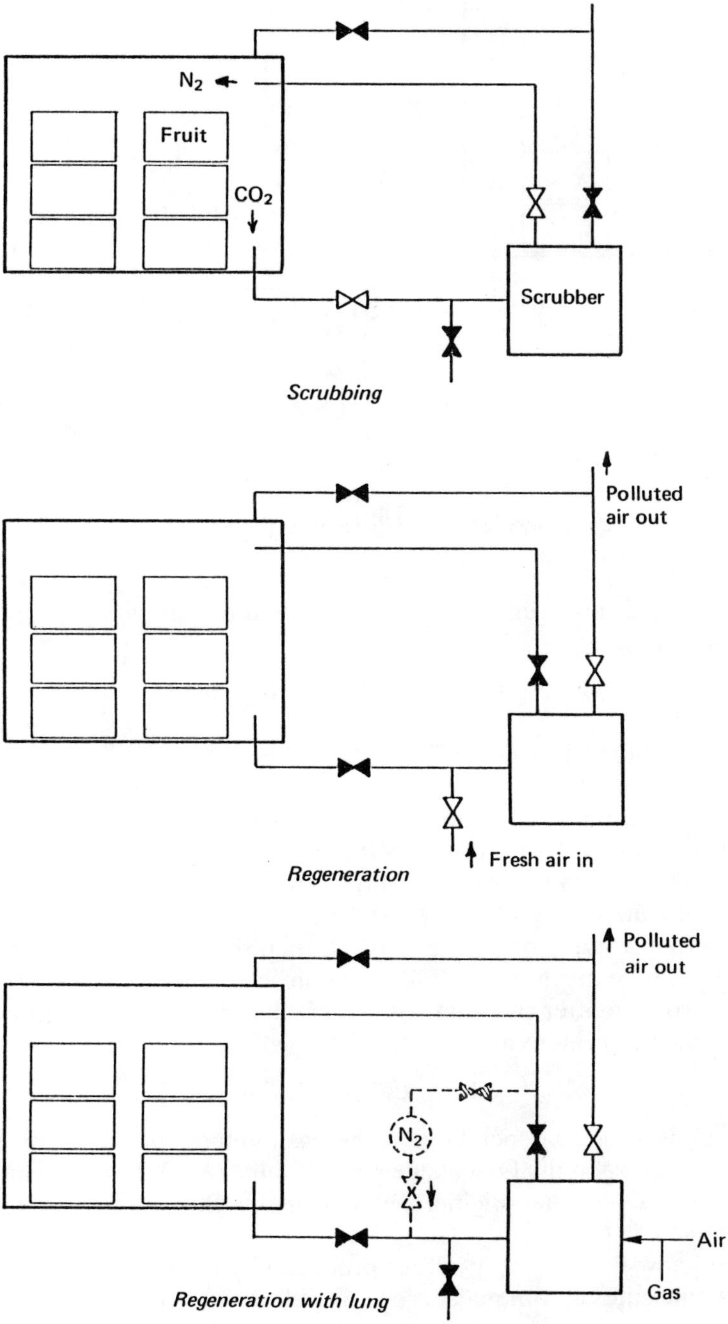

Figure 9.3 *Catalytic burner system*

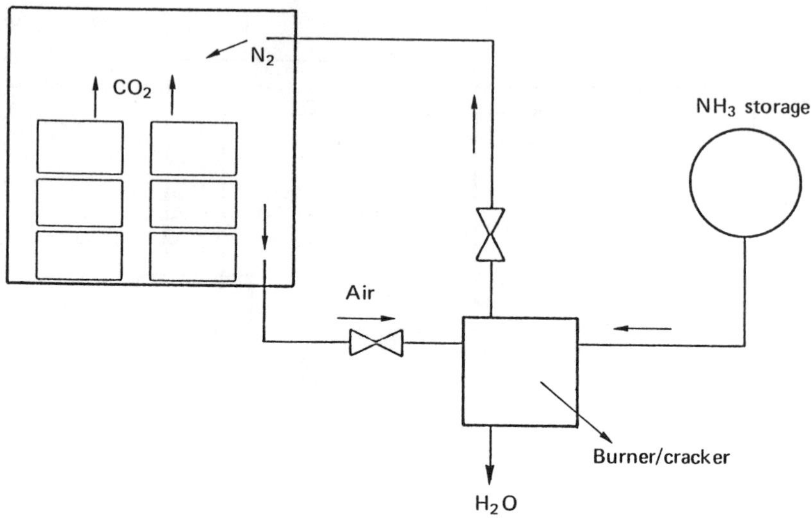

Figure 9.4 *Ammonia cracker burner system*

The mixture of nitrogen and hydrogen is then burned with the oxygen of the coldstore air:

$$2N_2 + 6H_2 + 3O_2 \rightarrow 2N_2 + 6H_2O$$

Calculation of the Pull-down Time

The pull-down time is the time necessary to bring a room with an oxygen concentration of 21% O_2 down to the desired percentage. Assuming the fruit density is 500 kg/m^3 and total quantity of fruit is T tonnes, then the volume of fruit V_2 is given by $T/500$ m^3.

If the total volume of the room is V_1, then the volume of the air in the room V_3 will be given by $V_3 = V_1 - V_2$ m^3.

If the volume after pull-down is V_4 m^3, this value can be calculated by the law of replacement as:

$$V_4 = C_0 V_3 \ln(C_1/C_2)$$

where C_0 is a factor depending on the gas-tightness of the room, the air circulation rate and the O_2 leakage rate. A value of 1.3 for C_0 is considered reasonable. C_1 is the original percentage of O_2 and C_2 the required percentage of O_2.

If we know the volume V_5 of air produced by the burner per unit time, we can calculate the time necessary for pull-down from V_4/V_5.

Note that a system has recently been introduced which uses hollow fibres to produce pure N_2 by 'filtering' the outside or coldstore air. The fibres are made from polydimethylpenylene-oxide (PPO). Air pressurized to 7 bar is fed into the fibre-ends. The oxygen easily passes through the fibre wall (while the nitrogen does not) and carries on to the end of the fibres, leaving as almost pure nitrogen. Depending on the operating pressure, the gas leaving the fibre-ends is 99.5% N_2. Another new system uses molecular sieves. The process is based on the separative properties of carbon molecular sieves for nitrogen (CMS-N_2). One member of a pair of absorbers is filled with CMS-N_2 while the other is fed with compressed air. While one absorber is producing nitrogen, the other is being regenerated by depressurization to ambient pressure. After approximately 60 seconds the pressure in the two absorbers is equalized, which means that the compressed air is used to maximum efficiency. It is then the turn of the second absorber to produce nitrogen, while the first is regenerated. This method is known as pressurizing absorption or PSA.

Because of its simple construction, this system will probably replace the burner systems in the near future. Its advantages are that it operates with the outside air, it does not create underpressure in the cold room, it brings in cooled air and it produces a very low O_2 percentage (although the percentage is only brought down to 3% by the system, and the fruit itself then reduces the percentage further, by respiration). During storage the O_2 percentage can be quickly lowered after thc cold room has been ventilated to remove excess ethylene.

9.1.3 Controls

There are a lot of values to measure, to control and to be recorded in a fruit coldstore.

(a) Temperature control:
- of the coldstore
- of the product
- of the outside air
- dT over the air cooler
- final temperature of the air cooler in order to interrupt the defrosting period when there is no longer need for defrosting.

(b) Humidity control.

(c) Composition of the coldstore atmosphere: oxygen, carbon dioxide and ethylene concentrations. There will also possibly be small concentrations of NH_3 or CFC which require measuring in order to detect leakage of the refrigerant.

(d) It is useful to know how much water evaporates from fruit. So the defrosting water leaving the air coolers is collected and the amount measured every 24 hours.

(e) Timing of the following functions must be controlled:
- period of rapid chilling
- commencement of rapid chilling
- commencement and period of defrosting
- period when free-defrosting with coldstore air is applied
- commencement and period of ventilation
- commencement and period of scrubbing
- commencement and period of regeneration
- commencement and period of burning oxygen
- commencement and period of lung filling
- time relays of compressors and oil differential pressurestats
- checking of the control equipment, testing the O_2, CO_2 and ethylene sensors.

(f) In addition, there are a lot of safety controls that must sound an alarm if anything goes wrong.

(g) Measuring and checking of pressure drops is carried out across filters to measure their states of pollution.

To control and register all these items is only possible with the help of a computer. Various other possibilities then arise: for instance, a program for energy saving can be utilized. This is done by a so-called 'may–must' system. The computer checks if the coldroom has reached a temperature at which cooling becomes necessary again; if so, it gives the necessary signals to the refrigeration equipment and at the same time checks if there is a compressor running on part-load.

The computer also checks if there is another coldroom close to the set-point which at that moment may already be cooled. If that is the case, cooling action in that particular coldroom is started too, and in other coldrooms as well until the compressor is working on full-load. Note that compressor efficiency is maximum at full-load.

In a similar way scrub-action, regeneration, defrost action and ventilation action can be programmed.

In the program a 'menu' for ventilation capacity of the air coolers can be presented. See table 9.3. Depending on whether the room is in a cool-down period or in a storage period, either the cooling capacity or the ventilation capacity can be adapted. Using air coolers with multiple fans and fans with two rotational speeds, many different air circulation rates can be realized. These are all energy-saving measures. Another possibility for energy saving is using a computer to monitor electric current consumption.

The maximum allowable current can be programmed for every day of the week, taking into account day and night tariffs or avoiding peak-load periods. By these means the capacity can be reduced during the day or at peak hours when energy is expensive, and full capacity can be restored at time when energy is cheaper, such as at weekends.

Table 9.3 Ventilation menu for two fans

	Fan 1	*Fan 2*
0	Off	Off
1	Low	Off
2	High	Off
3	Off	Low
4	Low	Low
5	High	Low
6	Off	High
7	Low	High
8	High	High

In the future there will be even more possibilities for using computers, such as:

- Calculating the superheat based on the measured suction temperature and the saturation temperature.
- Measuring the oil pressure and temperature and, with the help of this data, variations in viscosity of the oil can be compared with the desired and programmed values. At the same time, the CFC absorption in the oil must be measured and taken into account.
- Recording the results of acid tests in oils and checking them against the desired pH value.
- Logging the data for allowable internal leakage of screw compressors and checking values with the actual leakage as a measure for wear; in this way, preventive maintenance is possible.
- Measuring and registering 'hot-spots' inside the compressor.

9.1.4 Ethylene Scrubbing

At the time of writing, ethylene scrubbers are still in their development and research stages. Attempts have been made using calcium permanganate and platinum oxide, but reliable and economic ethylene scrubbers have not yet been produced, although the necessity of ethylene scrubbing has been known for many years. Even as early as the 1950s attempts were made to find a way to develop this type of scrubbing; wet-air coolers were still widely used and it appeared that even the water flow was able to absorb quite a lot of ethylene.

The present method of dealing with this problem is to ventilate the room from time to time when the ethylene concentration becomes too high. Naturally, after this the oxygen concentration needs to be restored.

9.1.5 Vacuum Cooling

We have seen how refrigeration stops the growth of micro-organisms and the enzymatic processes. Some products, especially vegetables, may spend a long time between harvest and arrival into storage. During harvest and transport to the coldstore there will inevitably be a large loss in quality. Fortunately this problem does not exist for other products; for example, meat coldstores are provided adjacent to the slaughterhouse, milk is refrigerated in milk tanks on the farm and fish is either stored in ice or frozen on board ship. For vegetable products, however, there is a special technique, called vacuum cooling, which can be applied in the field. By combining a decrease in temperature and pressure, a satisfactory result is obtained. In fact, because of the sudden change in osmotic pressure in the cells and the expansion of the capillaries, the outer surface of the product improves. The vacuum which is applied simultaneously with the cooling is not so deep as to harm the cell walls of the products, and only 5% of the moisture evaporates. A second advantage is that only part of the sensible heat needs removing when the product enters the coldstore. Using this vacuum cooling system, transport over short distances taking no more than 5 hours makes the use of refrigerated containers unnecessary.

The basic principle is the spontaneous evaporation of water when the pressure is decreased. By controlling the vacuum with the vacuum pump, the refrigerating capacity is controlled, and in 30 minutes the product can be cooled down to +1°C. See figure 9.5 for a comparison between normal and vacuum cooling processes.

The main components of the plant are:

(a) a vacuum space
(b) a vacuum pump installation
(c) a refrigerating installation.

After the product enters the vacuum cooler, the pump and the cooling starts until the desired pressure is obtained, between 2000 and 2666 Pa (1933 Pa corresponds to +17°C, 2640 Pa to +22°C). The moment cooling starts – the so-called 'flash-point' – an enormous quantity of water vapour is released, but the pump is unable to absorb this. At 2000 Pa, for instance, 57.5 m^3 of water vapour per kg of vegetables can be released and at 613 Pa about 200 m^3/kg; for this reason we need a refrigerating installation. At 653 Pa the pump stops. The water vapour condenses on the installed evaporator. When two vacuum cooling or other coldrooms are connected, the refrigerating equipment can be of the direct expansion kind; if not, a secondary medium must be used such as an ice tank.

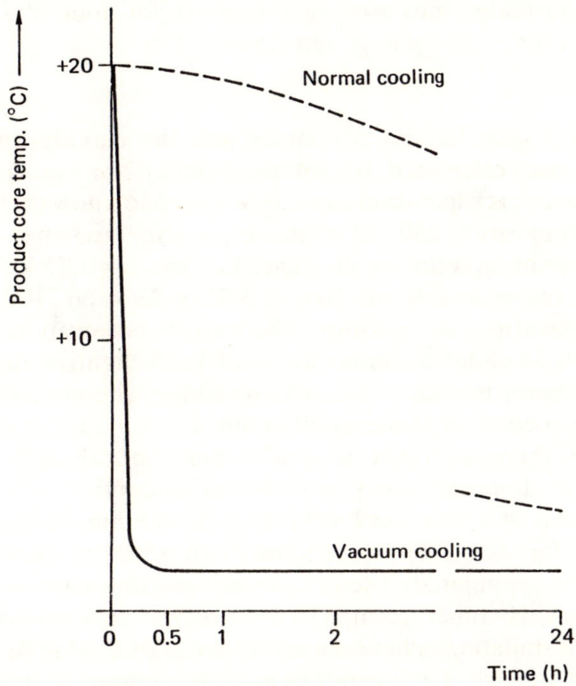

Figure 9.5 *Comparison of cooling curves for normal and vacuum cooling*

How do we select the correct size of the vacuum pump? The capacity of the pump is determined by:

(a) the dimensions of the vacuum space
(b) the time in which the pressure drop must be realized.

If S_{eff} = effective suction volume, without counting losses in tubes and valves, V = space volume, p_1/p_2 = pressure ratio, where $p_1 > p_2$ and t = time to achieve low pressure then

$$S_{eff} = [V/t]\ln(p_1/p_2)$$

The product will reach the saturated temperature at pressure p_2; this means that in order to get to +1°C, p_2 must be 653 Pa. The pumping time must be divided into two periods:

1. Pumping time t_1 from atmospheric pressure to saturation pressure at the initial temperature of the product. During this time there is no cooling.
2. Pumping time t_2 from flash-point to p_2. During this time, product moisture evaporates and the product is chilled. The total pumping time

is calculated taking into account two loads per hour, the loading and unloading time, and opening and closing of doors.

t_1 is about 5 minutes, t_2 25 minutes and the cooling time about 20 minutes. S_{eff}, once calculated, is then multipled by 2 or 3 as a practical rule.

To calculate the refrigeration capacity we need to know that for every kg of evaporated moisture 2500 kJ of thermal energy must be extracted. To cool down 1 tonne of lettuce with a specific heat $c = 3.77$ kJ/(kg K) from +25 to +2°C, one must subtract $1000 \times 3.77 \times 23 = 86\,710$ kJ of sensible heat. The quantity of moisture that must be evaporated is then $86\,710/2500 = 34.68$ kg or about 3.5% of the weight of the lettuce. In practice this figure is found to be 4% or 40 kg. The air coolers must be calculated for a cooling period of 20 minutes.

Remember then we must also add the other heat losses to the above-mentioned necessary negative thermal energy.

At a pressure of about 533 Pa the specific volume of water vapour is about 200 m^3 kg. For 1000 kg of lettuce, $40 \times 200 = 8000$ m^3 of water vapour must be eliminated. The realistic volume that must be taken out is between 8 and 10 times greater. The evaporating temperature of the refrigerating installation is between −6°C and −10°C. As the refrigerating installation is started at the same time as the pump, it runs during the period Z_1; the real cooling time though is somewhat longer than 20 minutes, with the maximum about 30 minutes.

To prevent the produce from freezing it is better to work with a somewhat higher evaporating temperature of between +0.5 and −1°C, which will obviously limit the weight loss. For that reason, vegetables are sometimes sprayed with water before being stored in the coldroom.

9.1.6 Wet Cooling System and Flow-through Cooling

Another method has been introduced recently, the so-called wet cooling system. This method removes the field heat very quickly from horticultural produce and at the same time prevents a reduction of weight and quality loss by moisture loss. This new system is used more and more in the flower and vegetable trade. In chapter 6 we have already dealt with the wet air coolers of this system. The system gives excellent results when flowers or vegetables have to be transported over long distances and it increases the shelf life of the produce. Good results are obtained, however, only when using this system in combination with a new method of air circulation called 'flow-through cooling'. Cold air is made to flow through all the packaged produce by means of pressure difference.

There are two systems:

(a) Pressure cooling, in which the air is pushed through the spaces between the product. Pressure cooling was developed at the University of Davis in California.
(b) Suction cooling, where the air is sucked through the space between the products.

The first system can only be used for products stored in more or less closed cartons, the second for products stored in more open cartons or cases, as used for tomatoes and mushrooms. Pressure cooling results in a more uniform produce temperature, and because of the investment costs, it is mainly used for large plant and for produce to be sold at auctions for instance.

In long rows, 12, 13 or 14 pallets may be put behind one another, but when using carton boxes, only 1 pallet high is possible. The height of a pallet is between 1.2 and 1.65 m. When wooden or plastic boxes are used, pallets can be arranged two high.

As figure 9.6 shows, the rows of pallets are placed on both sides of a suction channel and closed at the top by a roll-type curtain at one end; at the other end, the suction channel is connected to the space behind a separation wall. The air leaving the air coolers is then forced through the pallets, and between all the packages, at the same time and temperature and with the same velocity of 2 m/s; the air then returns to the coolers via the suction channel and the space behind the separation wall. Two rows of pallets form one section. Between every section there should be a corridor of 60 cm.

To achieve an adequate air velocity between the packages an airflow of at least 25 000 m^3/h per section is needed, and can even reach 50 000 m^3/h. The air passage between the boxes on the pallets is equal to about 5% of the surface. The suction opening in the separation wall is about 60 cm wide, but the height depends on the height of the pallets. The pressure drop over the suction opening is about 100 Pa.

For a refrigeration capacity up to 75 kW per section, cross-stream air coolers are used; above this capacity, counter-flow coolers are used; and above 95 kW per section, these coolers are fed with chilled water from a separate water chiller. The unit may be an ice-bank machine. An ice bank can be built up during the hours when the refrigeration capacity demand is low and the price of electricity is also low.

For very small capcities of less than 50 kW per section, dry air coolers with a large cooling surface, small dT, special suction pressure regulators and pump-circulation can be used.

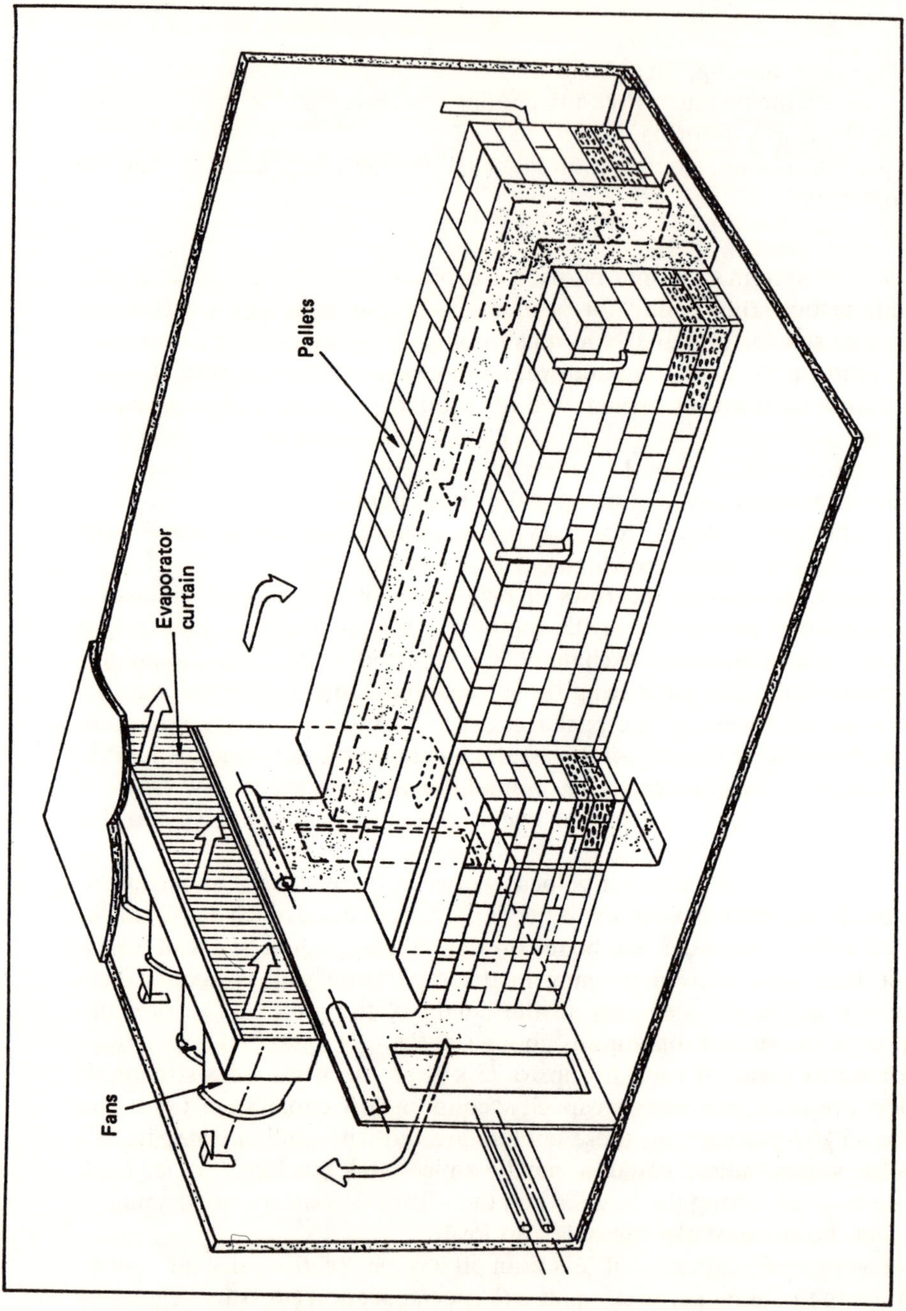

Figure 9.6 *Pressurized flow-through cooling system*

Some examples of cooling times are: tomatoes 1 hour, cucumbers 5 hours, peppers 2 hours. These products are cooled down to +14°C and even as low as +8°C.

The following products are cooled down to +4°C and sometimes down to +0.5°C: lettuce 12 hours, spinach 6 hours, cauliflower 8 hours, and Brussels sprouts 8 hours. Strawberries and mushrooms can be cooled in 2 hours.

9.1.7 Hypobaric Cooling

This kind of cooling is not now much in use. The principle is based on prolonging the storage time by lowering not only the temperature, but also the pressure in the room. The room must of course be reasonably gas-tight and also of sufficient mechanical strength to withstand the high pressure difference between the inside and outside air. The room is connected to a vacuum pump and a vacuum of between 1333 and 13 333 Pa is maintained; in addition a normal refrigerating plant is installed.

By means of an automatic device some purified fresh air is introduced from time to time. Before introduction this air is humidified by evaporating water under low pressure without adding much thermal energy.

The quantity of water vapour added is controlled with a dew point controller, which measures the relative humidity of the air taken out by the vacuum pump. The relative humidity is controlled between 80 and 100% to an accuracy of 3%.

Water for the humidification process is obtained by condensing the water vapour of the air handled by the vacuum pump. This water is filtered before re-use.

The advantages of this system are as follows:

1. In the room there is little air, so little oxygen. From the biological point of view this is equivalent to the controlled atmosphere storage method; this means less respiration in the case of fruit or vegetables, plants and flowers, and less bacterial growth.
2. As there is little air in the room, fresh air inflow is limited also resulting in reduced demand for refrigeration capacity.
3. CO_2, ethylene and other gases are continuously eliminated.
4. The air is replaced continuously without much energy consumption.
5. Water vapour is removed and then returned to the room.

Typical results are as follows:

Fresh meat can be kept for 45 days instead of for between 12 and 14.
Lettuce for 42 to 49 days instead of 14.
Flowers for 60 days.

Tropical and other fruits can be transported over longer times and distances, and expensive air transport is unnecessary.

9.2 Banana Storage

A special technique is used for the storage and ripening of bananas since they are harvested when green and have to be stored and transported that way. Like apples and pears, this fruit continues to live after harvesting, absorbing oxygen and producing carbon dioxide, ethylene, other volatile gases, water vapour and thermal energy, with a maximum production at the climacterium. At +12°C one tonne of bananas absorbs 2.4 m^3 of oxygen and produces 0.33 m^3 of carbon dioxide every 24 hours.

The chemical composition of the bananas depends on their origin and the time of harvest. Jamaican bananas harvested between the 86th and 90th day of growth contain 78.8% water and 21.2% dry matter, of which 16.7% is starch and 0.3% sugar.

The banana contains two pigments: chlorophyll that gives the green and carotene that gives the yellow colour. During the ripening process, most of the chlorophyll is destroyed by an enzyme action and the bananas become even more yellow. Ethylene both starts the ripening and continues it.

Temperatures

It is not possible to give general conditions for storing and ripening, as they depend largely on the origin and the time of harvesting; there are however some general rules.

For instance, temperatures below 11.5–12°C create damage by blocking the ripening process, while temperatures above +23°–24°C cause loss of aroma.

Transport of the green fruit is done in the pre-climacteric period, about 12 days after harvesting. The ship's hold, or any other means of transport, is pre-cooled to between +6°C and +8°C and the bananas are chilled down to +12°C in 36–72 h.

The ship's hold receives an amount of fresh air twice in 24 hours in order to eliminate CO_2 and C_2H_4. A circulation rate of between 60 and 80 air changes per hour is maintained, which is decreased to 50% of these values later on during the voyage.

Air leaves the air coolers at a temperature of 0.1°C. The relative humidity is kept at 85%. The respiration is about 1.4 kW/tonne at +12°C. During ripening, which takes 4–8 days, the temperature is kept at +18°C.

Ripening is started artificially by injecting ethylene into the room. The first day, the refrigerating equipment is not used, but heating is applied, either using hot water batteries at +70–75°C or electric heaters, to provide a warming-up rate of approximately 1.5K/h.

The second day the refrigeration process is started in order to eliminate the respiration heat. The evaporation temperature is about +4°C with a dT about 14K. When glycol is used as a secondary refrigerant, its temperature will be between 0 and +3°C. The normal cooling-down rate is 1K/h, although some processes specify a rate that does not exceed 0.5K/h.

This cooling-down rate is maintained only for a few hours, and then only when necessary; nevertheless the air coolers must be designed to accommodate it.

Tables 9.4 and 9.5 contain information about ideal time/temperature schemes for bananas on the stem and in boxes.

Humidity

During storage the relative humidity must be 85%, but during ripening humidifiers are used to create relative humidities of 95–100%.

Table 9.4 Time/temperature schedule for bananas on the stem

	Temperature of air in ripening room (°C) when the number of days before removal for selling is as stated								
Ripening schedule (days)	*1*	*2*	*3*	*4*	*5*	*6*	*7*	*8*	*9*
4	17.8	17.8	16.7	15.6					
5	16.7	16.7	16.7	16.7	15.6				
6	16.7	16.7	15.6	15.6	15.6	14.4			
7	15.6	15.6	15.6	15.6	15.6	14.4	14.4		
8	14.4	14.4	14.4	14.4	14.4	14.4	14.4	14.4	
9	14.4	14.4	14.4	14.4	14.4	14.4	14.4	14.4	14.4

Table 9.5 Time/temperature schedule for bananas in boxes

	Temperature of air in ripening room (°C) when the number of days before removal for selling is as stated						
Ripening schedule (days)	*1*	*2*	*3*	*4*	*5*	*6*	*7*
4	17.8	17.8	16.7	13.3			
5	17.8	16.7	16.7	15.6	18.9		
6	17.8	15.6	14.4	13.3	13.3	13.3	
7	16.7	14.4	14.4	14.4	14.4	13.9	13.3

Respiration

It is difficult to give exact data about the production of thermal energy, as again this depends on the origin, quality and time of harvest. Here is some average data:

Green	+15°C	58–133 W/tonne
bananas	+20°C	88–154 W/tonne
Ripe	+15°C	88–167 W/tonne
bananas	+20°C	96–240 W/tonne

1 tonne = 1000 kg

The climacteric point lies at about +22°C, where respiration is double that during the pre-climacteric period.

For calculation purposes, figure of about 125 W/tonne can be assumed for a ripening room. Specific heat is about 3.5 kJ/(kg K).

Ethylene Injection

During the first 24 hours ethylene must be injected into the room; afterwards the fruit itself develops enough ethylene to maintain the ripening process. It has been proved that extra ethylene injection does not accelerate the ripening. For the first 24 hours a C_2H_4 concentration of 1% must be maintained, and as this gas is explosive even at small concentrations, a mixture of 5.5% C_2H_4 with N_2 is injected at a ratio of 18:20 litres per m^3 volume. A row of gas bottles with a tube system and solenoid valves is then installed for automatic regular injection.

(e) Fresh Air Intake

Carbon dioxide and other gases inhibit the ripening process, therefore a complete change of air is necessary for a 20-minute period in every 24 hours; this maintains the proper O_2 concentration simultaneously. Automatic systems are used with a timer, a fan and an air grill. The air change rate inside the room is usually greater than 150 per h, but during storage only 40 per h.

To ensure that all the air passes over the bananas, even in a half-filled coldstore, a movable ceiling is used that can be lowered down to the highest stacked box of bananas.

In order to gain space, sometimes vertical ripening rooms are built. Their advantage is a more equal spread of temperature over all the boxes, as the air stream passes only through one pallet. The air stream circulates from one wall to the other, and the air cooler is mounted against the ceiling. This system also results in a lower fan energy consumption because the air stream follows a shorter route.

(f) Temperature Control

As it is very important to control the ripening process, heating and refrigerating equipment must be controlled by electronic thermostats to an accuracy of 1K. The position of the temperature sensor between the products is of great importance and the optimum position can only be found by a skilled specialist, usually by a trial-and-error process. The sensor can be placed into any box of bananas.

Nowadays temperature control is achieved by microprocessors. The air-coolers are equipped with a valve which regulates the evaporation temperature.

(g) Storage Methods

The bananas are seldom stored while still on the stem; they are mostly packed in boxes, the size of which depends on their origin.

Some examples are: 45 × 34 × 25 cm containing 17.5 kg and 53 × 40 × 23 cm containing 20.4 kg, with 8 boxes stacked on a pallet. Note that storing conditions during ripening for bananas on the stem and in boxes are different, as detailed in tables 9.3 and 9.4. A coldstore of 500 m^3 contains 100 tonnes of bananas and needs a refrigeration capacity of about 75 kW.

Boxes are placed 5–8 layers high; they are stacked that way so that air can pass easily inbetween and in all directions. Usually pallets are stacked 3 across and 6 behind each other. Pallets are not stacked one upon another in traditional horizontal rooms, whereas in vertical rooms 3 or more pallets are placed on top of each other, but then racks are used.

9.3 Citrus Fruit

Lemons are stored at a high temperature, namely +14.5°C, and at between 86% and 88% RH. The air circulation rate is about 12 times the room volume per hour. At +15°C the respiration rate is about 30 W/tonne of fruit. They also produce CO_2 at a rate of about 24 l/tonne day. The concentration of CO_2 in the coldstore must be kept under 0.1%. Fresh air should be introduced into the store at a rate of 5.8 m^3/h tonne of fruit.

Lemons are packed in crates measuring 85 × 35 × 35 cm, each containing 35 kg.

Oranges are stored at between +1°C and +4°C, and 85% and 90% RH, depending on their origin.

The air circulation rate is 230 m^3/h tonne during the cooling-down period, and 93 m^3/h tonne during the storage period. Fresh air intake is between 0.77 and 1.0 m^3/h tonne. The concentration of CO_2 should be about 0.2%.

Respiration at +4°C is 18 W/tonne and CO_2 production rate 24 l/tonne day.

Grapefruit is stored at between +10°C and +15°C, and 80% and 90% RH, also depending on their origin. The rate of fresh air intake is about 4 m^3/tonne h and the air circulation rate in the room is 25 per hour. The fruit has a respiration rate of 30 W/tonne at +15°C.

9.4 Bakery Products

With bakery goods, a special factor is the limited storage time because the product quickly becomes stale. This process starts as soon as the product leaves the oven. It is better to cool down bakery products before freezing them, as we will see in the following explanation. The process of becoming stale starts to be an important factor at +60°C, but is most significant between +30°C and −7°C. This temperature range must be passed as quickly as possible; it is not good practice to stop the freezing process at −7°C and continue the freezing process in the holding rooms.

When using air temperatures higher than −30°C, the crust will separate itself from bread; fruitcakes also need very low air temperatures: −35°C or lower. Small cakes can be frozen at about −25°C air temperature.

The maximum freezing time for bread is about 4 hours; for fruitcakes and ricecakes 3 hours, and for rolls and small cakes 1 hour. These values are for unpacked products. To freeze packed products, freezing times are longer and the air temperature in the freezing tunnels must be −40°C or lower. In general the air velocity must be high, between 2 and 3 m/s, but not as high as for other products.

Storing is best done without air circulation or with a very low air velocity and at a temperature between −15 and −18°C; when defrosting is required it must be done in an oven at +250°C.

When freezing bread, which contains 35% water, from +30 to −20°C, 176 kJ/kg must be subtracted.

The above values refer to products which are already baked; but often the product is frozen unbaked, that is, as dough. Refrigeration can be used to slow down the fermentation process; for this reason, special air conditioners are available.

The fermentation period normally precedes the baking process by at least 2 hours. By prolonging this process artificially, by between 10 and 12 hours using refrigeration techniques, the delay can be timed for overnight operation, so that the baker need not work a nightshift, the baking process then begins the following morning. Fermenting times can sometimes be extended by up to 18–40 hours; relative humidity must be held between 75 and 88%, with a dT of between 8 and 10°C.

The evaporators of the air-conditioning system are usually of the natural convection type. If forced convection is used, the RH should be kept between 85 and 90%, with a dT of 4 to 5°C. In each case, ventilation should be kept to a minimum.

It is necessary to fill the room completely and to have a good overall air circulation and temperature spread. The plant should consist of several small rooms or a big room divided into several sections.

Some extra considerations of the heat load are:

1. Cooling down the dough from +25 to +10°C. Specific heat is 2.72 kJ/(kg K), and the cooling time is 4 hours.
2. Fermentation heat 0.84 kJ/kg, to be eliminated in about 10 hours.

9.5 Eggs

A thousand eggs of average size represent about 50 kg of product weight. The size of a box needed for 1000 eggs is about 1.5 × 0.2 × 0.25 m. The recommended storage temperature for eggs is +0.5°C at 80–85% RH.

Egg-yolk is usually frozen in tunnels or rooms between −25 and −30°C and stored between −18 and −20°C; the yolk is packed before freezing in metal cans (diameter 0.26, height 0.22 m), each of which contains about 10 kg of yolk. This gives a storage capacity of 1000 kg of yolk per m^3. Twenty eggs provide about 1 kg of yolk. The freezing time is about 10–12 hours in air of −30°C at 3 m/s velocity. The defrosting is done in rooms at +10°C.

9.6 Ice Cream

Ice cream is one of the oldest known foodstuffs, so it is interesting to take a look at its history. In ancient times Arabian kings consumed a kind of water and fruit ice. The Arabian word for drink is 'sjerbat', from which comes the word 'sorbet', a kind of fruit ice much appreciated in our day too.

In AD 55 the emperor Nero served his honoured guests sorbet made from rosewater, honey, fruit, raisins and snow brought down from the mountains by slaves.

Marco Polo reported that the Chinese told him they had made ice cream from milk, water and other food products long before the Egyptians built their pyramids. In 1292 he was able to bring back from China the first recipes for ice cream making. The Chinese put the ice-cream ingredients into pots over which they poured a solution of saltpetre in order to freeze them. The Sicilian de Catane used this technique in 1530 and even improved it.

The notorious Catherine de Medici brought an Italian maker of ice cream with her to the French court. This man knew how to prepare ice cream with the flavours of lemon, orange and various berries. In Europe during this time, milk or cream was not used in the preparation of ice cream, just water and fruit.

The first person to prepare real 'cream' was the French cook Tirsain who served at the English court of king Charles I. He added fresh cream and milk to his recipe. When Charles I was beheaded in 1649, Tirsain fled back to France where he opened, shortly after his return, the first ice-cream parlour in Paris. He called the ice cream 'Neapolitan' ice. This in a way marked the start of ice-cream production all over the world.

Deliciously flavoured ice cream has become a very popular item on the family shopping list throughout the world. In the USA consumption has reached about 22.5 l/year per inhabitant; the figure for Sweden is 12.4 and for Belgium 5.4; Portugal still has the lowest ice-cream consumption of Europe, but this is increasing rapidly; consumption in the USSR is very high.

There are three main types of ice cream, depending on the ingredients: ice cream, milk ice and water ice, the last usually made with fruit-flavouring.

The components are fat, sugar, egg-yolk, water and a stabilizer; these constitute the 'mix'. The mix is pasteurized, homogenized and then cooled down to +3°C; this is done in a plate cooler and a period of ripening follows. Then, during the following freezing process, air is injected which is called the 'over-run'. At the finish, air is added to the mix to give it a better structure, flavour, quality and more volume.

The percentage of over-run on the original volume of the mix is either

$$\left(\frac{(\text{volume of ice cream} - \text{volume of mix})}{\text{volume of ice cream}}\right) 100\%$$

based on 1 kilogram, or

$$\frac{(\text{weight of mix} - \text{weight of ice cream})}{\text{weight of ice cream}}$$

based on 1 litre. Normally the over-run varies from 60% to 100%.

The freezing is done in a drum with double walls, between which the refrigerant evaporates. Inside the drum a scraper rotates, serving as a mixing-screw and beating the air through the mix. The evaporation temperature is between −28 and −30°C. Figure 9.7 shows a schematic arrangement of a typical ice-cream freezer.

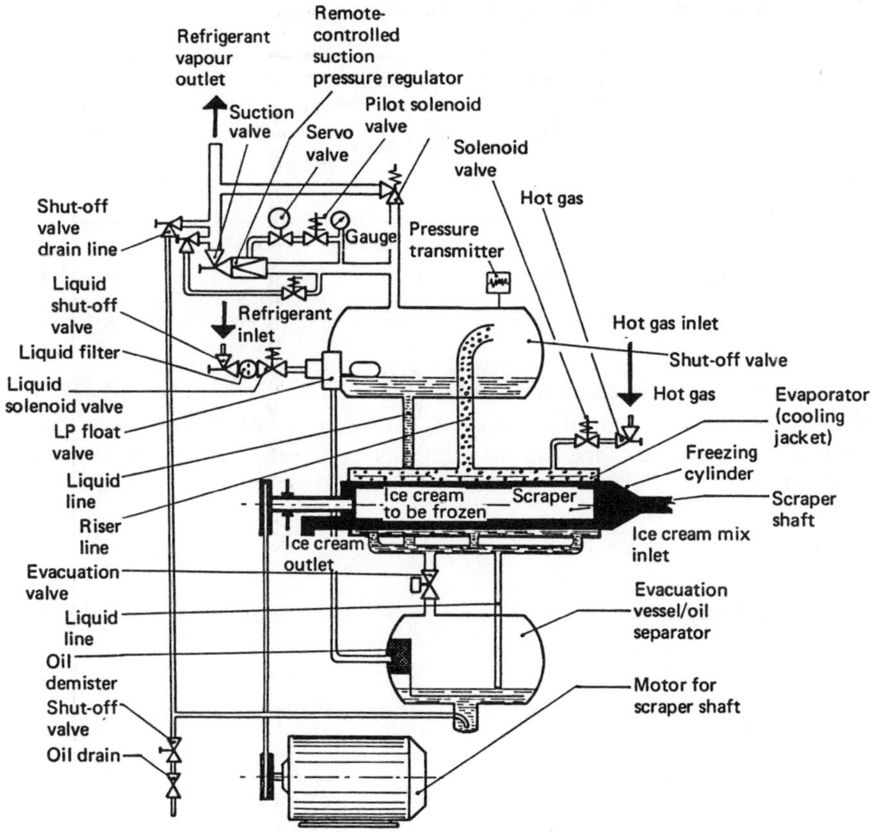

Figure 9.7 *Schematic diagram of ice cream freezer (Courtesy of Gram, Denmark)*

The ice cream is now ready for consumption and in this particular state and quality is called soft ice. It has a temperature of between −5 and −10°C but unfortunately it cannot be stored. Freezing can continue without adding more air. The ice cream is 'hardened' till it reaches a temperature of −28°C and is stored then at this temperature. After distribution it is consumed at −13°C.

Hardening is done in tunnels or rooms, but in many cases automated tunnels are used. As for all freezing processes, hardening must be done at a high speed in order to keep the ice crystals small. The hardening room temperatures are kept between −30 and −35°C and tunnel temperatures between −40 and −45°C; the evaporation temperatures are then between −40 and −45°C for rooms and about −50°C for tunnels. Hardening times vary from between 6 and 12 hours. Because ice cream is a mixture of liquid and solids, the freezing point will be lower as the percentage of solids is increased. The composition of the mix varies for different varieties of ice cream, as table 9.6 shows.

Table 9.6 Constituents of different ice creams

Quality	*% mass of ice-cream constituent*				*Freezing point (°C)*
	Fat (butter, cream, oil, etc.)	*Sugar*	*Stabilizer (gelatine)*	*Water*	
1	8.5–16	15	0.28–0.4	60–85	−2.5
2	1.2	20	0.5	67.3	−3.4
3	0	32	0.5	67.5	−3.5

Ice cream with more than 12% milk fat is 'real' ice cream; the others are milk ice, or water ice which contains no fat at all. Besides the above-mentioned ingredients, ice cream contains egg-yolk, flavour and colour additives.

Example

Ice cream quality 1: 61.7% water and 38.3% solid matter of which 12.5% is fat.

Frozen water (%)	*Mix temperature (°C)*
0	−2.47
5	−2.59
10	−2.75
15	−2.9
20	−3.11
25	−3.31
30	−3.50
35	−3.87
40	−4.22
45	−4.65
50	−5.21
55	−5.87
60	−6.79
70	−9.79
80	−14.94
90	−30.20

Refrigerating capacity: It is complicated to determine the necessary refrigerating capacity for this product, as we are dealing here with a mixture of constituents. We therefore have to know the specific heat, specific mass and solidification energy of more than one constituent. Also, fat behaves differently during a solidification process from items such as water and so on.

First the average values of these characteristics have to be calculated:

$$\text{average specific weight} = \text{specific weight of the mix}$$

$$= \frac{100}{\left(\dfrac{\%\text{ butter oil}}{\text{specific weight [0.93]}}\right) + \left(\dfrac{\%\text{ other solid matter}}{\text{specific weight [1.58]}}\right) + \left(\dfrac{\%\text{ water}}{\text{specific weight [1]}}\right)}$$

Depending on the chosen over-run and the desired end-volume V of the ice cream, the necessary quantity of mix (the weight M_a) can now be calculated:

$$M_a = \frac{100}{100 + \%\text{ over-run}} \times \text{average specific weight} \times V$$

The average specific heat is about 3.35 kJ/(kg K).

The heat loads for the freezer can be summarized as follows:

$M_a dT_{mix}$ 0.95 kJ.
Max % water in mix × % of water to be frozen × 335 kJ
$M_a(T_{\text{freezing point}} - T_{\text{end1}})$ 1.8J where T_{end1} refers to the value after the freezing process
Motor power of the scraper
10% general heat-losses

The heat load for the hardening tunnel:

Max % water in mix × % water to be frozen × 335 kJ
$M_a(T_{\text{freezing point}} - T_{\text{end2}})$2 kJ where T_{end2} refers to the value after the hardening tunnel
10% general losses

Generally 100% over-run is used and an average value for the heat load for hardening is 120 kJ/kg.

9.7 Chocolate

Chocolate is put into steel moulds in a liquid and warm state, and must be cooled down to become solid and hard. The starting temperature is +32°C and the specific heat 1.25 kJ/(kg K). Solidifying takes place at +25°C and the solidification heat is 92 kJ/kg. The chocolate is then cooled down to +5°C in a cooling tunnel. Normally the weight of the moulds is about 50%

of the weight of the product. The specific heat of the mould is 0.47 kJ/(kg K). General losses have to be added, resulting in an average heat load of about 210 kJ/kg chocolate.

9.8 Cheese

Part of the cheese-making process is the so-called ripening period. As far as solid cheese varieties such as the Dutch, Flemish or Swiss are concerned, ripening is carried out in rooms at a temperature of +13°C. Since this temperature must be maintained throughout the year, both refrigerating and heating equipment has to be installed.

In addition to temperature, humidity must also be controlled accurately at a value in the range 85–90%. Air movement must be limited; therefore air ducts equipped with adjustable grilles are most frequently used.

Air velocity is limited to 0.1 m/s, or the air change rate to 15 per h. The discharge duct is placed in the middle of the ceiling; suction ducts are provided on the ground and on the ceiling and against the side walls.

The duct air velocity is about 6 m/s and the grille air velocity 2.5 m/s. The ripening energy for solid cheeses such as Gouda, Gruyere and Edam is given as 5225 kJ/tonne day.

Normal calculations produce heat load values of 23 W/m^3 for refrigeration and 35 W/m^3 for heating. Fresh air intake must be about 5 times the room volume per 24 h. In a room of 200 m^2 about 90 tonnes of Gouda cheese can be placed on racks.

In the storage room the temperature must be kept at +6°C; at the same time a low air velocity and a well controlled humidity are advisable. Instead of ducts, false ceilings can be fitted or air coolers using small air volumes. Temperatures and other conditions are however different for other varieties of cheese, such as soft French cheeses, for example.

9.9 Beer

Thanks to refrigeration techniques, modern breweries are able to produce good beers. Throughout the whole process of beer-making, cooling is necessary. In the past, up to the previous century, natural ice was collected during winter and stored in ice cellars. When low fermentation processes were introduced, the need for cooling became especially necessary, so the breweries decided to use artificially made ice blocks. In 1876 Professor Linde developed the first ice-block-making machine. The first ice-block-plant was located in a brewery. This marked the end of the period during which natural ice blocks were used worldwide and the beginning of the development of modern refrigerating techniques. The application of such techniques to the brewing process starts with the conservation of the prime ingredients like hops and barley, and continues later with conditioning of the malt.

As we all know, beer has been around for a very long time. Babylonians, Assyrians and the ancient Egyptians all brewed a drink very similar to beer. So did the ancient Germans and Celts – although without proper cooling methods, the quality must have been poor. The ingredients of ancient beer were barley and water, just as they are today. The process of changing barley into malt and the fermentation process would have taken place naturally, without any human control.

Nowadays, the temperature is kept between 0 and −2°C, and the RH at 75%, in the hop storage rooms. The average heat load is 58 W/m^2. During the process it is important to eliminate the germination energy from the malt house. The barley, which is simply dumped on the floor, has refrigerated air at +10°C blown through it and is then kept at +12°C.

The germination process is the same as the respiration process of fruit and vegetables and, similarly, thermal energy is also produced during this process. As there is not enough oxygen for respiration inside the bulk, alcohol is formed and the germ is killed. The amount of thermal energy developed is about 1086 kJ/kg but this varies during the process; if it takes 6 days, thermal energy production is at the maximum during the third and fourth day, and the calculation is based on a heat load of 2 W/kg including general losses. About 10–15% of fresh air is included in the calculation of these figures, as well as the cooling-down of the barley in 5 days over 4K with a specific heat of 1.7 kJ/(kg K) and 3% moisture extraction. The circulating air volume is 600 to 800 m^3/h tonne.

Using direct expansion air coolers with a close fin spacing of about 3 mm, the air-inlet conditions to the air coolers are +18 to +23°C and 75% RH, the air-outlet conditions +11 to +15°C at 90–95% RH. Evaporating temperatures between +1 and +5°C are used. When designing the fans, a pressure drop over the barley must also be taken into account. A 75 cm layer of barley with a density of 1.357 m^3/tonne placed on a perforated metal floor results in a pressure drop of between 0.4 and 0.6 mbar.

It is important to note that air coolers must be corrosion-resistant since the atmosphere in the malt house is very aggressive.

Wet-air coolers are often used in place of dry coolers, in order to achieve a high relative humidity. The water chiller feeding these wet-air coolers operates at an evaporating temperature of +5°C, a water outlet temperature of +8°C and an inlet temperature of +11°C. Now a brief summary of the remainder of the brewing process will be given.

The next step in the refrigerating process of beer takes place inside the brewery. First of all the wort, which has a temperature of +50°C or more after brewing, is cooled down to +20°C. During the cooling process, water at about +20°C is used. The water volume flowrate is 1.5–2 times that of the beer volume flowrate. To cool it down to +15°C and lower, water from a water-chilling unit is need. This can be ice-water or glycol/water at about −5°C.

The specific heat of the beer wort is 3.8 kJ/(kg K) at 12° Balling or 4 kJ/litre, and the specific weight is 1.05 at 14° Balling.

The next step is the elimination of the thermal energy produced during the fermentation process. In the so-called 'low yeasting' process, the temperature in the room is +5°C. Some breweries work at a temperature between +15 and +17°C; the process is then called 'high yeasting'. Copper or stainless steel coils are placed in the yeasting tanks to achieve the cooling, or the tanks are provided with double walls, in which ice-water or glycol/water solution is passed as secondary refrigerant; this is termed jacket cooling. The thermal energy developed during the fermentation process is 2100–2600 kJ/Hl day. The water volume flowrate is 2.8 l/h Hl with a d*T* of 2K.

The heat load to cool the room alone is between 48 and 60 W/m^3.

In beer cellars, the beer is kept at an ambient temperature of between 0 and −2°C, although some beers are stored between +1 and +3°C. The heat load is 85 kJ/m^3 24 h.

Today, most tanks used are cooled individually by means of ice-water inside the double wall jacket. Only the front end of the tanks are placed in a corridor between two rows of tanks that must be cooled.

Other figures for the heat load are:

Storage cellar: 30–35 W/m^2
Fermentation rooms: 50–60 W/m^2

When calculating the heat load for both kind of rooms it should not be forgotten that thermal energy is introduced by the clean water used for rinsing the tanks. At each cleaning operation an amount of water equal to one-quarter of the tank volume is used.

In order to chill the beer, some refrigerating capacity must be added. In fermentation rooms the radiation of the tanks must be added too, which is roughly 70 W per tank of 30 Hl.

Using open tanks, the amount of fresh air intake has to be considered too. For each 100 Hl about 25 m^3/h fresh air enters the room during an 8 hour period each day. Nowadays, however, fermenting is mostly done in closed tanks in order to recycle the CO_2 that develops during the fermentation process.

As the last step in the refrigerating process of beer manufacture there are coldstores for yeast storage at +3°C and 85% RH, and the beer chiller in which the beer is chilled from +5 to 0°C by means of ice-water or glycol brine. In the large breweries, which operate continuously day and night, ice-water combined with ice storage is replaced by glycol/brine systems.

These systems operate at lower temperatures, thereby reducing the necessary heat exchange surfaces.

One of the ingredients of beer is hops. The heat load of the storage room for hops is 4000 kJ/m^2 24 h.

9.10 Deboned Meat and Prepared Dishes (TV Dinners)

Deboned meat is first wrapped in PVC film, packed in carton boxes and then completely frozen. Each box contains 25–30 kg of meat. Because of the packing, the freezing time is relatively long. In a blast freezing tunnel with air at −35°C and high air velocities it takes between 15 and 20 hours to achieve a core temperature of −8°C. At this temperature the blocks are rigid enough to be piled for storage.

The condensers of the refrigerating system must be oversized in cases where the tunnels are loaded with warm processed meat. Cooked meat is often put straight into the tunnels. So at the start of the freezing process the compressor will work at high evaporating temperature and high capacity. This is the case when freezing all kinds of cooked prepared meat such as ham.

Red meat should be frozen in liquid nitrogen, especially if it consists of small individual portions of beef etc. (portion control). On the other hand, minced meat must be completely frozen with low temperature gases; CO_2 is most commonly used for this special application in order to reduce the multiplication rate of the bacteria. Temperatures can rise to very high values in the meat during the butchery process.

Prepared dishes, stored in positive temperature cold rooms for no more than 6 days, must be cooled down quickly immediately after preparation from +50 to +60°C to +5°C. Individual portions normally have a weight of 450 g, are 3–4 cm high and must be chilled individually. This chilling process should be completed in special rooms with a high air velocity and in less than 1 hour. Deep-freezing to a final temperature of −18°C should take no longer than this. Sometimes the freezing process in the tunnels is stopped between −5 and −10°C, and further freezing is done in the cold store. This is not recommended, as the quality will suffer.

The above-mentioned chilling and freezing times can be obtained with mechanical refrigeration as well as with cryogenic refrigeration – liquid gases such as N_2, R12 and CO_2. The mechanical refrigeration will need to operate with very low evaporating temperatures (in the range −45 to −50°C) otherwise a quality product might not be achieved.

The above-mentioned products must be frozen as near as possible to the eutectic point.

For luxury products, such as lobsters and shrimps, only cryogenic freezing is used. Such systems consume between 1 and 1.2 kg of cryogenic gas for each kg of product.

9.11 Poultry

Chilling poultry in snow ice or ice water after slaughtering is no longer permitted for hygienic reasons, and chilling now takes place in refrigerated air in the cold room. The poultry are placed on racks or small trolleys with racks which usually measure 1.9 × 0.68 × 1.69 m, each holding 200 chickens or 250 kg of meat. The air circulation rate in the cold room is 250 to 300 per h, the air temperature between 0 and −1°C, the air velocity more than 2 m/s and the chilling time 3 hours. The final temperature at the centre of each piece of meat will then be +3°C.

Hannan and Shephert give the following data:

Air velocity: 2.75 m/s
Weight: 1 kg, chilling time 0.4 h
Weight: 2 kg, chilling time 1 h
Weight: 3 kg, chilling time 1.7 h
Fresh air intake rate: 4–6 per day is standard practice
Weight loss: between 1 and 2%

After the chilling process the poultry are packed in PVC film and placed into cartons: 20 kg in boxes of 665 × 445 × 145 mm and 10 kg in boxes of 560 × 370 × 110 mm. These are then stored in coldstores at between 0 and −1°C ambient temperature. Other data:

Freezing tunnel air temperature: −35°C
Air velocity: 3 m/s
Freezing time: 3 h in boxes without top cover

9.12 Fish

Fish is very similar to meat in its composition. It is composed of proteins, lipids, some traces of glucids, vitamins B, PP, A, and D in flat fish, mineral salts, phosphorus, iodine, chlorine, potassium, magnesium, and 70–75% water. Fish can sometimes contain 10% lysin, which means 40% more than the reference protein. The mineral salts and vitamins in fish compare closely with those of meat. The main difference betwen fish and meat is that fish contains less fat: just 0.5% in the leanest fish such as cod and haddock. This allows fish to be stored in the dry state. The fatter varieties of fish contain between 10 and 20% of fat; the fish in the latter category are still less fat than meat of a similar description.

Rigor mortis is the physical manifestation of a series of complex biochemical reactions that occur in animal tissue after death. The first stage is 'onset', the next stage 'hardening' or 'stiffening' and the third stage 'relaxation'. The times required for onset, hardening and relaxation depend on a number of factors and vary from a few hours to several days.

Trawled fish are usually exhausted by the time they are landed on deck and in consequence pass through the stages of *rigor mortis* very quickly in a few minutes. In general, the higher the temperature the shorter the duration of the three stages. On the other hand, the onset of *rigor* and the other stages can be delayed by rapidly cooling the recently caught fish in plenty of ice and keeping it close to 0°C.

When fish are filleted immediately after death the fillets themselves will pass through the *rigor* changes. However, because the muscular tissue is no longer supported, the fillets will actually shrink. This shrinkage is only slight at temperatures close to 0°C, but increases at higher temperatures.

It is essential that freezing be sufficiently fast to prevent the development of any adverse quality in the product. The freezing process must therefore be applied so as to minimize physical, biochemical and microbiological changes, all of which together affect the flavour and texture. If this is done, the product immediately after freezing compares very favourably in all respects, particularly organoleptic, with fresh unfrozen fish.

Fish fillets may be frozen packaged or unpackaged, individually or in multiple packs. Packaged products, particularly fish blocks for retail sale, should be frozen so as to prevent distortion. This can best be done in a pressurized freezer, such as the multiplate design. Individual unpackaged fillets are usually frozen in an air blast. Care must always be taken to minimize dehydration or discolouration during freezing.

Freezing rates for fish fillets of between 0.6 and 4 cm per hour are indicative of established commercial practice in the major fish-producing countries. For technical as well as practical reasons, it is recommended that these rates are used as a guide in freezing fish fillets.

When freezing whole or dressed fish, the rates should approximate to those for fillets. In the case of large fish or with packages of whole or dressed fish, a freezing rate of 0.6 cm/h may be difficult to achieve with existing commercial practice. Since this might be detrimental to the quality of fish for reprocessing, attempts should be made to improve the freezing techniques where needed.

Care must be taken to minimize dehydration and oxidative rancidity when storing unpackaged frozen products. This can be accomplished by using a water or other suitable glaze after freezing and by renewing this glaze as necessary during storage. It is recognized that certain fish such as tuna, herring and halibut are frozen in sodium chloride brine. In order to minimize salt penetration and because it is impractical to work with this brine at a temperature lower than −18°C, fish so processed should be frozen rapidly to approximately −12°C, to −15°C in the centre, and the temperature finally lowered to −18°C in cold storage.

Fish is susceptible to deterioration of quality by micro-organism attack, perhaps to a greater degree than most foods and food products. Furthermore, fish are readily contaminated through poor handling and processing procedures, and dirty equipment during processing. The fact that fish may be processed at temperatures considerably higher than that of its natural environment tends to accelerate the attack by microbes and so hasten quality deterioration.

Low-temperature storage prevents fish from deteriorating, but since fish contains much unsaturated fatty acids the greatest problems are oxidation and hydrolytic separation. As long as fish is kept moist it will maintain its quality better. For this reason, ice is mixed with the fish in the containers from the moment they are caught. Some of the ice melts and this water removes the heat very quickly and prevents the fish from drying out.

The ice is produced by spraying water over a drum with a jacket in which the refrigerant is evaporated. The ice film is continually scraped from the drum by knives or broken from it by means of a roller containing sharp pins. The ice is subcooled a little down to a temperature of about −0.5°C so that it lasts longer in the refrigerated storage hold. The room temperature is maintained at a value between −3 and −5°C, with an evaporating temperature in the ice-maker of between −7 and −12°C for small production units on board the vessels. For big industrial plants ashore, evaporation temperatures of about −20°C and lower are used depending on the thickness of the desired ice layer. This ice layer can be between 1 and 4 mm and subcooled to between −6 and −8°C. Big ice-bunkers have special rotating chain systems to prevent the ice blocks from freezing together.

In the far and middle East and African countries there are still many block-ice plants which produce ice in cans of 12.5, 25, 50 kg or more. The cans are placed in a brine tank where the water is frozen. After they have been taken out of the cans, the blocks are stored at temperatures between −6 and −8°C. Five tons can be stacked on an area of 9 m^2. The heat load of the coldstore is about 20 W/m^3. In order to produce clear ice during freezing, air is blown in the cans.

For further details on ice production and storage, see section 9.14.

The cold room for the fish/ice mixture is kept at a temperature of between 0 and +1°C and an RH between 90 and 98%; not lower because it is necessary that the ice is able to melt as discussed above. Average storage time is 7 days, but some kinds of fish can be preserved up to 20 days this way. The ratio ice:fish is about 2:3.

For a longer storage time, fish must be frozen; nowadays this is mostly done on board ship in vertical plate freezers or blast tunnels. Before freezing, water is added to the fish surface to produce the glazing. Storage time is between 4 and 6 months at −25 to −30°C. The best results are obtained using a low air circulation rate or even natural convection cooling.

Tuna fish is cooled in water tanks. First the fish is cooled in seawater at −1.5°C. After two days the fish reaches a temperature of −1°C and is then frozen in brine tanks with the brine at a temperature of −12°C. The heat load is 800 W/ton of fish and the freezing time between 2 and 3 days.

9.13 Refrigerated Cargo

A ship's hold above 0°C has a heat load of between 70 and 80 W/m^3. The fresh air intake rate is 10 times the hold volume a day and the air circulation rate in fruit holds is 60 per day, in meat holds 45 per day. The heat load of holds for frozen products is about 50 W/m^3. Holds in small coasters sometimes have small coldstores at only −10°C. The heat load for such a store is about 90 W/m^3. Lloyds require a dT for the air coolers of not more than 4.5K.

9.14 Ice Plants

9.14.1 Block Ice

As we saw in Chapter 1, the use of blocks of ice is an old tradition in refrigeration technology. After natural ice came the use of artificially frozen blocks. Although the production of such blocks has almost disappeared in the Western world, we will give this subject some attention as it is still of interest in developing countries.

To produce ice blocks, steel moulds or cans are filled with water. The moulds are placed in a brine tank by means of racks containing several rows of blocks, as shown in figure 9.8.

The brine is kept at a temperature of between −6 and −8°C by means of an immersed evaporator placed in the brine tank. The brine is kept in

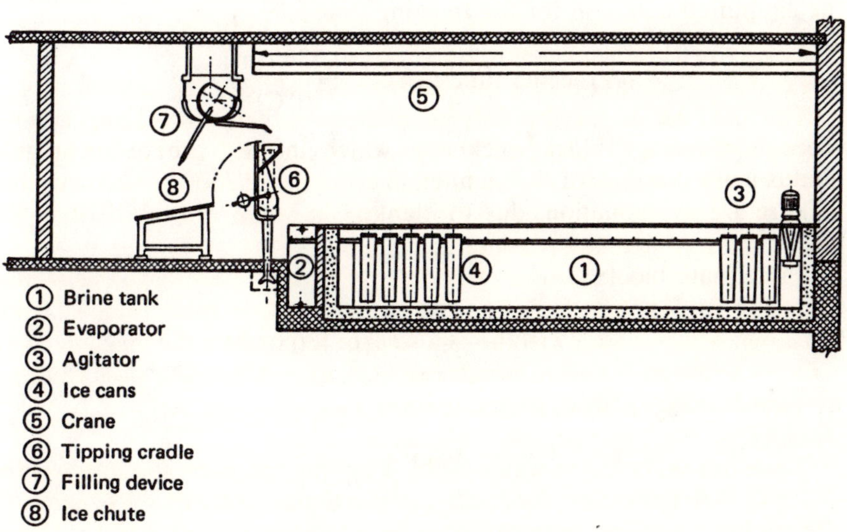

Figure 9.8 *Block ice plant*

circulation over the evaporator and around the moulds by means of motor-driven agitators. The brine velocity is 0.15 m/s.

In order to obtain clear ice, the water in the moulds is agitated by means of a mechanism of steel bars, or by means of compressed air blown in the moulds through a connection nipple at the bottom of the moulds.

The moulds are made of welded galvanized steel plates of 1.5–2 mm thickness. After welding they are coated with tin or lead. Typical block sizes are 25 and 12.5 kg, although there are many other sizes, such as 50 kg, 100 kg and larger. The smaller sizes are of course the most easy to handle.

Table 9.7 shows the main dimensions of the two smaller sizes.

Table 9.7 Dimensions of ice moulds

Capacity (kg)	*Dimensions (mm)**				
	a	*b*	*c*	*d*	*l*
12.5	190	110	160	80	1100
25	190	190	160	160	1100

*See figure 9.9 for use with this table.

The number of moulds per brine tank can be calculated from the formula:

$$\frac{\text{daily production in kg}}{\text{weight per block in kg}} \times \frac{\text{freezing time in h}}{24\text{ h}}$$

The simplified equation for the freezing time is

$$\text{Freezing time } t\text{ (h)} = \frac{0.33\delta^2\text{ (m)}}{t(°\text{C})}$$

where δ = average block thickness, which in the case of rectangular moulds is the average of the smaller size.

There are two equations due to Plank:

(a) For square blocks

$$t\text{ (h)} = \frac{3130}{T_{\text{brine}}}[b(b + 0.036)]$$

(b) For rectangular blocks

$$t\text{ (h)} = \frac{4540}{T_{\text{brine}}}[b(b + 0.026)]$$

Use values of b in metres from table 9.6.

A block of 25 kg needs about 18–22 hours of freezing time. The refrigeration capacity for a given supply of ice can be obtained from table 9.8.

Table 9.8 Refrigeration capacity requirements for differing ice production rates in tropical countries

Ice production (kg/h)	*Refrigeration capacity (kW/kg)*
5	75
10	165
15	155
20	150
50	145
200	140
300	135
500	125
1000	120
2000	119
4000	117

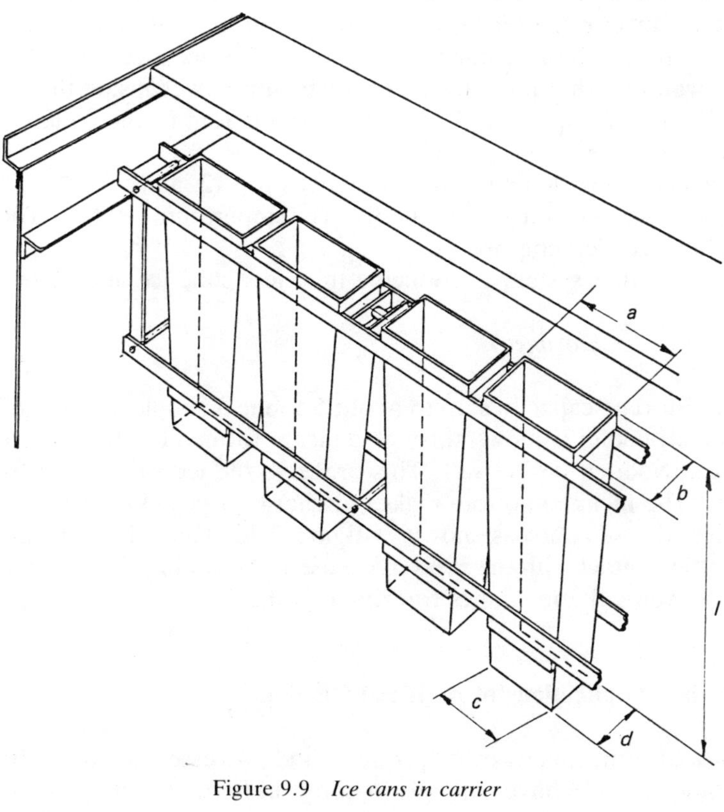

Figure 9.9 *Ice cans in carrier*

9.14.2 Block Ice Storage

The blocks are stored in an insulated room kept at between −3 and −5°C. The average density of storage is about 900 kg/m^3. One pallet 1 × 1.2 m can carry 1 tonne of ice made up of 8 layers of 5 blocks of 25 kg. The height of the pallet and load is about 1.75 m. Between every layer of blocks wooden slats are placed, 2 for each block.

9.14.3 Slice or Flake Ice

Where water ice is still needed in Western Europe and the USA, flake ice is now used, mainly in the fishing industry.

A typical flake-ice machine is shown on figure 9.10. Ice is frozen by spraying water over a refrigerated rotating drum. The thin ice layer is scraped from the drum by fixed knives. The evaporating temperature inside the double wall of the drum is selected between −15 and −30°C depending on the slice thickness and the production capacity wanted.

As shown in figure 9.11 the ice production capacity depends on the drum speed, the ice thickness and the evaporating temperature.

Another method of slice ice production is shown in figure 9.12. A cylinder is surrounded by an evaporating coil so that water fed to the cylinder is frozen on the inside of the cylinder. The ice is removed from the cylinder wall in a thin layer by a worm which revolves inside the cylinder and pushes the ice upwards. It is pressed and frozen further, then broken off and ejected.

The evaporating temperature is between −12 and −30°C, the ice thickness between 6 and 7 mm, and the ice temperature −0.5°C. This kind of flake ice is called chip ice.

There are other systems, producing tube ice, cube ice and plate ice.

9.14.4 Slice Ice Storage

For small storage capacities up to about 5 tonne a simple silo is used. The air of the air cooler is blown through a jacket formed by the insulated silo wall and a wooden inside wall. This prevents the ice slices from freezing together. The inclined bottom makes discharging easy. For bigger storage capacities, the silos are as shown in figure 9.13. Here the circulating air makes little contact with the mass of ice used. A rotating chain, driven by a motor, prevents the ice flakes freezing together.

9.15 Other Applications of Artificial Cooling

There are of course many other products and processes that use refrigerating techniques. We have only given some important examples in which

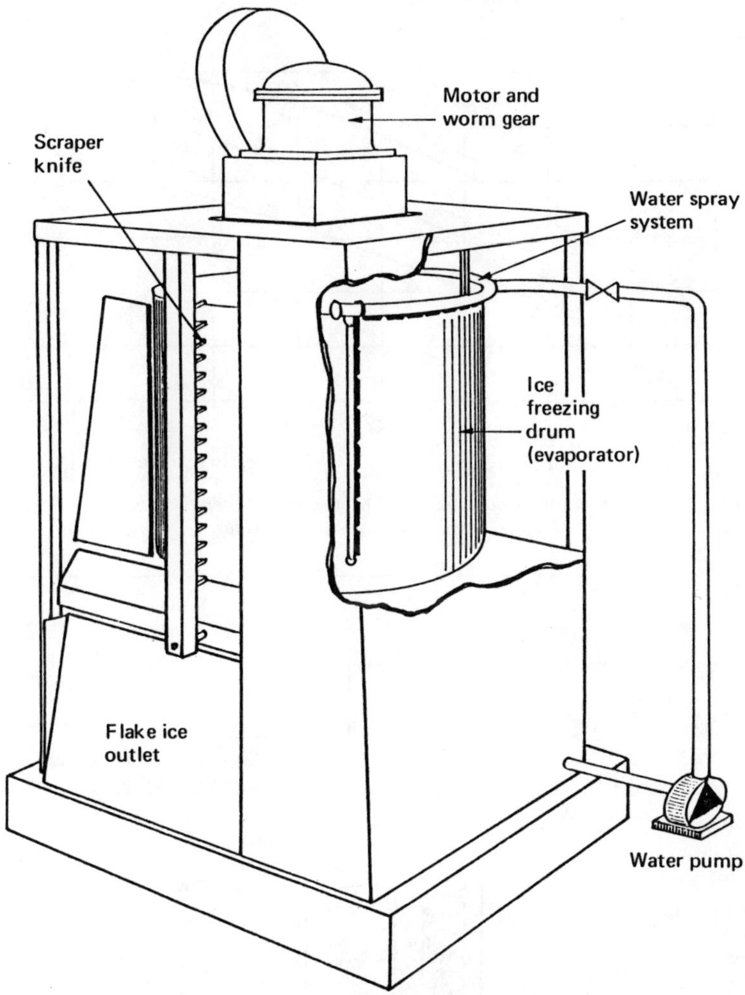

Figure 9.10 *Rotating drum flake ice maker*

special chilling or freezing systems with particular characteristics are used. In this book we have dealt only with those systems which have application in the food industry.

Refrigerating techniques used in chemical plants, civil works, the plastics and metallurgical industries do not have standard designs. They differ from case to case. Some other specialities are artificial ice-rinks and mine-cooling; open air ice rinks, for example, use a heat load of 380 W/m^2 and covered rinks 260 W/m^3, and the average evaporating temperature is -7°C depending on the sizes of the tube used. The centre to centre distance between the tubes is between 100 and 110 mm. The dimensions of an Olympic rink are 30 × 60 m, and for such a rink 21 km of tube of 25 mm

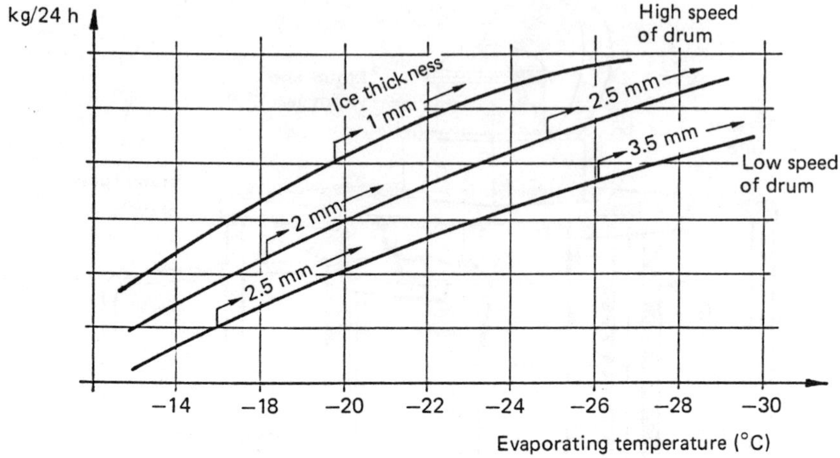

Figure 9.11 *Graphs showing dependence of ice production capacity on drum speed, ice thickness and evaporating temperature*

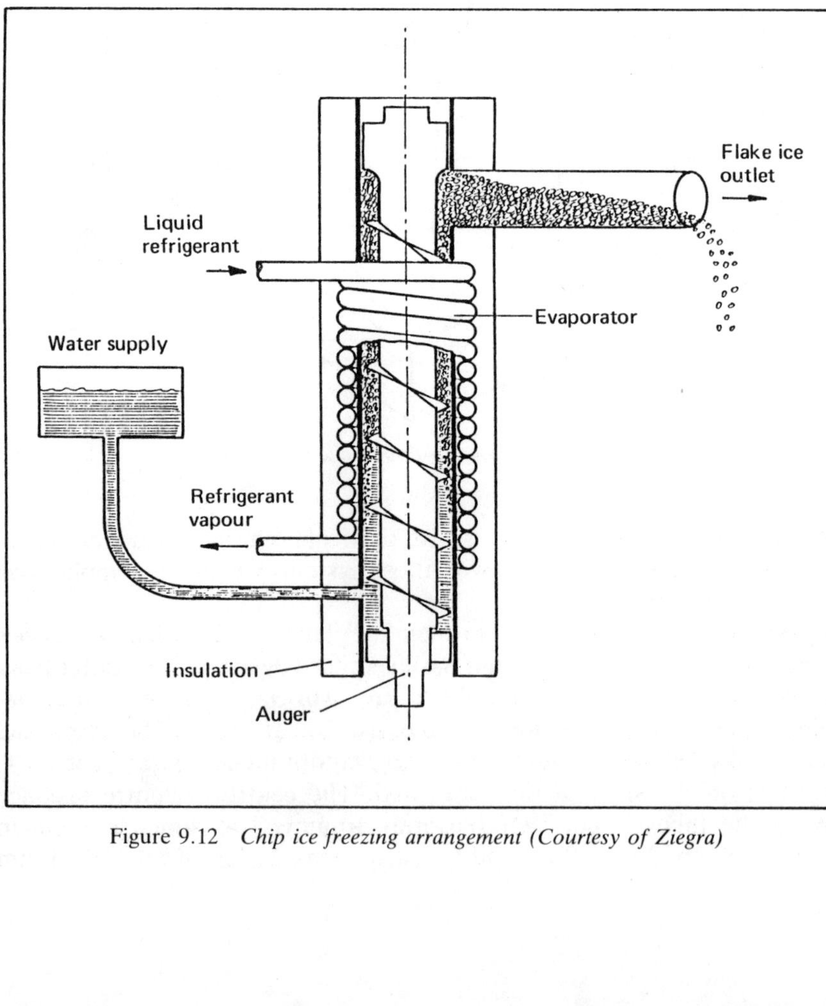

Figure 9.12 *Chip ice freezing arrangement (Courtesy of Ziegra)*

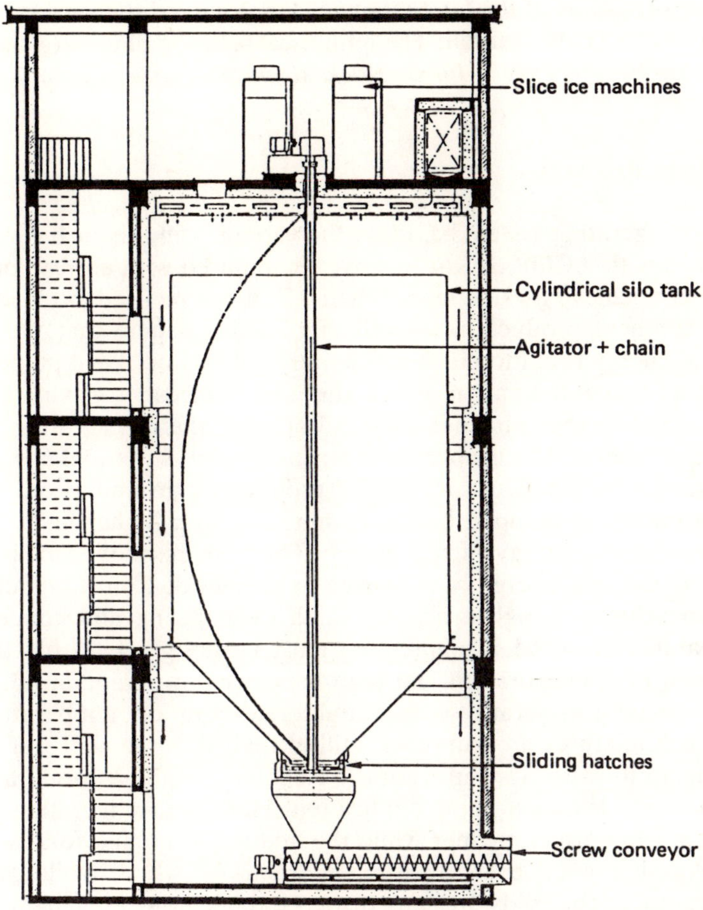

Figure 9.13 *Vertical ice silo*

or 35 mm diameter is necessary. The ammonia content in the plant is between 3.5 and 4.5 tonne and that of CFC refrigerants between 7.5 and 10 tonne. The low-pressure separator must be able to hold all the refrigerant. Sometimes the tube circuit is fed by means of an ammonia or CFC liquid pump system; sometimes brine is used.

Since mine cooling is such a specialized field, we can here only refer the reader to the literature. For example: Joachim Vosz, *Grubel Klima*, Verlag Gluckauf GmBH, Essen, West Germany; R. M. Stroh, The refrigeration systems on Western Deep Levels, *J. Mine Vent. Soc. S. Africa*, 27 (1979) No. 1, pp. 7–18; P. Wenthen (J. C. Ince and Son (Pty) Ltd), Air coolers in mines with moist and warm climatic conditions, *Proceedings of the International Mine Ventilation Congress, Johannesburg, 1975*; J. Voss (J. C. Ince and Son (Pty) Ltd), Control of the mine climate in deep coal mines, *Proceedings of the International Mine Ventilation Congress, Johannesburg, 1975*; D. W. Jordan, The numerical solution of underground heat transfer problems, Part 3, *Int. J. Rock Mech. Min. Soc.*, Volume 2.

9.16 Heat Recovery

As the refrigerating system based on the vapour compressor has very high efficiency, with a COP of 3 or just over, compared with absorption plants that do not achieve a value higher than 0.7, it can be used as an economic system for heat production as well. This does assume that the cost of electrical energy is not too high compared with fuel oil. Heat pumps are in fact refrigerating systems, in which the goal is to use a maximum of the heat rejected by the condenser, instead of using the refrigerating capacity of the evaporator. The evaporator absorbs low-level thermal energy from for instance the ambient air, or the ground, or groundwater, and this thermal energy is pumped by the compressor to a higher level together with the electric energy of the motor. The condenser then delivers this high-level thermal energy to the space or subject that must be heated. In order to achieve a high efficiency, high evaporating and condensation temperatures are used. Compressors must be designed so that they can work with high evaporation and high condensation temperatures; sometimes, special refrigerants with variable evaporation temperatures are used. Such mixtures are, however, still in their development stage.

Refrigerants with a steep saturation characteristic are preferable. An ice rink is an ideal heat-pump application. Here the refrigerating capacity is used in the normal manner while the heating is applied to a swimming pool. Another good example is a supermarket, where while providing refrigeration to the coldstores and displays, the heating is available for use in the main building. Unfortunately the need for maximum heating and maximum refrigeration seldom coincide.

Heat pumps are used mainly in the field of air conditioning, but there are also industrial applications, as for instance in the agricultural industry.

In fact, every refrigerating plant can be considered as some kind of heat pump. The only problem is that the thermal energy of the condenser is in most cases at such a low-temperature level that it is useless for heating purposes. In some cases it is interesting to invest in a gas/water heat exchanger in which the superheat of the discharge gases is pumped out before it enters the condenser. In this way, clean water of about 50°C can be provided free of charge. In such cases, refrigerants with a high discharge temperature give interesting possibilities for heat-pump applications; their only disadvantage, though, is that the superheat is only available in relatively small amounts.

When using the condensation energy the temperature level in the best case is about +45°C, so water temperatures of only +35°C can be obtained.

It is always possible to increase the condensing temperature artificially to a higher level, for instance to between +50 and 55°C which is the limit for a normal industrial compressor. This will lower the efficiency of the refrigerating plant which of course is not desirable. The main goal of a refrigeration installation is to deliver 'cold' at a high rate of efficiency.

As the consumption of warm water is mostly out of phase with heat production, storing water in a tank becomes necessary. As the local bylaws normally prescribe that heat exchangers connected to the water distribution system may not contain a refrigerant, the use of a secondary heat exchanger is often necessary. This in turn lowers the efficiency because of the involvement of two temperature differences, and results in a higher investment cost. As corrosion problems must be prevented, the cost of investment can be driven up to a very high level.

All things considered, heat recovery systems are not as useful as might be expected. A thorough cost–benefit calculation must be made before any decision about investment in such a system can be made.

10

Insulation Techniques and Coldstore Construction

10.1 Background Information

10.1.1 Heat Transfer

The overall heat transfer coefficient or U-value, and its inverse R, the thermal resistance, is usually determined in a relatively simple way when calculating heat loads for industrial or commercial refrigeration applications.

In air-conditioning projects, where walls, doors, windows, roofs and floors of different and often complicated material combinations are used, the calculation is more complex. According to the theory of resistance in electrical theory, one can add the heat-transmission resistances of the different layers of the construction. So

$$1/U = 1/\alpha_1 + \Sigma(\delta/\lambda) + 1/\alpha_2 \ (\mathrm{m^2\ K})/\mathrm{W}$$

where α_1 is the surface heat transfer coefficient from the air to the wall on the warm side in $\mathrm{W/(m^2\ K)}$

α_2 is the surface heat transfer coefficient from the air to the wall on the cold side in $\mathrm{W/(m^2\ K)}$

λ is thermal conductivity of any layer of solid material in the structure in $\mathrm{W/(m\ K)}$

δ the thickness of any layer in m.

Values of α depend on the velocity of the air. For example, at a velocity of 5 m/s inside a coldstore, α_2 varies between 8 and 10 $\mathrm{W/(m^2\ K)}$ for vertical walls and 6 and 9 $\mathrm{W/(m^2\ K)}$ for horizontal walls. For normal work, α_2 is taken to be 8 $\mathrm{W/(m^2\ K)}$. Similarly the value of α_1 for the outside air is taken to be 29 $\mathrm{W/(m^2\ K)}$.

Modern coldstores are now generally built using prefabricated insulation panels composed of two galvanized steel, aluminium or sometimes polyester cladding sheets, each 0.6 mm thick. The space between the sheets is filled by foaming polyurethane or by placing a polystyrene layer therein.

As the sheet thickness is always small, only δ/λ of the insulation material is taken into account; when stone, concrete or other construction material is used to build the walls, the values of δ/λ for the individual layers must be calculated and added together.

The most important value in the calculation of insulation heat transfer is that of λ, and in this respect the best insulating material is non-moving dry air or gas. λ for non-moving air at 273 K is only 0.024 W/(m K) and that of R11 0.016 W/(m K).

Metals have the highest values of λ. For instance, copper has a value of 380 W/(m K). λ for most commonly used construction materials varies between 0.58 W/(m K) for brick to 2.3 W/(m K) for heavy concrete.

Until the invention of plastics and their derivatives, cork was the most commonly used insulation material. Being a natural product, cork was readily available in big quantities in southern Europe and America, where it is still widely used, but in most countries it has now been replaced by polystyrene, polyurethane, foamed glass, glass wool, rock wool, perlite and other artificial insulation materials. Pure and expanded cork slabs with a density of 120 kg/m^3 have a λ value of 0.04 W/(m K) at 293 K.

Modern insulation materials are based on the principle of creating a material consisting mostly of air or gas, enclosed in cells of a material with a low density; the smaller the cells and the thinner the cell walls, the closer the value of the material and gas comes to the ideal value of air or gas.

Standard polystyrene cells are filled with air, steam is used in the production of the white polystyrene and the blue variety is produced by extrusion.

Polyurethane (PUR) or the variation Polyisocyanurate (PIR) until recently had their cells filled with R11. Because of its high ozone depletion potential, R11 can no longer be used in the production of polyurethane and new chemical processes are required. Some manufacturers have already replaced 50% of the R11 with water, a technique which has been used before but discarded because of its drawbacks. New methods of production will use the replacement refrigerants such as R123 or R141b. Further research will be necessary in the future. Foamed glass cells are filled with H_2S.

For calculation purposes the following λ values are used:

Expanded polystyrene: 0.03–0.035 W/(m K) {5–10 kg/m^3}
Extruded polystyrene: 0.029–0.038 W/(m K)
PUR: 0.029 W/(m K) {40 kg/m^3}
PIR: 0.023 W/(m K) {40 kg/m^3}

Actual values depend on the following:

(a) The kind of insulation material used.
(b) The degree of humidity. If a porous material is used, it is possible that the material may contain water. As the λ of water is 0.58 W/(m K), its negative effect on the overall value of a material can be imagined. Figure 10.1 shows this effect on PUR.
(c) The temperature of the material. Values given by manufacturers are generally laboratory figures measured at temperatures between 273K and 293K. Figure 10.2 shows the influence of the temperature on λ for

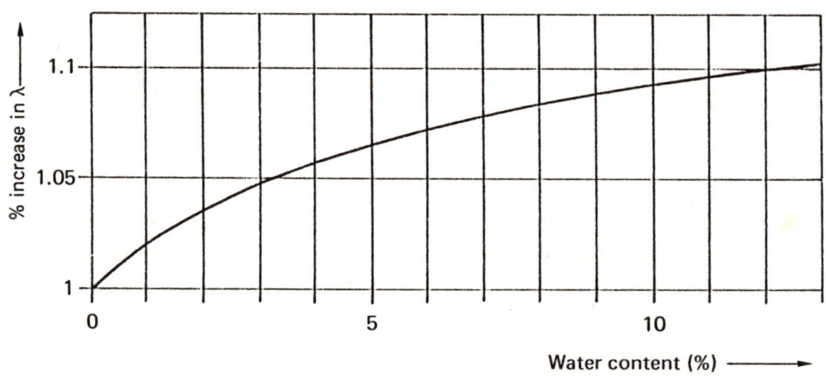

Figure 10.1 *Effect of water content on the coefficient of thermal conductivity of PUR*

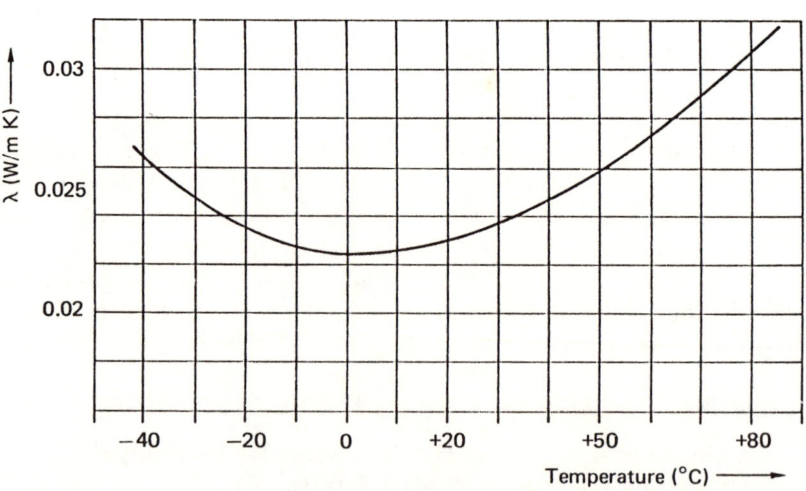

Figure 10.2 *Effect of temperature on the coefficient of thermal conductivity of PUR and PIR*

PUR and PIR. Because an insulation material has a very low density, internal radiation has a big influence on the heat transfer, and as radiation is highly influenced by temperature, λ increases at higher temperatures.

PUR and PIR behave differently from the other insulants. In figure 10.3, λ shows an increase of between 273 and 233K owing to condensation of the R11 in the cells.

(d) The density of the material.

(e) The age of the material. λ does not always remain stable over the years. PUR and PIR cells, especially those exposed to the open air, will alter; they tend to break open and lose their gas contents. Therefore it is better always to keep polyurethane surfaces covered with a layer of steel sheeting or paper, or wrapped with PVC etc.

When choosing the insulation material that is to be used for a project, many characteristics have to be taken into consideration. These are:

1. The λ value, which must be lower than 0.058 W/(m K).
2. The linear expansion coefficient. This value varies from one material to another. It should not be too high; if it is, the space between the

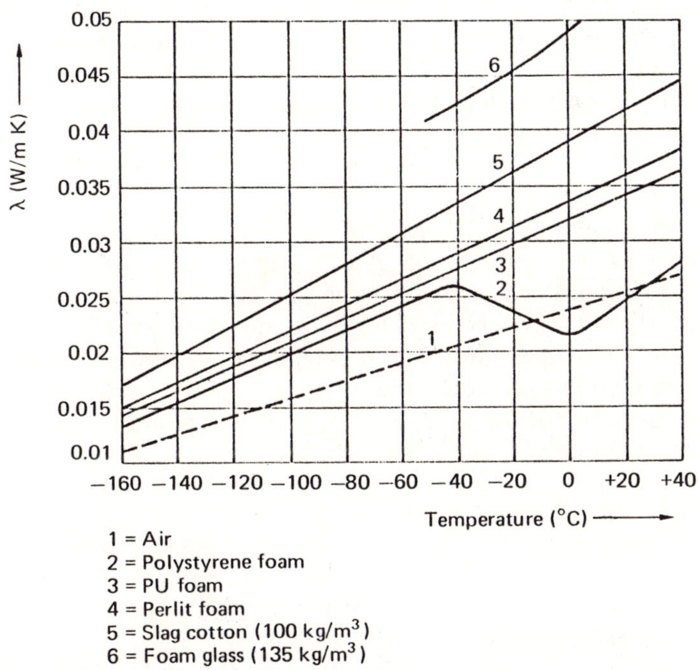

Figure 10.3 *Effect of temperature on the coefficient of thermal conductivity of various insulants*

cladding sheets will get larger after a certain period of time. When using materials with high linear expansion coefficients, such as expanded polystyrene, it is advisable not to use these materials within 3–6 months after the date of manufacture, in order to allow time for the expansion to reach its maximum.

3. Stability. As λ and expansion values vary over a time period, one should use materials that are as stable as possible.
4. Permeability. Some materials absorb water; others do not, such as extruded polystyrene. Most insulation materials are not water-vapour tight and do, to varying degrees, absorb the water vapour in the air or the gas. It is therefore important to compare the water vapour transmission resistance coefficients, discussed later in this chapter.
5. The mechanical strength is another characteristic of special importance for floor insulation. Polyurethane, cork and foamed glass have the highest mechanical strength, and extruded polystyrene is stronger than expanded polystyrene.
6. The density. For reasons of mechanical strength it is inadvisable to use polystyrene with a density of less than 20 kg/m^3 and polyurethane with less than 35 kg/m^3. However, the higher the density, the greater the ratio of material/gas volume, and the higher the value of λ.
7. Some insulation materials have distinct disadvantages. Polystyrene is sometimes eaten by rats. Since glass wool has no mechanical resistance, it may collapse when used in walls and when it is not properly supported.

   Foamed glass has a high λ value; also when blocks are broken, H_2S is released from the cells, creating a bad smell.

   Previously manufactured polyurethane contains R11, which, in case of fire, will form phosgene, a lethal gas.
8. The only commonly used insulation materials that are fire-proof are foamed glass and glass wool. All the others may have more or less fire-retarding or low fire-transmitting characteristics, but once they start to burn they do so very quickly. Such materials are classified by the insurance companies into several categories on which their premiums are based. Unfortunately, there is as yet no worldwide or even European standard for these materials. Some countries, though, have set certain standards, as tabulated below.

United Kingdom	BS 476 Part 7
United States	ASTM D84 or ASTM D1692–S9
Germany	DIN 4102
France	CSTB PV 65,2259 A
The Netherlands	NEN 1076
Belgium	NBN 713–020 (1968)

9. Price. Prices vary according to the world economic situation. They may be classified from the cheapest to the most expensive, as follows:

   Glass wool
   Expanded polystyrene
   Extruded polystyrene
   Cork
   Polyurethane (PUR)
   Polyisocyanurate (PIR)
   Foamed glass

Sometimes, for very large projects and especially in plants that store liquefied gas at extremely low working temperatures, insulation thickness is calculated on an economic basis.

In specialized literature one can find equations for these calculations based on: cost-price of material and labour, investment cost of insulation, building and machinery, maintenance costs, running time per year, interest rates, and the ratio of insulation surface area to store volume.

In daily refrigeration practice for coldstore tunnels' pipework and vessels, λ is chosen in such a way that the losses by heat transmission are limited to between 8 and 9 W/m^2. This means that $\delta = dT\lambda 100/8$ cm or $dT\lambda 100/9$ cm.

The insulation thickness for pipework and vessels is normally calculated based on the prevention of condensation of the outside surface of the insulation. See below.

Note that the U-value for non-insulated floors of concrete lies between 1.7 and 2.2W/(m^2 K), depending on the λ value of the ground underneath (see DIN 4701).

10.1.2 Water Vapour Diffusion, Vapour Barrier and Material Characteristics

Just as important as the value of the heat transfer coefficient of the insulation material is the value of the water vapour transfer coefficient of the vapour screen or barrier covering the insulation. Where heat transfer is created by temperature difference across the separating wall, water vapour diffusion is created by vapour pressure difference across the wall as shown in figure 10.4.

This partial water vapour pressure in the mixture of air and water vapour in the atmosphere depends on the temperature (T) of the air and the relative humidity (RH or ϕ). Outside the coldstore or tunnel, the temperature will be higher. The partial vapour pressure will be higher when RH and T are higher.

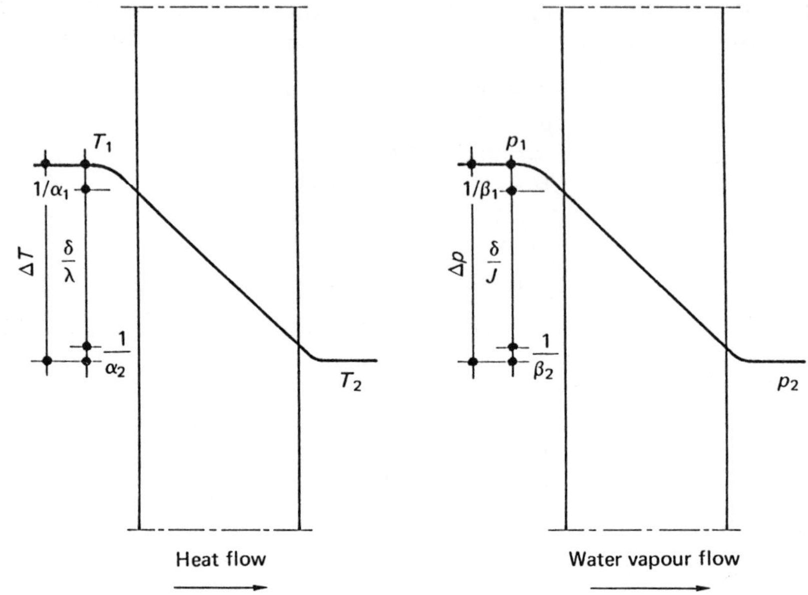

Figure 10.4 *Comparison of heat transfer and water vapour diffusion through an insulating material*

In the cooled space, the vapour pressure generally will be lower. Of course, during winter, especially in insulated rooms at temperatures above 273K, the opposite may occur.

In freezing rooms and freezing tunnels, in particular, there will be a low water vapour pressure, which results in the tendency of water vapour to diffuse from the outside to the inside (figure 10.5). Most insulation materials, except foamed glass and phenol-based products, have a low resistance against water vapour diffusion. That is why we have to add an extra layer of material with a high resistance against water diffusion to the warm side of the insulation layer. Besides having a good diffusion resistance, the material must be:

(a) not too expensive,
(b) easy to fix,
(c) stable over the years,
(d) of a sufficient mechanical strength.

The value of the diffusion resistance is expressed by the symbol μ.

What happens when water vapour enters the insulation material?

(a) It can pass through it without leaving a trace, and then enter the coldstore and sublimate on the evaporators.

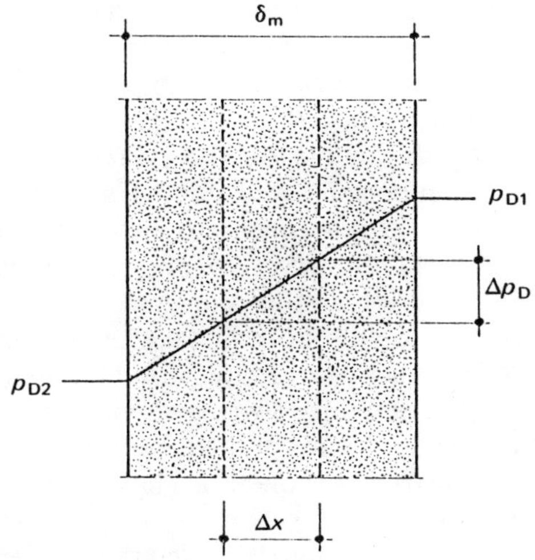

Figure 10.5 *Water vapour diffusion caused by pressure difference*

(b) It can condense inside the insulation and have a negative effect on the λ value. The value of λ for water vapour is 0.018 W/(m K), but remember that for water it is 0.58 W/(m K).
(c) Depending on the temperature inside the insulation material, this condensed water can freeze. Note that the λ value for ice is 2.3 W/(m K). Ice also has a destructive effect on the insulation layer, because of the expansion between two insulation layers or a layer and a wall, ceiling or floor. Later we will see the effects of this on floor insulation and on the foundations of a building.

As stated above, the barrier is placed on the warm side. If it is not certain that a 100% vapour tight barrier is being used, it is better not to place a barrier on the cold side, because the vapour entering is not absorbed by the insulation material, but is allowed to find its way to the evaporators.

For rooms that operate at high temperature levels, such as banana ripening rooms at 285K in countries with a cold or moderate climate, it is better to put a vapour barrier on each side if material is being used that is 100% vapour tight; if not, no vapour barrier at all should be provided, as in winter periods water vapour diffusion works in the opposite direction, that is, from the inside to the outside.

The importance of a good vapour barrier will be evident from the following figures. A split 0.5 mm wide in a vapour barrier, at $dp = 0.01$ bar, lets in water vapour at a rate of 0.08 g/(h m) throughout the length of the split. Every gramme of water/h m entering at $dT = 20$K increases the U-value of the insulation by 0.035 W/(m^2 K).

Note that like λ, the water vapour diffusion value of a construction is not constant with time. In general, the vapour barrier should be of such quality that, after 10 years of operation in an environment of 38°C and 90% RH, it has not increased by more than 10%.

We will now develop the theory of water vapour diffusion. As water vapour diffusion is very important, we will provide some more theoretical background. This theory is not meant to be used for calculations in everyday practice, but mainly serves to give an understanding of the motive power behind diffusion and its dangers.

Because nowadays, more and more metal vapour barriers sealed with silicone kits are used, which are 100% water vapour tight, it is not necessary to calculate the amount of water vapour entering the construction when calculating the heat load. In the case of non-metal barriers, and especially when poor construction conditions have to be contended with, it could be useful to do the calculation.

In disputes about bad construction, it is sometimes necessary to calculate the humidity situation inside the insulation or the construction materials concerned in order to judge the condition of the construction.

Definition of the Water Vapour Diffusion Resistance Value

The water vapour diffusion resistance of a material, μ, is defined as the ratio: (quantity of water vapour D_1 transferred through a layer of air of thickness $\delta = 1$ m, per unit time, per m^2, at temperature T, total pressure p and a concentration of 1 kg water vapour/m^3 air) to (quantity of water vapour D transferred through a layer of the material concerned of the same thickness per unit time, per m^2 at the same conditions). Thus

$$\mu = D_1/JR_DT$$

where J is the amount of water vapour in kg/(Pa s m) that passes per unit time per m through a layer of the material with thickness of 1 m at a constant temperature and a pressure difference of 1 Pa.
R_D is the gas constant for water vapour (461J/kg K)
T is the average temperature of the layer.

Table 10.1 gives values of μ for different materials.

The value of D_1 for air has been calculated by R. Schirmer for different conditions. Since by definition D varies with the pressure p and the temperature T, D_1 can be calculated from the equation:

$$D_1 = 0.083(10\,000/p)(T/273)^{1.81} \text{ m}^2/\text{h}$$

At 1 bar and 273K this gives a value for D_1 of 0.185×10^{-9} m^2/h, a figure mostly used in German literature.

For a better understanding of this subject it is easier to use the concept of J (water vapour flow rates). A list of values of J is given in table 10.2.

Table 10.1 Water vapour diffusion resistance values μ for different materials

Material	*Density (kg/m³)*	μ
Impregnated/expanded cork	100–200	10–50
Polyurethane	30–50	30–50
Expanded polystyrene	15–25	20–90
Phenol foam	30–60	30–50
Extruded polystyrene	40	200
Foamed glass	130–150	∞
Aluminium (0.2 and 0.1 mm)		7×10^5–∞
Bitumen/rubber-based emulsions		4×10^4
Bitumen-based emulsions		760
Waxed paper		2.5×10^5
Aluminium roofing (0.05 mm)		2.7×10^5
Special bitumen		10^6
Double roofing		10^4
Polythene sheet (0.2 mm)		10^5

Table 10.2 Water vapour flow rates (J) for different materials

Material	*Density (kg/m³)*	J *(kg/Pa s m)* $\times 10^{-12}$
Insulation corkboard	100–200	18–4
Extruded polystyrene	40	0.9
Expanded polystyrene	15–35	9–2
Polyurethane	40	5
Stone	1900	13
Low-density concrete	1000	26
Medium-density concrete	1900	12
High-density concrete	2200–2500	5–18

Stefan gives the following equation for water vapour diffusion:

$$\text{water vapour flux } m = J(p/\{p - p_D\})(dp_D/dx)x \text{ kg/(m}^2\text{ s)}$$

where p = total pressure of the mixture air/water vapour

p_D = partial water-vapour pressure, varying linearly from p_{D1} to p_{D2}.

Or in simplified form:

$$m = J(dp_D/dx) \text{ kg/(m}^2\text{ s)}$$

where p is in kg/m².

Integrating over the total thickness and taking into account the fact that the temperature in every point of the layer is T_m gives:

$$m = J/\delta(p_{D1} - p_{D2}) \text{ kg/(m}^2\text{ s)}$$

We can now compare the process of water vapour transfer with that of heat transfer. Heat transfer by conduction is given by Fourier as $C = dQ/dT = -\lambda A dT/dx$, which in practical form simplifies to $Q = (\lambda/\delta)A dT$ (W), and water vapour diffusion is given by $m = (J/\delta)A dp$ kg/s.

In heat transfer, there is not only transfer by conduction through the material but also exchange with environmental air by convection. In vapour transfer there is also an exchange with environmental air; the exchange has a coefficient given the symbol β. Compare α_1 to β_1 and α_2 to β_2. U is the overall heat transmission coefficient, and the overall water-vapour transmission coefficient is called U_w.

This is similar to $1/U = 1/\alpha_1 + \Sigma(\delta/\lambda) + 1/\alpha_2$. Now we can state:

$$1/U_w = 1/\beta_1 + \Sigma(\delta/J) + 1/\beta_2$$

Finally the equation can be written:

$$m' = U_w dp_D \text{ kg/m}^2 \text{ s}.$$

The heat flux equation $q = UdT$ compares here with the water vapour flux equation $m' = U_w dp_D$. $\Sigma(\delta/J)$ is once again derived by adding the individual resistances over the different layers. One of the layers may be a layer of air, but this is negligible.

With the help of this theory we can construct a graph showing the evolution of the moisture transfer in an insulated wall. The total vapour flow is

$$m_f = U_w dp_{Dtot} \text{ kg/m}^2 \text{ s}$$

where $dp_{Dtot} = p_{\text{outside the room}} - p_{\text{inside}}$
and $p_{\text{outside}} = \phi_{\text{outside}}\, p_{ws}$ where ϕ is the relative humidity and p_{ws} is the saturated water vapour pressure at the temperature concerned. We can find these values in the Mollier or Carrier graphs for humid air. U_w is found from

$$U_w = 1/\beta_1 + \Sigma(\delta/J) + 1/\beta_2$$

We can calculate the pressure drop in each layer as follows. The pressure drop from the outside air to the wall surface is given by

$$dp_1 = (1/\beta_1)U_w dp_{tot}$$

The pressure drop in the first layer of material is given by

$$dp_f = (\delta_1/J_1)U_w dp_{tot}$$

and in the next layer by

$$dp_s = (\delta_2/J_2)U_w dp_{tot}$$

and so on.

Using these values of pressure drop we can now locate the intermediate values of p on a graph, as is shown in figure 10.6.

In the same way we can calculate the temperature drop in each layer by following the same procedure but using the total heat flux $q = U\mathrm{d}T_{\mathrm{tot}}$, where $\mathrm{d}T_{\mathrm{tot}} = T_{\mathrm{outside}} - T_{\mathrm{inside}}$.

U we find from $U = 1/\alpha_1 = \Sigma(\delta/\lambda) + 1/\alpha_2$. Then the temperature drop from the outside air to the wall surface is given by

$$\mathrm{d}T_1 = (1/\alpha_1)U\mathrm{d}T_{\mathrm{tot}}$$

The temperature drop in the first layer of the material is given by

$$\mathrm{d}T_2 = (\delta_1/\lambda_1)U\mathrm{d}T_{\mathrm{tot}}$$

and so on.

These values of temperature drop may be used to determine the intermediate temperatures on the same graph. At points where the actual water vapour pressure is higher than p_{ws}, then condensation should definitely occur as shown on figure 10.6.

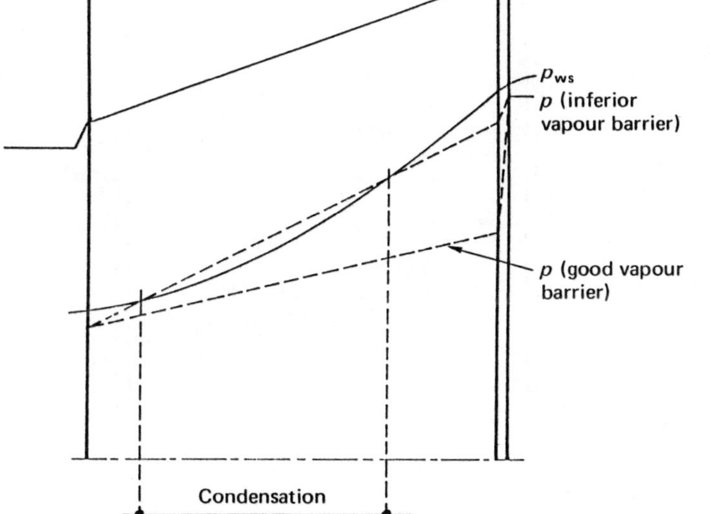

Figure 10.6 *Water vapour pressure and temperature gradients through an insulated wall*

The Glaser Method

Using this theory and procedure, Glaser developed a method by which it is possible to determine the necessary thickness of the vapour barrier, given a certain value of the material used. He stated that the above-mentioned situation in which the actual vapour pressure lies above the saturated vapour pressure can actually never exist. In other words these lines cannot cross each other, but the line of actual vapour pressure will at the maximum be a tangent to the line of saturated vapour pressure at the actual temperature.

He also stated that the vapour flow or flux either remains constant through the composite wall, or will eventually decrease. But it will never increase if we do not add water artificially, and it decreases when the vapour condenses on certain spots; however, as long as $p_D < p_{ws}$, the flow is constant. As long as there is no condensation in the material, dp_D is proportional to the water vapour diffusion resistance δ/J.

The method involves marking on a graph all the δ/J values of the layers that make up the wall on the *x*-axis, as shown on figure 10.7. On the *y*-axis

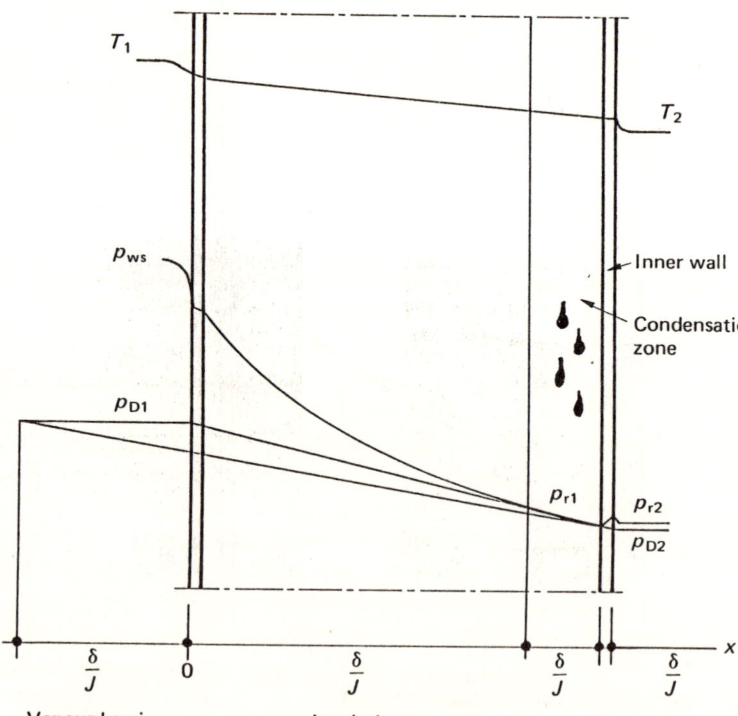

Figure 10.7 *The Glaser method of selecting a vapour barrier*

we mark the calculated temperatures of each point and next we put in at each point the p_{ws} corresponding to the temperature at those points; having got a p_{ws} line we now put the values of $p_{outside}$ and p_{inside} on the graph. As stated above, the connecting curve cannot cross the p_{ws} line.

A tangent has to be drawn from each point to the p_{ws} line and we can see that the condensation zone lies between the points of contact p_{r1} and p_{r2}.

In fact, there is diffusion from the outside to the inside from p_{D1} to point p_{r1} of quantity $m_1 = (p_{D1} - p_{r1})J_1/\delta_1$ and diffusion from inside to outside from point p_{r2} to p_{D2}, being $m_2 = (p_{D2} - p_{r2})J_2/\delta_2$. The amount $m_2 - m_1$ remains as condensed water vapour inside.

If we want to know what thickness of vapour barrier is needed to avoid condensation, in other words to create a situation in which $p_{r1} = p_{r2}$ and there is only one contact point, so that m = constant, then we have to draw a tangent through p_{r2} until it crosses the line p_{D1}. On the x-axis we read the required value of δ/J of the vapour barrier, being the distance between the point of intersection and the initial point O on the x-axis.

The engineers Wagner, Seiffert and Eickler have tabulated all the results of these calculations. These tables can be found in the specialist literature only.

Note that in the English literature one can find another coefficient for water vapour diffusion, the Perm. 1 Perm/inch is 1 grain of water vapour per hour, per square foot, per inch thickness, at 1 inch mercury pressure difference.

10.2 Insulation of Coldstores

10.2.1 Floor Insulation

A particularly important function of the coldstore floor is its need to carry heavy loads. Except for very small commercial coldstores, the floor is almost always insulated in the traditional way and not with prefabricated insulated floor panels of the sandwich design.

Here again the vapour barrier is very important; an inadequate barrier can cause serious problems to the foundations of the building. As we know, even with insulated floors there will be more or less 8 W/m^2 heat flux through the floor, which can of course be limited by increasing the floor insulation. Owing to the relatively low average ground temperature compared with the ambient air, there is, however, a tendency to provide a thin floor insulation. In cases of relatively high coldstore temperatures, floors are sometimes not insulated at all.

However, for low temperature coldstores, depending on the building construction and the composition of the ground, the 273K isotherm tends to descend into the ground year after year; in other words, the ground is 'frozen-up' as is shown in figure 10.8.

The soil delivers a certain quantity of thermal energy, between 1 and 2.5 W/m^2 depending on its composition; however, this is far too small in relation to the 8 W/m^2 'heat flux' created by the conditions in the coldstore.

Ward and Sewell developed an equation to calculate the penetration 'p' of the 273K isotherm as follows:

$$T_p - T_m = \frac{(T_0 - T_m)}{\left(1 + \frac{2\delta}{r_0} - \frac{k_a}{k_i}\right)}\left(1 + \frac{2p^2}{r_0^2} - \frac{2p}{r_0}\sqrt{\left[1 + \frac{p^2}{r_0^2}\right]}\right)$$

where T_p = minimal temperature at depth 'p' expressed in feet
T_m = average temperature of the ground
T_0 = coldstore temperature
δ = insulation thickness in feet
r_0 = radius in feet of the equivalent area of the coldstore floor
k_a = heat transfer coefficient of the ground
k_1 = heat transfer coefficient of the insulation.

Depending on the values of the different coefficients mentioned above, 'p' can have values of 10 feet after 7 years, as was calculated and measured in −25°C coldstore in Ottowa, Canada. The deformation of the floor of this particular store showed an elevation of 27 cm in the middle of the

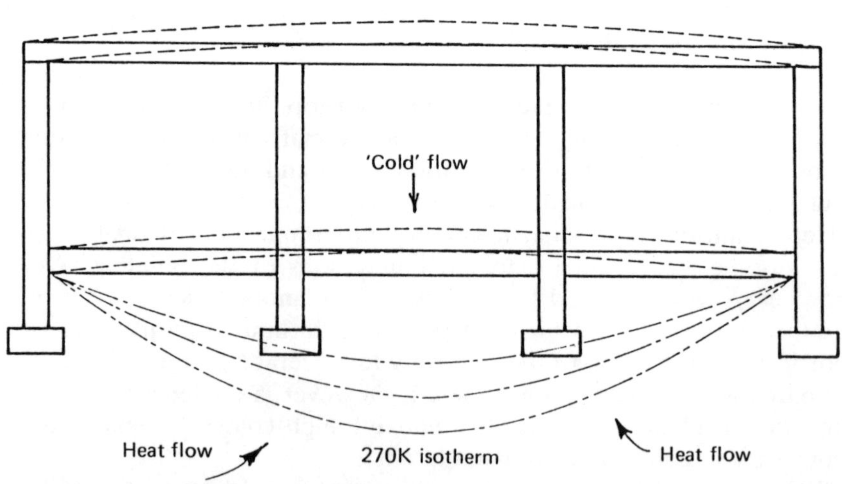

Figure 10.8 *Lowering of the 270K isotherm into the ground beneath a coldstore*

floor, Thus, the development of the situation depends very much on the composition of the soil. Non-binding material, such as gravel, freezes as one solid layer. In other typical soil types, layers of ice crystals are formed; however, certain sliding layers are very dangerous because their expansion can cause serious damage to the foundations.

A first solution in cases where there are doubts about the composition of the soil is to remove a layer of 0.5–1 m and replace it with a non-binding material. But still there might be the risk of a high water table.

Another possible way of avoiding problems is to build a cellar under the coldstore floor. Such a construction is expensive since it must be combined with the creation of a loading-bay.

The best and most frequently used solution is underfloor heating which can be achieved in several ways:

(a) By means of 4-inch diameter asbestos tubes put into the first layer of concrete at distances of about 1 m, connected on both ends to an air-duct, preferably placed with one end in a warm environment such as in the machinery room. If necessary, the air circulation can be forced by small centrifugal fans connected to the duct, preferably on the warm side.

(b) By means of electrical resistance cables manufactured by specialists and available in mats or rolls placed on the concrete layer. A floor surface of 1000 m^2 requires about 25 W/m^2 (40 V).

(c) By means of a floor heating system consisting of $\frac{3}{4}$-inch diameter plastic tubing at 600 mm distance set in the concrete layer (as shown in figure 10.9), and through which a water/glycol solution is pumped. The solution passes over a heat exchanger, acting as a small water-cooled condenser, to recuperate surplus energy from the discharge gas.

Where heat is needed when the plant is at standstill, electric heating resistance is provided. Everything is controlled by thermostats, the ground being kept at about +6°C by circulating the glycol solution at +10°C. Several circuits are available in case one should fail.

To illustrate what happens below the surface, we will now give some calculations which can be followed on figure 10.9:

$$1/U_{tot} = 1/\alpha_1 + \delta_{c1}/\lambda_{c1} + \delta_i/\lambda_i + \delta_{c2}/\lambda_{c2} = \delta_s/\lambda_s + 1/\alpha_2$$

where c = concrete layer (1 or 2)
i = insulation layer
s = soil.

Also: heat flux $q = U_{tot} dT_{tot}$ W/m^2

$dT_{tot} = T_{soil} - T_{coldstore} = 40K.$

$\alpha_2 = 8\ W/(m^2\ K).$

$\lambda_c = 1.5\ W/(m\ K)$

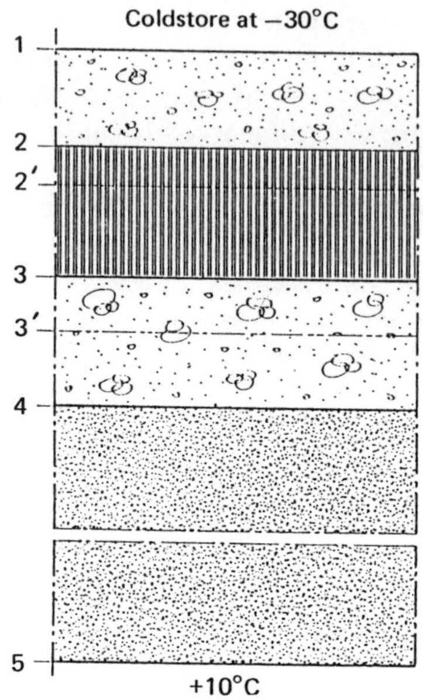

Figure 10.9 *Section through a coldstore floor showing insulation and underfloor heating*

$\lambda_i \quad = 0.037$ W/(m K)
$\lambda_s \quad = 1.2$ W/(m K).

For calcareous soil with between 10 and 28% water, λ varies between 1.15 and 2.3 W/(m K). Then ignoring α_2:

$$1/U_{tot} = 1/8 + 0.15/1.5 + 0.1/0.37 + 0.15/1.5 + 1.2/1.2$$
$$= 0.125 + 0.1 + 2.7 + 0.1 + 1 = 4.025 \text{ (m}^2\text{ K)/W}$$

So $U_{tot} = 0.248$ W/(m^2 K)
and $q = 0.248 \times 40 = 9.94$ W/m^2.

Now the temperature in each layer can be calculated:

$$\mathrm{d}T \text{ across a layer} = \text{resistance of the layer} \times q$$

To determine point 1: $\mathrm{d}T = 0.125 \times 9.94 = 1.24\text{K}$ so $T_1 = 244.24$K
point 2: $\mathrm{d}T = 0.1 \times 9.94 = 0.994\text{K}$ so $T_2 = 245.23$K
point 3: $\mathrm{d}T = 2.7 \times 9.94 = 26.84\text{K}$ so $T_3 = 272.07$K
point 4: $\mathrm{d}T = 0.1 \times 9.94 = 0.994\text{K}$ so $T_4 = 273.06$K
point 5: $\mathrm{d}T = 1 \times 9.94 = 9.94\text{K}$ so $T_5 = 282.99\text{K} \approx 283$K.

The 273K isotherm is situated just under the first concrete layer, point 4. As this soil is rather humid at this point, the situation is rather dangerous since the isotherms have a tendency to descend over the years. If it were required to compensate for this by increasing the insulation thickness, 1 m of insulation would be needed. If we wanted to keep point 3 above 273K then we need to supplement the heat flow q.

Then considering only the layers above point 3 and allowing for a $dT_{1\text{-}3}$ of 30K, the local q is given by:

$$q = U_{1\text{-}3} \times dT$$

where

$$1/U_{1\text{-}3} = 1/8 + 0.15/1.5 + 0.1/0.037 = 2.93\ (\mathrm{m}^2\ \mathrm{K})/\mathrm{W}$$

Thus

$$U_{1\text{-}3} = 0.34\ \mathrm{W}/(\mathrm{m}^2\ \mathrm{K})$$

and

$$\text{local } q = 0.34 \times 30 = 10.2\ \mathrm{W/m^2}$$

We could reduce this by taking the heat flow of the soil into account. However, as we saw at the beginning of this chapter, this flow is rather small and it is safer to neglect it. Taking into account a reasonable safety factor, a floor heating system should be installed with a capacity of 12–15 $\mathrm{W/m^2}$ immediately below the insulation layer.

10.2.2 Roof Insulation

In contrast to the floor situation, roof insulation is subjected to the highest ambient temperatures, depending on the aspect of the roof, its construction and the reflection of solar energy. This is especially true with prefabricated constructions where the insulation panels are only protected by a thin layer of roofing, or are immediately exposed to solar radiation, so that surface temperatures of the panels become very high.

Very often, as we will see later on, the ceiling panels are protected from direct solar radiation by a roof and an airspace which may be ventilated. Even then the surface temperature can reach high levels, but not to the extent as in the case of direct exposure.

A calculation will provide some idea of the surface temperatures involved. The panel receives an amount of solar energy Q_{tot}, part of which is absorbed and part reflected by radiation and convection, and as a result the surface temperature T_x will increase:

$$Q_{tot} = \alpha_0(T_x - T_0) + U_i(T_x - T_i).$$
$$1/U_i = 1/\alpha_i + \delta/\lambda$$
$$1/U_{tot} = 1/\alpha_i + 1/\alpha_0 + \delta/\lambda$$

So

$$1/U_{tot} = 1/U_i = 1/\alpha_0$$

and thus

$$U_i = U_{tot}\alpha_0/(\alpha_0 - U_{tot})$$

We can find the intensity Q of the solar energy for each day, hour and situation from table 10.3. Depending on the composition of the roof, the penetration of solar energy will be delayed by some hours, figures for which are given in table 10.4.

Since only a part of the solar energy falling on a surface is absorbed, this should be taken into account. Table 10.5 lists values of α', the absorption coefficient, for different surfaces.

The heat flow into the surface will be $\alpha' Q$. Thus:

$$\alpha' Q = \alpha_2(T_x - T_0) + U_i(T_x - T_i)$$

or

$$T_x = (\alpha' Q + \alpha_2 T_0 + U_i T_i)/(\alpha_2 + U_i)$$

Example

$T_0 = +25°C = 298K$ and $T_i = -30°C = 243K$.
Q on July 1st at 14 h is 843 W/m^2 and $\alpha' = 0.9$:

$$\alpha' Q = 0.9 \times 843 = 759 \text{ W/m}^2$$

$$U_i = 0.124 \times 29/(29 - 0.124) = 0.125 \text{ W/(m}^2\text{ K)}$$

$$T_x = (759 + 29 \times 298 + 0.125 \times 243)/(29 + 0.125) = 323.8\text{K} = 51°\text{C}$$

So even if it is only +25°C outside, the roof has a temperature of +51°C. When using insulation material with a relatively high expansion coefficient, one must provide enough space for expansion between the plates, which is filled up with a sealer of flexible insulating material. Imagine a coldstore with an inside temperature of −25°C; in this case the temperature drop across the panel is more than 75K. But be aware that many insulation materials shrink with time, so the expansion space increases after some time; sealant must therefore be pre-pressed and maintenance staff will need to observe the situation very closely and eventually take measures if necessary.

10.2.3 Gas-tight Insulation

Special insulation techniques are necessary in the case of coldstores that provide a controlled atmosphere as, for example, in the storing of fruit and vegetables. The insulation must not only prevent thermal energy leakage and water vapour diffusion, but also gas leakage. In addition to a vapour barrier, a gas barrier must be provided.

Table 10.3 Intensity of solar radiation (W/m^2) on walls and roofs in northern Europe

Time of day / *Date*	*0400* *2000*	*0500* *1900*	*0600* *1800*	*0700* *1700*	*0800* *1600*	*0900* *1500*	*1000* *1400*	*1100* *1300*	*1200* *1200*	*1300* *1100*	*1400* *1000*	*1500* *0900*	*1600* *0800*	*1700** *0700†*	*Day total*	*Surface*	
I Jan						48	130	185	204						930	FLAT ROOF	
1 May		3	94	238	393	534	654	733	762						6071	FLAT ROOF	
1 July		49	170	309	456	602	725	812	843						7059	FLAT ROOF	
1 Sept			29	150	293	433	545	623	651						4757	FLAT ROOF	
1 Jan						306	533	638	670						3675	VERTICAL WALLS	SOUTH FACING
1 May					159	300	423	506	534						3326	VERTICAL WALLS	SOUTH FACING
1 July					72	207	317	398	428						2396	VERTICAL WALLS	SOUTH FACING
1 Sept				57	192	335	461	548	577						3745	VERTICAL WALLS	SOUTH FACING
1 Jan						194	379	479	512						2652	INCLINED ROOF	SOUTH FACING
1 May			43	207	420	612	778	887	927						6780	INCLINED ROOF	SOUTH FACING
1 July			65	241	430	625	785	901	943						6990	INCLINED ROOF	SOUTH FACING
1 Sept			12	158	350	542	704	814	854						6036	INCLINED ROOF	SOUTH FACING

Time of day / Date	0400 2000	0500 1900	0600 1800	0700 1700	0800 1600	0900 1500	1000 1400	1100 1300	1200 1200	1300 1100	1400 1000	1500 0900	1600 0800	1700* 0700†	Day total	Surface
1 Jan						404	577	566	473	336	177				2582	Vertical walls, SE and SW facing
1 May		42	270	437	570	616	604	520	378	195					3629	
1 July		100	248	372	481	530	513	437	301	124					3070	
1 Sept			159	407	548	618	613	543	408	231	38				5559	
1 Jan						242	401	444	414	328	201	21			2082	Inclined roof, SE and SW facing
1 May		24	216	426	625	770	868	894	848	743	569	366	169		6606	
1 July		92	271	454	634	786	884	921	882	765	594	401	205	28	6920	
1 Sept			105	334	528	683	778	812	769	656	492	301	115		5559	
1 Jan						265	284	164							735	Vertical walls, East and West facing
1 May		98	458	619	645	571	430	230							3128	
1 July	2	291	513	601	608	544	409	222							3233	
1 Sept			254	519	584	538	407	220							2466	
1 Jan						173	255	242	177	78					930	Inclined roof, East and West facing
1 May		52	315	504	663	748	782	749	659	520	351	176	17		5513	
1 July	1	188	405	568	698	793	832	814	730	592	422	250	91		6397	
1 Sept			151	413	545	643	676	650	544	430	270	105			4373	

Date / Time of day	0400 2000	0500 1900	0600 1800	0700 1700	0800 1600	0900 1500	1000 1400	1100 1300	1200 1200	1300 1100	1400 1000	1500 0900	1600 0800	1700* 0700†	Day total	Surface	
1 Jan																VERTICAL WALLS	NE AND NW FACING
1 May		95	379	437	345	193	5								1465		
1 July	3	313	478	479	379	240	64								1965		
1 Sept			199	327	277	144									933		
1 Jan						27	16								43	INCLINED ROOF	
1 May		51	271	424	512	558	558	536	471	374	265	155	55		4245		
1 July	2	199	386	507	584	641	658	640	578	484	370	256	155	81	5582		
1 Sept			124	293	392	447	452	424	361	269	165	65			2966		
1 Jan																VERTICAL WALLS	NORTH FACING
1 May		37	78	0											247		
1 July	2	151	163	53											785		
1 Sept			28												38		
1 Jan																INCLINED ROOF	
1 May		22	121	207	261	312	355	381	393						3687		
1 July	1	117	229	295	358	416	468	504	516						5257		
1 Sept			38	102	158	207	242	266	276						2303		

*Top row of times of day applies to S, SE, E, NE and N facing surfaces.
† Bottom row of times of day applies to S, SW, W, NW and N facing surfaces.

Table 10.4 Heat gain time lag of building materials in hours

Material	*Time lag*
Flat roof	2
Stone, 22 cm	2
Concrete, 15 cm	3
Wood, 5 cm	1.5

Table 10.5 Absorption coefficient α′ for surface materials

Material	*Absorption coefficient*
Black surface	0.9–1
White surface	0.3
Aluminium sheet (rough)	0.4
Aluminium sheet (painted)	0.25–0.3
Polished aluminium	0.1
Red, yellow and green paint	0.5–0.55
Blue paint	0.7
Asbestos concrete (new)	0.42
Asbestos concrete (grey)	0.71
White coated stone	0.26
Natural stone	0.76

By using prefabricated sandwich panels of 60–120 mm PIR covered with metal sheets up to a thickness of 0.6 mm, one part of the problem is solved, because metal or polyester sheets are gas-tight. The problems of gas-tightness are confined to joints and to non-prefabricated, traditionally insulated floors. Joints should be carefully filled up with gas-tight silicone kits and a layer of 100 mm fibre glass tape. Every passage of pipework and cables should also be carefully sealed with silicone. Doors should have special gaskets that can be expanded with compressed air. The floor, which should preferably be insulated with 6 cm foam glass or PUR, should be provided with a gas and vapour barrier of special gas-tight bitumen 110/30 with a thickness of 2.5 mm. In corners connecting with the walls, a gas-tight polyvinyl film of 0.75 mm should be installed up to a sufficient height and finished off with silicone.

Afterwards the rooms are put under an overpressure of about 1332 Pa. A pressure gauge is then used to measure the time it takes for the pressure to fall to about 933 Pa. If this takes more than 10 minutes, the room can be considered to be gas-tight. Figure 10.10 shows degrees of acceptability for various leakage rates.

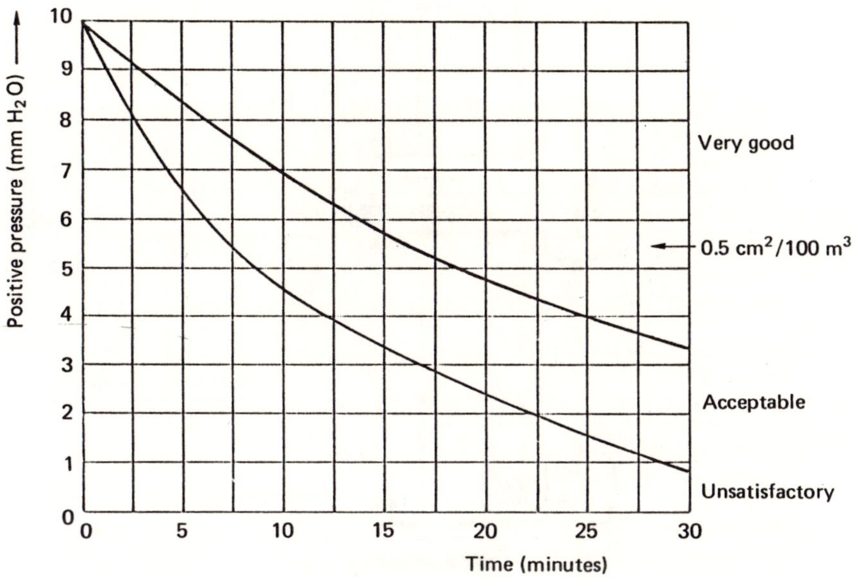

Figure 10.10 *Pressure/time testing curves for CA storage*

Sometimes small smoke-bombs are also used for leak detection. A leakage is measured in $cm^2/100\ m^3$. A leakage of 0.2 $cm^2/100\ m^3$ is permissible.

10.2.4 Overpressure Protection

When the difference in pressure inside and outside a coldstore is 1333 Pa for instance, there is in fact a pressure of 1333 Pa on the walls. Such pressures on the construction must be avoided. In a gas-tight coldstore and even in normal prefabricated constructions there are no normal means of equalizing air in order to deal with differences in air pressure between the inside and outside atmosphere. These pressure differences, especially when starting up the plant, may be very big indeed.

Therefore, when starting up a plant it is necessary to keep the door slightly open at the beginning. If this is not done the pressure on the walls may exceed the permitted construction tolerances. For normal operation of the plant, sufficient equalizing valves must be installed. Typical examples are shown on figure 10.11a, b and c. Electric heaters must be incorporated in valves where the coldstore temperature goes below 273K.

We give below two causes of pressure differences in coldstores.

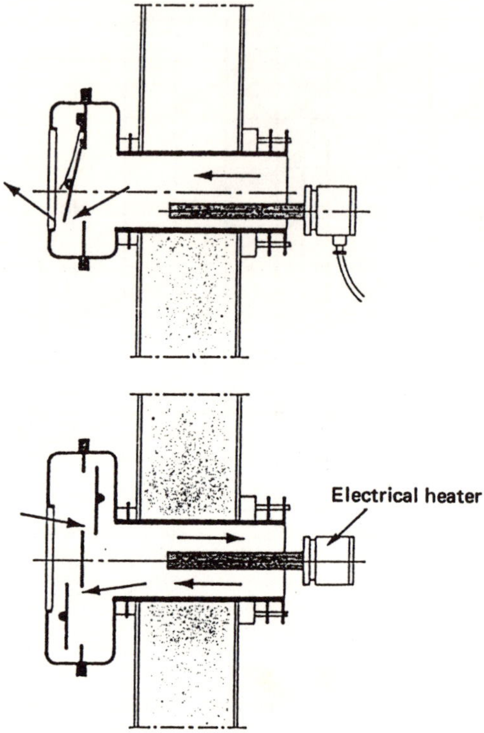

(a)

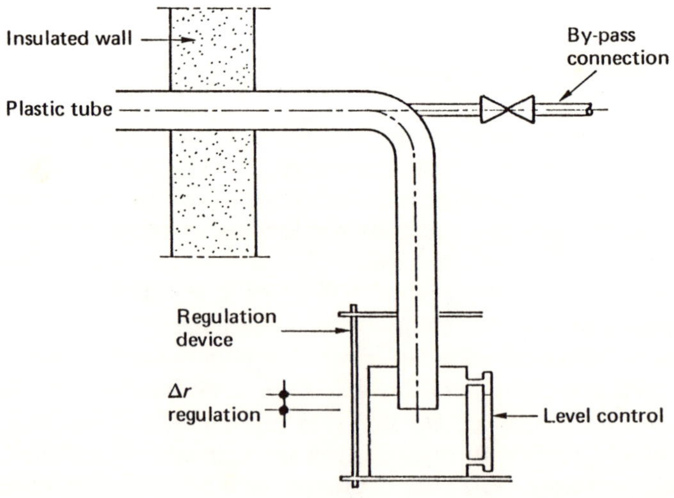

(b)

Figure 10.11 *(a), (b) and (c) Pressure-equalizing devices. (d) Pressure/flowrate characteristics of a typical pressure-equalizing device*

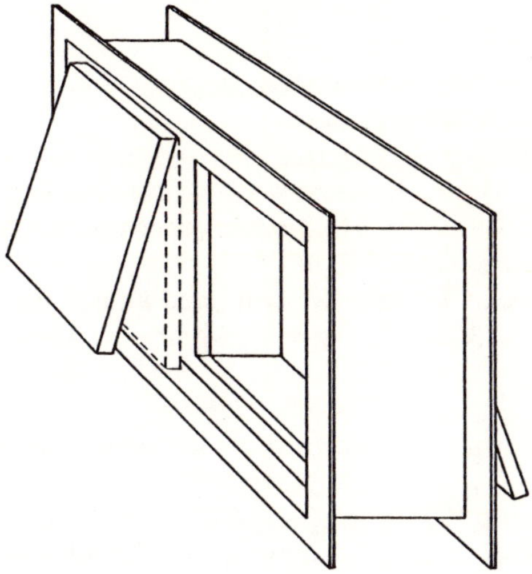

(c)

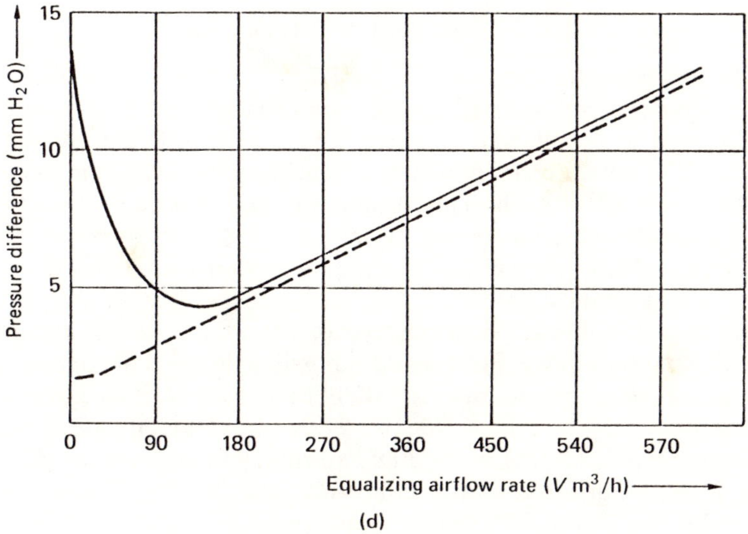
Pressure difference (mm H_2O)
15
10
5
0
90
180
270
360
450
540
570
Equalizing airflow rate (V m^3/h)

(d)

Example 1:

Suppose that the outside atmospheric pressure changes from 97.3 kPa to 101.3 kPa over a 12 hour period, and assume a coldstore volume of 8800 m^3 at +4°C, 90% RH, with outside conditions of +40°C, 60% RH or +19°C (80% RH). Under these new conditions the apparent air volume would be 8800(101.3/97.3) = 9162 m^3, in other words the volume increase of air in the coldstore $dV = 362\ m^3$.

At an air velocity of 1.5 m/s, we will need an equilibration tube with a section of 362/(12 × 3600 × 1.5) = 0.056 m^2 or a 4-inch tube.

Example 2: Pull-down of a coldstore avoiding open doors

Ambient temperature +30°C (303K). Coldstore final temperature +4°C (277K). A dT of 26K and a pull-down time of 6.5 h results in a dT/dt of 4K/h or 1K per 15 minutes. Atmospheric pressure is 103.320 kPa.

With a temperature at start-up T_1 = 303K, and a temperature after one hour T_2 = 299K, then the internal pressure becomes:

$$103.320(299/303) = 101.960 \text{ kPa}$$

resulting in a pressure drop dP of 1360 Pa.

So the apparent volume will be decreased to 8800(299/303) = 8684 m^3, and will need to be allowed to enter the store in order to equalize the pressures. The resulting dV/dt will be 116 m^3/h or at outside air conditions 116(303/299) = 121.4 m^3/h. This volume flowrate can be provided either by 3 tubes of 4-inch diameter or by the special devices shown in figures 10.11a and b. These devices are based on a certain allowable pressure on the panels of normally 300 Pa.

In any case it is always better not to cool down the coldstore too quickly during the first pull-down. However it is possible during daily operation for a change of temperature rate of 4K/h to occur. If, for example the temperature changes from 276K to 272K then the apparent volume changes to 8800(272/276) = 8672 m^3, resulting in a value of dV/dt of 128 m^3/h. With outside conditions of +40°C or 313K, then outside air at a rate of 128(313/272) = 147.3 m^3/h should enter.

The French engineers Professor Duminil and M. Patin have studied the problem. Professor Duminil proposed the equation below:

$$dV/dt = T_o(V_1/[T_s]^2)(p_1/p_o)(dT_s/dt)$$

where s denotes store and o outside.

M. Patin simplified this by eliminating the term (p_1/p_o) because a pressure difference of 300 Pa is allowed and under these conditions (p_1/p_o) = 103 620/103 320 = 1.002. He suggested a normal temperature change of 4K/h as we used above.

The equation now becomes dV/dt in $m^3/h = 4T_o(V_1/[T_s]^2)$.

In our example: $4 \times 313 \times 8800/276^2 = 144.6$ m^3/h, which compares favourably with our solution above.

It is good practice to allow some safety tolerances and allow for lower pressure differences across the walls.

10.2.5 Practical Guidelines to Coldstore Dimensions

Local circumstances sometimes make it necessary to use traditional insulation techniques by applying insulation slabs on to stone or concrete coldstore walls. In each case, as already mentioned above, floor insulation is generally applied in this traditional way.

So before dealing with the subject of modern prefabricated constructions, we will consider the traditional methods of fixing insulation slabs and vapour barriers.

Traditional insulation installation involves 100% manual application and, as on the building site, quality control is difficult; the results depend much on the local conditions, the skill of the operators and the quality of the materials used. The operation will also be a lengthy one. It is for these reasons that today most installations are built from prefabricated insulation panels.

The best way to apply traditional insulation is as follows:

Insulation material

- Extruded polystyrene, $\geqslant 25$ kg/m^3 density, $\lambda = 0.03$ W/(m K), and mechanical resistance to 0.75bar; or
- Polyurethane $\geqslant 30$ kg/m^3 density, $\lambda = 0.023$ W/(m K), and mechanical resistance to 0.35 bar.

Linear expansion coefficient: both about 7×10^{-5} cm/(cm °C).

Refer also to the other insulation material characteristics mentioned above.

The floor insulation should have adequate mechanical strength, and before placing the final concrete slab, the insulation must be covered with a layer of polyethylene film to prevent water from the wet concrete penetrating the insulation. The final concrete slab or finishing slab must have the following characteristics:

- mechanical strength capable of bearing the static and dynamic loads
- suitable expansion joints
- no smell, no dust and ease of cleaning for hygienic reasons
- non-flammability
- slip resistance for safety reasons
- durability.

In small rooms and stores in which the product hangs on rails, as in the case of meat, the floor can be made of a lighter construction, such as stone. On spots where point loads will occur, caused for instance by the use of steel beams, the load must be spread by placing an insulation material locally with a high resistance against pressure. Also, locally reinforced concrete may be used. See figure 10.12 for a section through a typical floor and the effects of a severe point load. Nowadays, better precautions must be taken, as the loads are becoming greater, with piles of stacked product becoming increasingly high, thereby adding to the load per m^2. The insulation must have adequate mechanical strength. Foamed glass, cork, extruded polystyrene and polyurethane (PUR) are the best materials; expanded polystyrene must be of high density, of more than the normal 25 kg/m^3.

When using pallets, the load is relatively well spread. When stacking 3 or 4 pallets high, loads of up to 0.35 kg/cm^2 may occur. In palletized stores of

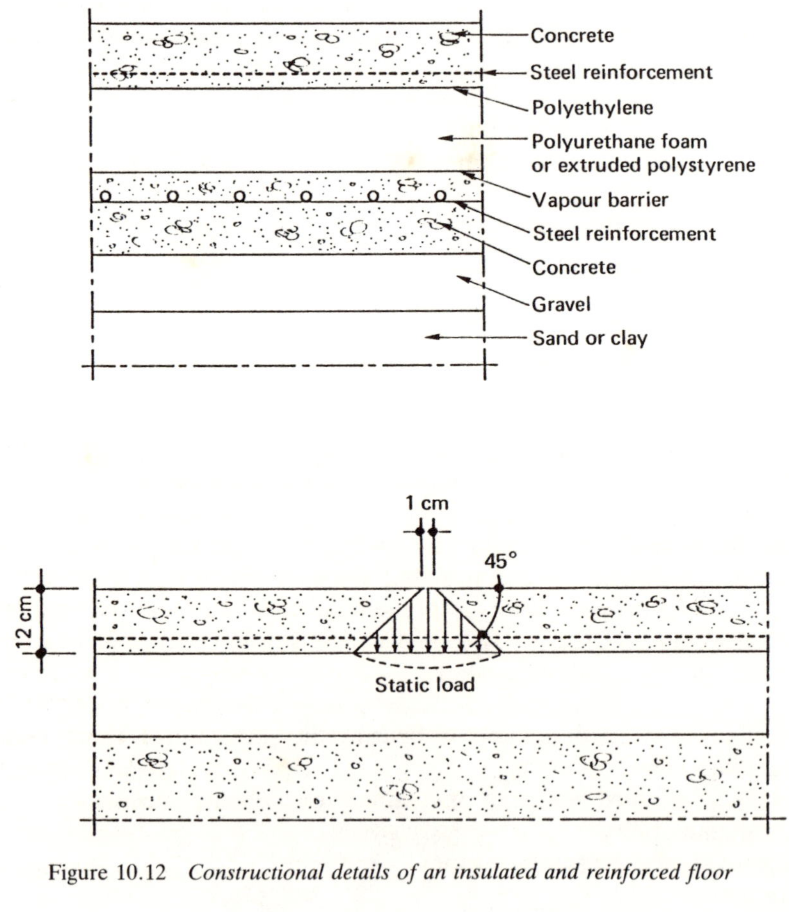

Figure 10.12 *Constructional details of an insulated and reinforced floor*

6 m height, loadings of between 2 and 3 tonne/m^2 may occur. In addition we have to take into account the dynamic loading of the forklift and pallet trucks. Forklift trucks may add to these figures an additional loading of 1200–1500 kg per wheel. A forklift truck may have a total weight of 6–8 tons, spread over 4 wheels, with a small contact area. As the load is transferred to the ground at an angle of 45°, a 2000 kg wheel load imposes a 1.8 bar load on the insulation, as shown in figure 10.13, when using a concrete slab of 12 cm thick. The thicker this slab, the larger is the loaded surface on the insulation. As 1.8 bar is unacceptable for any kind of insulation steel reinforcement or a thicker concrete slab must be used. There is also the friction load of the wheels to be considered. Because of the movement of the wheels, the insulation is subjected to a wave movement with a small amplitude, resulting in a local bowing of the concrete slab as illustrated in figure 10.14. In order to obviate this

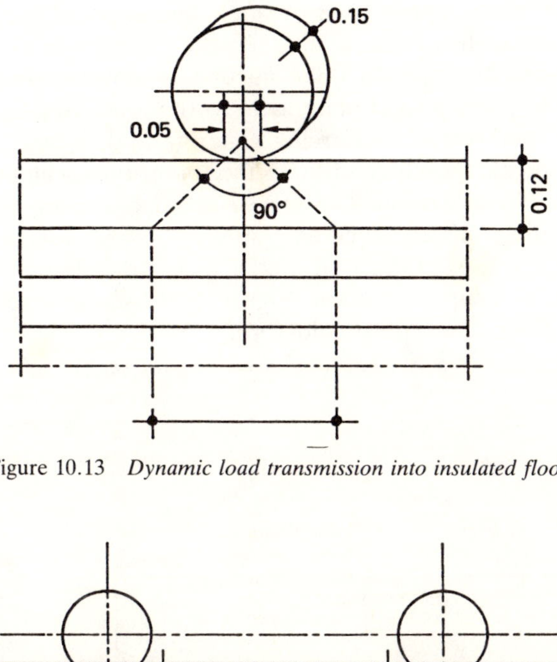

Figure 10.13 *Dynamic load transmission into insulated floor*

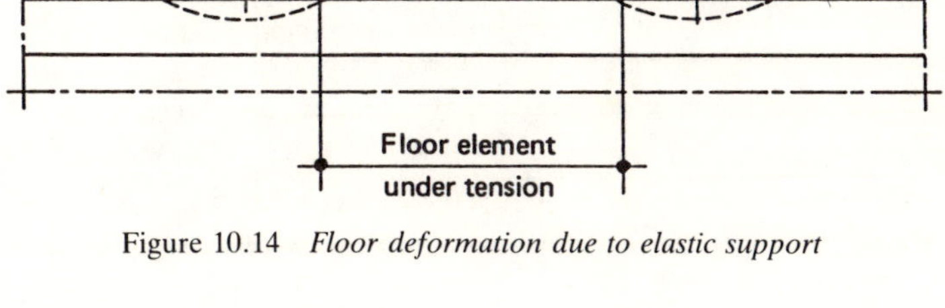

Figure 10.14 *Floor deformation due to elastic support*

problem, the reinforcement must be set in the upper-side of the concrete slab. An insulant with little tendency to wave movement should be used; the maximum deformation allowed is 1–1.5% or 3 mm on 20 cm. This may lead to expensive insulation material, so steel reinforcement of the concrete slab is a cheaper solution. The lower layer of reinforcement should be placed 2 cm above the insulation surface, and the upper layer of reinforcement 2 cm below the concrete slab surface. The upper reinforcement layer is lighter then the lower one. There are also forces due to the settling of the building during start-up. In freezer stores, temperature differences can reach 50°C and may cause the concrete slab to contract quite a few centimetres.

At low temperatures, concrete has a very low elasticity. To calculate the exact details of the steel reinforcement, a specialist in this field must be employed since this work is not the job of a refrigeration engineer.

Between two slabs of concrete an expansion joint must be provided every 5 × 5 m. In most cases it is enough to make a joint in the upper layer and only interrupt the upper layer of steel. The edges of the joint must be protected mechanically against damage by the store traffic. The joints must be closed with an elastic product like silicone as shown in figure 10.15. It is good practice to place a steel plate local to the joint. See figure 10.16 for this and other methods of reinforcing concrete floor joints.

In small rooms the polythene film in the top of the insulation provides a means for possible expansion. This can be avoided, if necessary, by adding a layer of a sliding material such as graphite.

The final concrete slab must be mixed with a thin layer of small metal particles as an anti-slip protection. Figure 10.17 shows a section through the walls and floor of a typical coldstore.

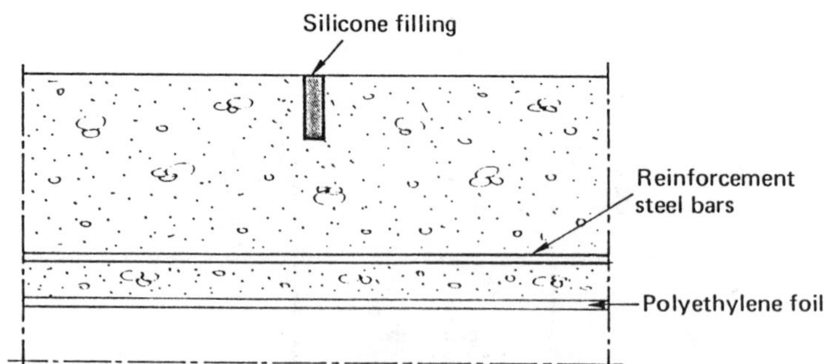

Figure 10.15 *Reinforced floor showing expansion joint and polythene*

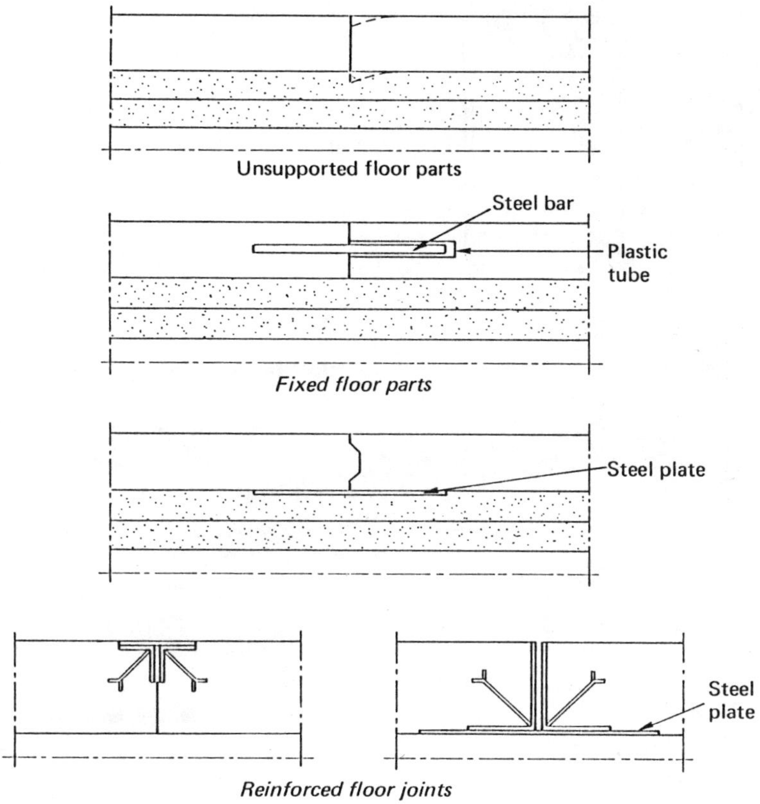

Figure 10.16 *Typical joint reinforcement methods for coldstore floor*

Ceiling and Wall Insulation

After placing the vapour barrier, the first layer of insulation is fixed to the ceiling or wall by means of mechanical devices such as steel pins with galvanized wire and plates. The second and subsequent layers are fixed to the previous layer by means of wooden pegs inserted at an angle using at least 8 pegs per m. A non-clay-based bituminous emulsion is used as the adhesive between the layers.

The ceiling is made of wooden beams placed 50 cm apart, calculated on a sag of 1/500.

All insulation can be covered with epoxy paint, plaster, aluminium or other sheets. A one layer brick wall is a very effective but also very expensive method of protection. As a vapour barrier roofing felt, bitumenized paper of 3 mm thick or alu-kraft-paper, reinforced with fibreglass, may all be used. Polyethylene film can be used for the floor, but it must have a thickness of at least 0.2 mm.

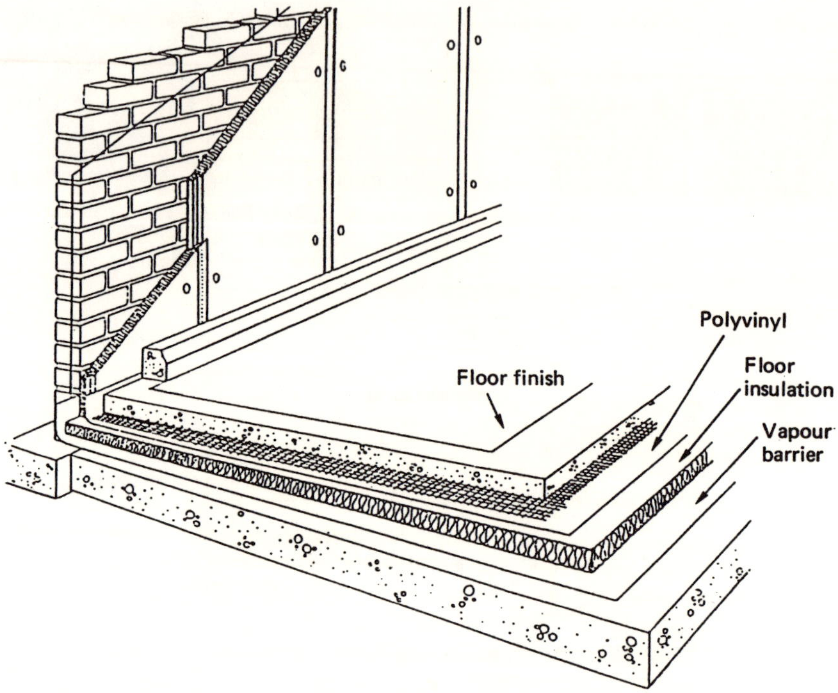

Figure 10.17 *Section through wall and floor of insulated coldroom*

Prefabricated Construction

A steel frame is built to which the prefabricated walls and ceiling panels are fixed. See figures 10.18 to 10.22 for details of the two methods of assembly.

The methods of assembly are:

(a) Placing the panels outside the steel structure. In this method the steel structure will be inside the cold room and can be used to fix piping, air-coolers etc. Solar radiation falls directly on the panel surface, creating great temperature differences across the outside and the inside surfaces of the panels. This method of construction is called the North European method.

(b) Placing the panels inside the steel structure. The walls and ceiling inside are smooth; thermal bridges are inevitable if we want to fix supports for the pipework of air-coolers which can be limited to small nylon bars, passing through the panels.

Steel structures inside the store, placed on the floor, may support air-coolers and pipework. Outside the store on the steel frame of the

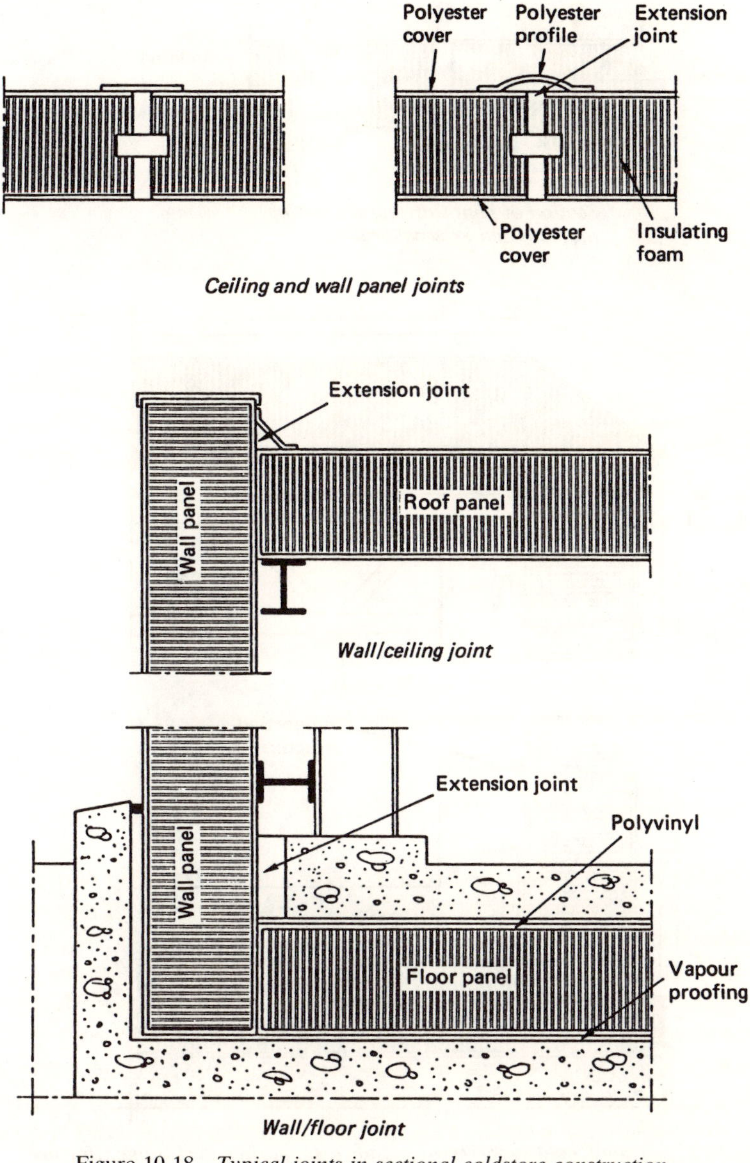

Figure 10.18 *Typical joints in sectional coldstore construction*

building, metal cladding is fixed in order to protect the walls and ceiling from solar energy penetration, cladding that can be chosen to meet local conditions. This construction method is the South European method and is to be advised for warm climates.

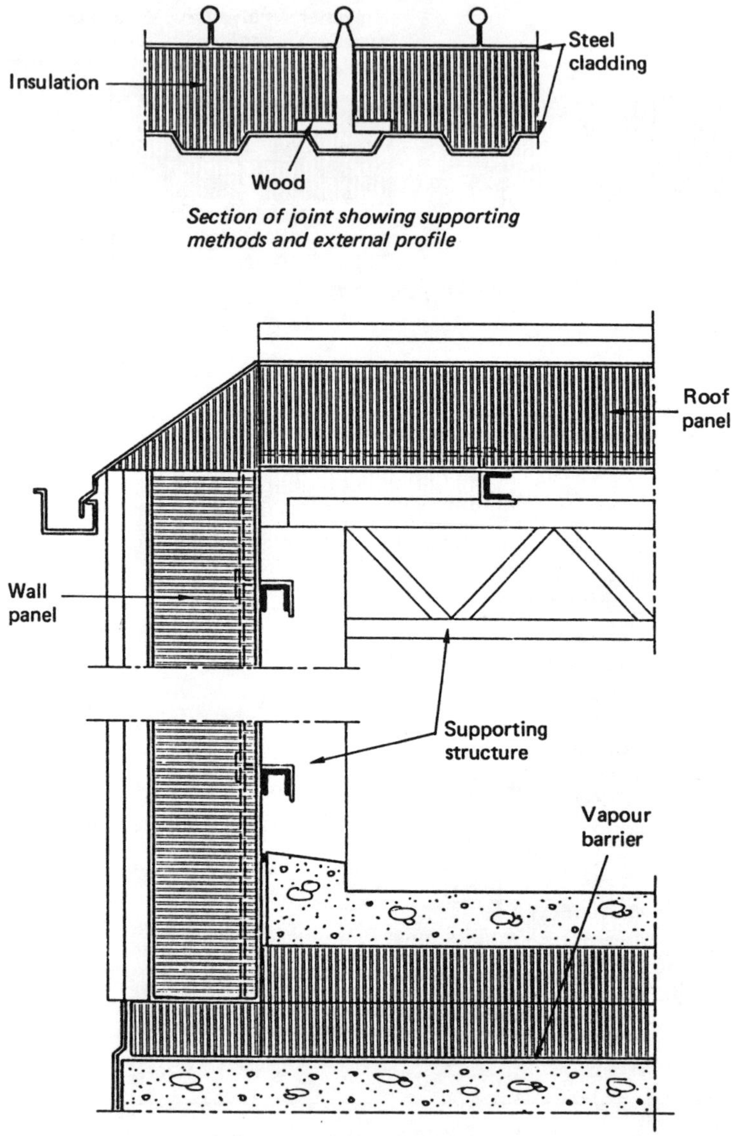

Figure 10.19 *Sectional coldstore erected outside a supportive steel-frame structure*

The South European method has, however, the disadvantage that the frame is subjected to expansion forces, which could be transferred to the insulation panels.

There are several dimensions of panels to choose from. With long panels there are more joints but they are more difficult to handle than short ones.

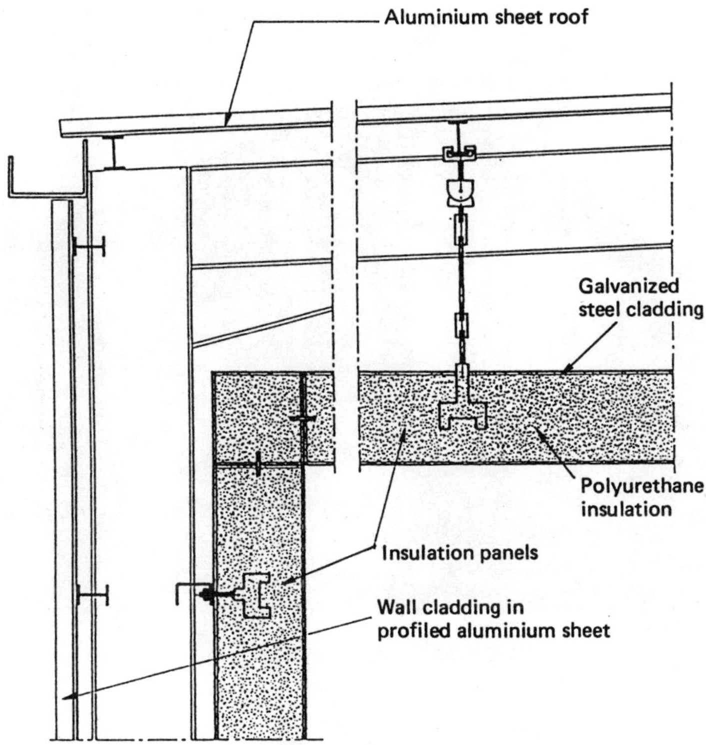

Figure 10.20 *Inside weather-proof building, showing suspension methods for wall and ceiling panels*

When making a choice between the many available types of panels, certain details must be carefully considered:

- the method of fixing them to the steel frame must avoid thermal bridges as much as possible, but still ensure that the panels are firmly held to the frame;
- the panels must be of adequate mechanical strength and must have a protective layer against corrosion;
- no wood should be used inside the sandwich panel, for deformation reasons as well as for fire resistance;
- the interconnection of the panels must be firm and solid.

Mechanical systems, such as those using steel or plastic fixing devices together with neoprene joints are the most reliable. The systems must not create thermal bridges; when not using mechanical connection systems, male–female connections should at least be used. Injection of polyurethane into the joints is not very effective as there is no control of the quality of the injection procedure: it could be inspected by means of infra-red cameras, but repair work after application is difficult.

Figure 10.21 *Sectional coldstore prior to final cladding*

Figure 10.22 *Erection of coldstore ceiling panels*

Coldstore Dimensions

In order to determine the dimensions of a coldstore and choose the type of construction, we need the following data:

- the kind of product to be stored – how it will be packed and stacked
- how much will be stored
- the cost of the land to be built on
- the availability of handling equipment; failing that, the availability of manual labour
- the rotation time for the goods
- the number and variety of products to be stored; for instance, only frozen products, only ice-creams, only fruit, or are there different kinds of products to store, subject to different storage conditions
- the standard dimensions of the insulation materials used
- the type of internal ventilation system required
- the nature of any special local conditions, such as aesthetic design requirements or rules, fire-protection, soil conditions, etc.

To determine the dimensions, the data given in tables 10.6, 10.7, 10.8 and 10.9 may be used to calculate the exact floor area; to this, space for corridors, for equipment and air-circulations must be added. Products must stand at least 30 cm from walls and ceilings. The gross volume should be 36% higher than the calculated net volume.

Empirical figures are available for a multipurpose coldstore: 200 tons per 1000 m^3, using stacks of 3 pallets of 4.5–5 m height; about 160 tons per 1000 m^3 for chilled products only; for frozen products about 300 tons per 1000 m^3; and about 225 tons per 1000 m^3 for apples and pears in stores with controlled atmosphere of about 6–7 m height.

Once we know the volume of the coldstore we need to determine the dimensions. As land is expensive everywhere, it might be more practical to build as high as possible rather than a building which spreads over a larger area.

But of course we have to consider the conditions of the type of soil we would like to build on; is there a chance the ground will move under the weight of the building and its contents, etc.? The height of the coldstore depends on the stacking method chosen.

For example if meat is to be hung on rails, the height of the coldstore is determined by the height of the rail system plus sufficient space for the air-cooling system. For pigs, calves, cattle quarters, goats, sheep and other smaller animals, the rail height is 2.3–2.5 m, the distance between them 0.65 m. Five pig-halves may be hung on each metre of rail. The rail height for beef is 3.3–3.7 m, the distance between the rails 1.2–1.5 m. It is permissible to hang 3 beef-quarters, 1–1.5 carcasses, 2–3 halves or 2–3

Table 10.6 Food storage conditions

Product	*Packing state*	*Mass (kg)*	*Daily rotation*	*Store temp. (°C)*	*RH*	*Storage time*
Fish	Covered with ice	300	20%	0/+2	90/100%	5–10 days
Fine fish	Covered with ice	500	Storage	−14	100%	2–3 months
Dough for bread	Baskets or plates	100	100%	+6/+8	80/85%	8 hours
Pastry	Plates	30	100%	+6/+8	80/85%	1–2 days
Beer	Barrels	3	30%	+5/+7	100%	a few days
Strawberries	Baskets	200	20%	+0.5/0	80/90%	8–10 days
Raspberries	Baskets	200	20%	+0.5/0	80/90%	8–10 days
Plums	Baskets	200	20%	+0.5/0	80/90%	8–10 days
Pears	Baskets or cases	200	Storage	+0.5/0	90/95%	1–3 months
Apples	Baskets or cases	300	Storage	+0.5|0	90/95%	3–6 months
Apricots	Baskets or cases	200	Storage	0/+2	90/95%	3–4 weeks
Peaches	Baskets or cases	250	Storage	0/+2	90/95%	15 days

calves on each metre of rail. A slaughtered cow normally weighs about 400 kg, a pig about 80–100 kg, a calf 45 kg, and a sheep 25–35 kg.

A greater height also has the advantage of minimizing the outside surface:volume ratio, which results in less heat transfer by walls, ceilings and floors, and reduced insulation costs.

A volume of 6400 m^3 at 8 m height has a 3560 m^2 surface. The same volume at 4 m height results in a 4000 m^2 surface. Where the floor is not insulated the difference in insulation surface is 1760 m^2 compared with 2400 m^2. To have a square floor would be most economical but would be difficult for stacking and handling of the product. A height of 15 m is the present economic limit; however, coldstores can be made higher using fully computer-controlled automatic stacking systems. Such systems only apply

Table 10.7 Products on pallets 120 × 100 cm

Product	*Packing*	*Dimensions (cm) and gross weight (kg/m)*	*Weight of products on pallet (kg/m)*	*Number of pallets*	*Staple height (m) and staple weight (kg/m)*	*Charge per m^2 (kg/m)*	*Charge per m^3 (kg/m^3)*	*Note*
Meat	Wooden cases	64 × 36 × 19 27.3 kg	955	3	4.5 m 2865 kg	2000	445	
Frozen meat	Cardboard box with 3 plates CEGF*;	52 × 35 × 22 28.5 kg	800	4	4.15 m 3200 kg	2300	550	Cardboard between 2 layers of tin cans
	Square tin 20 kg	25 × 25 × 35 11 kg	1000	3+	4.1 m 3300 kg	2450	600	
Fresh eggs	Canadian cases	70 × 35 × 40 27 kg	430	3	4.85 m 1180 kg	875	180	Basic weight per egg, 75g
Butter	Cubic cases	35 × 38 × 35 29 kg	780	3	3.6 m 2340 kg	1740	485	Cardboard between the baskets
	Baskets	10.3 kg	820	3	4 m 2460 kg	1760	440	
Frozen fish	Per 4 cases	24 × 32 × 25 11 kg	770	3	4.2 m 2310 kg	1710	420	—

(*continued overleaf*)

Table 10.7 (*continued*)

Product	*Packing*	*Dimensions (cm) and gross weight (kg/m)*	*Weight of products on pallet (kg/m)*	*Number of pallets*	*Staple height (m) and staple weight (kg/m)*	*Charge per m^2 (kg/m)*	*Charge per m^3 (kg/m^3)*	*Note*
Frozen fruit & vegetables	Cardboard box with 3 plates CEGF*;	52 × 35 × 22 27 kg	750	4	4.2 m 3000 kg	2150	510	Limited height because of nature of products
	Cardboard box for peas	38 × 30 × 19 16 kg	800	3	3.3 m 2400 kg	1780	540	
Fresh oranges		66 × 31 × 31 37 kg	815	2 + 1 palette of 18	4.8 m 2150 kg	1590	330	As above
Fresh apples	Cases	57 × 34 × 22 16.5 kg	600	3	4.4 m 1800 kg	1330	300	As above
Fresh peaches	Plates	57 × 34 × 8 5 kg	360	3	3.3 m 1800 kg	800	245	As above

*CEGF = French packing standard.

Table 10.8 Floor loading for food in coldstorage

Products	*Loading (kg/m²)*	*Packing methods*
Cows and oxen		
complete	550–600	Pile
half	600–650	Pile
quarter	600–650	Piles of 10–15 layers
Pork	500–550	Piles of 5–7 layers
Bacon	500–550	Piles of 5–6 layers
Canned meat	600–700	Cases with 5–6 layers
Salt meat	400–800	Barrels of 100–250 kg
Salt bowels	900–1000	Barrels of 250 kg, in 2 layers
Cooled fish	350	Cases, covered with crushed ice
Frozen fish	550–650	Between partitions
Caviar, granular	600–700	Cases
Caviar, pressed	600–800	Cases
Frozen poultry	300–400	Barrels
Eggs	550–600	Cases
Butter	650–700	Cases
Margarine	750–800	Barrels or cases
Soft cheese	750–800	Barrels or cases
Hard cheese	500–600	On racks
Milk	600–750	Bottles or pitchers
Fruits	300–400	On racks, unpacked
Strawberries	500–600	Cases
Vegetables	250–400	Baskets
	500–700	

Table 10.9 Storage on rails for meat carcasses

Product	*Distance between rails (m)*	*Capacity per m²*	*Capacity of floor surface (kg/m²)*
Cows and oxen			
(300 kg) on carts on	1	2 or 3 half-cows	250–300
bi-rail, on hooks	1.2	2 half-cows per 0.8 m	250–300
Calves (70 kg)	0.7–1	3 calves	180–200
Pork (80 kg)	0.6	4 half-pigs	200
Sheep (20 kg)			
4 hooks	1	4 sheep	70
8 hooks	1.2	8 sheep	120

to volumes of above 75 000 m^3, whereas standard volumes are those between 5000 m^3 and 25 000m^3.

There are international standard pallet sizes; the most frequently used have the dimensions of 100 × 120 mm, 120 × 80 cm and 200 × 100 cm. Alternative dimensions are 80 × 100 cm and 120 × 120 cm. 100 × 120 cm is the standard size pallet, which is now being used more and more.

The height of goods stacked on a pallet varies between 1 m and 1.77 m for different products. The pallet load also varies for different products and height of stacking, and ranges from 750 to 1000 kg.

Stacking Methods

From a volume/surface point of view, block-stacking is the most successful, although the maximum height is about 3 pallets. Height can be increased using special pallets and precise stacking methods. Block stacking can only be used under the following conditions:

(a) Where only the same kind of goods are stored, making corridors unnecessary, densities of 400 tonne/1000 m^3 can be achieved.
(b) Where only a few different types of goods are to be stored, and so one or two corridors are sufficient, goods can be placed in deep rows on both sides of the corridors.
(c) There is no need for the first-in first-out system; otherwise double the number of corridors is necessary.
(d) Fresh goods are brought in and the stored ones are taken out in one operation.

Block stacking, which requires an internal height of about 6 m, is still the most widely used; it can achieve an optimum when using static racking as shown in figure 10.23 or even drive-through racks as shown in figure 10.24. If block stacking is not possible for all the stock, it can be used for one part, for which the first-in first-out method is absolutely necessary. This way, a partial stock rotation is obtained.

When using racks, the stacking height can go up to 12 m which means a store height of 15 m. This, however, demands special forklift trucks, such as turret-trucks; and preferably those able to turn the forks in different directions as shown in figure 10.25. When using stacker-cranes a height of 35 m, 18 pallets, can be reached, but then the investment for cranes, rails and building is rather high. This system gives little gain in volume, but much gain in floorspace

Drive-through racks permit stacking up to 20 rows deep. However, it will often be necessary to provide several corridors in order to be able to reach the different products; in that case the choice should be for forklift trucks using narrow corridors.

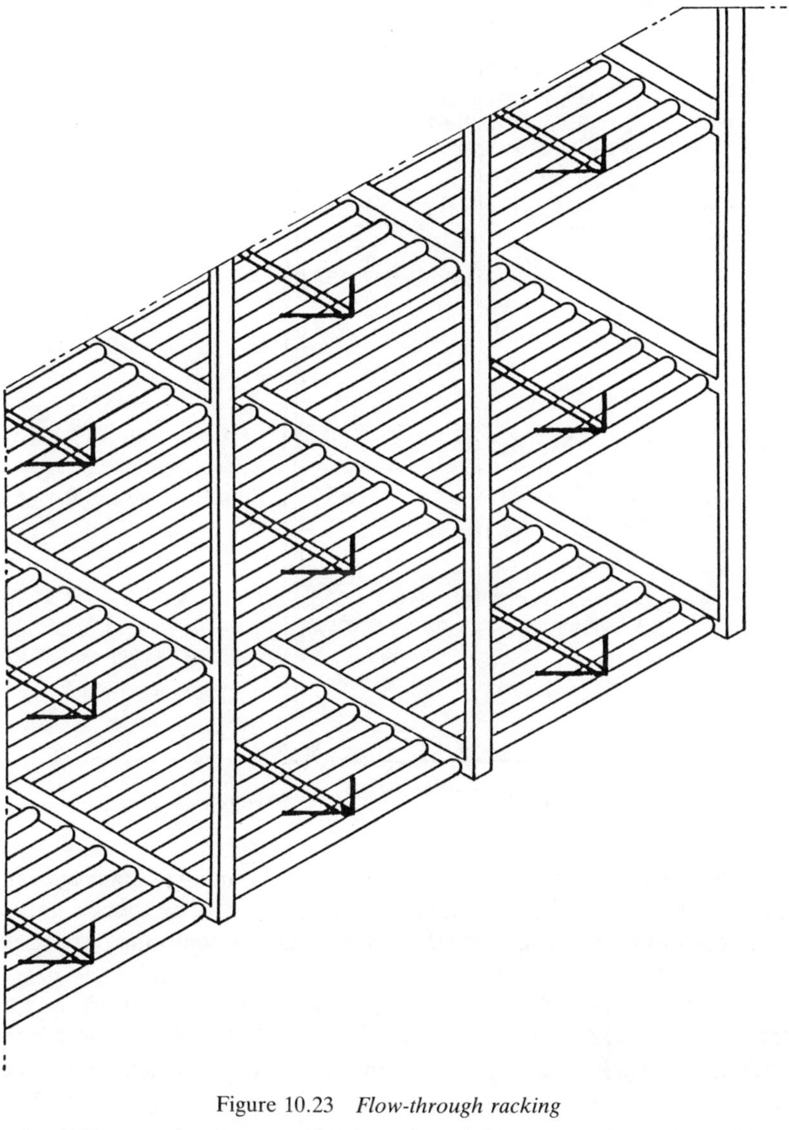

Figure 10.23 *Flow-through racking*

A normal forklift needs a corridor of 3.5 m, a so-called reach-truck needs 2.4 m, whereas a tri-directional turret-truck only needs 1.65 m and is able to reach as far as 7 pallets high.

The truck does not need rails, just a simple system between the racks, and it can work in different corridors. This type of truck is efficient in long corridors (about 15–20 m) and with piles of at least 4 m height.

A stacker crane needs a corridor of 1.4 m; it works on rails and is not able to work in different corridors, so each corridor needs its own crane,

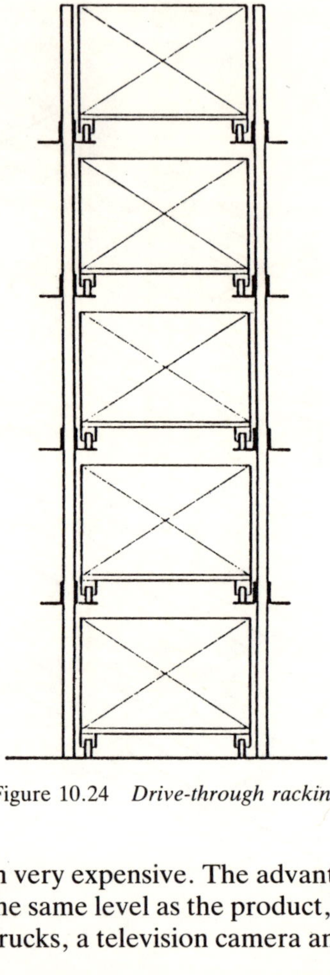

Figure 10.24 *Drive-through racking*

which makes the system very expensive. The advantage of stacker cranes is that the operator is at the same level as the product, so he sees exactly what he is doing. On turret-trucks, a television camera and monitor are required for that purpose.

In order to reduce investment costs, a combination of forklift, reach and pallet trucks is often normal practice. For each forklift truck there must be 1 or 2 pallet trucks. Another combination is turret-trucks and stacker-cranes.

A more efficient block system is the mobile rack system; by using mobile rack systems one can utilize 80–90% of the floorsurface. This is double the efficiency rate when stationary racks are used, in which case there must be a corridor between every two racks, whereas normally only one corridor is needed for the whole store. See figures 10.26, 10.27 and 10.28. The width of the corridor also depends here on the equipment used, and since we have only one corridor we can design it somewhat larger and so make handling easier.

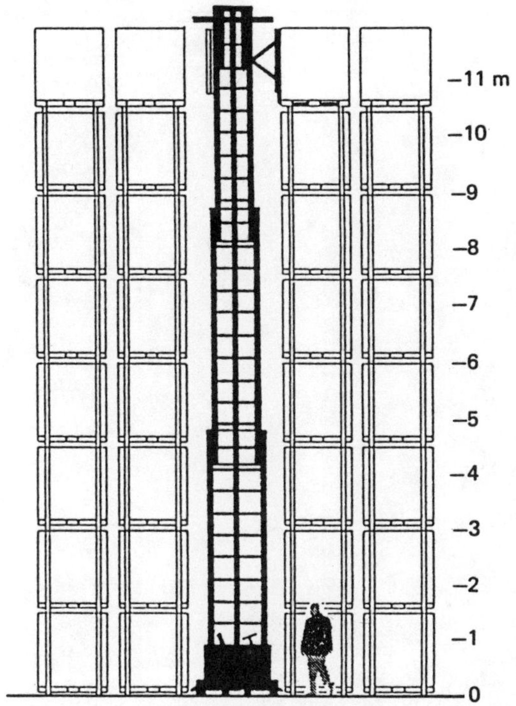

Figure 10.25 *Turret-truck*

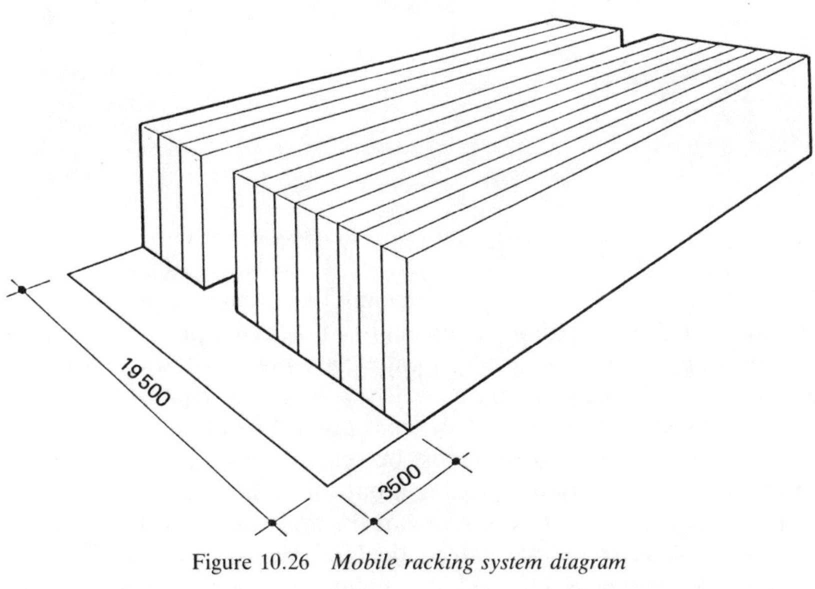

Figure 10.26 *Mobile racking system diagram*

Figure 10.27 *Actual mobile racking system*

The racks are placed on rails by means of a framework with wheels, driven by a small electric motor. Normally all racks are placed against each other, forming a block. Generally 4 pallets are placed on the racks, which gives a height of about 8 m and a coldstore height of 10 m. It is possible, though, to stack more pallets when using special forklifts.

If a particular pallet in a row must be reached, a corridor is created by moving the racks, simply by pressing a button at the beginning of the row concerned. For security reasons, a button must be activated inside the already existing corridor at that particular time. As long as this is not activated, the system is blocked, to prevent accidents involving people

Figure 10.28 *Another actual mobile racking system*

between the moving racks. In systems where the buttons are outside the coldstore, a photoelectric cell is placed at the entrance to the corridors; when people pass this cell the moving system is automatically blocked.

Every rack has a brake system activated by touching the baseboard of the rack. Activating the buttons outside will save time for the truck operator. While he is driving in, the racks will be moving to the desired position. The velocity of movement is about 10 cm/s.

It must be appreciated, however, that the efficiency of the loading and unloading systems depends on a logical organization being built-in at the design stage of the store. The floor must be designed after calculating the extra weight and the location of the rail-systems, and this may involve 10–20 % of the investment cost.

All the above-mentioned systems can be combined into one coldstore to obtain the highest efficiency. A layout of this type is shown on figure 10.29.

The system used in a particular section of the store depends on the product rotation. In order-picking areas, the highest degree of automation is desirable in order to gain time and save on labour.

Relationship Between Length, Width and Height of a Coldstore

The height depends on the stacking system. Length mostly depends on the throw of the air stream leaving the air coolers. With modern fans a throw

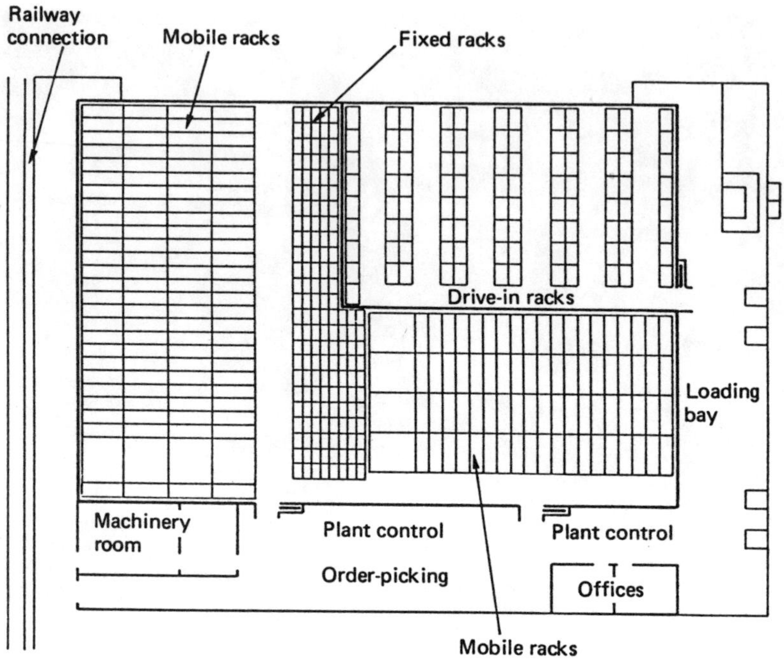

Figure 10.29 *Coldstore with different racking methods*

of 25–80 m can be obtained without using ducts or false ceilings. See chapter 6 on air coolers. It is not always necessary to blow air over the whole length of the store. However, when air is blown over the full width, more air coolers must be used.

For a long jet of more than 30 m, a smooth ceiling is necessary, which means a coldstore construction with an external steel frame. With this construction there are no obstructions to the air stream. The width is also limited by the maximum economical allowable span between the columns over the width of the steel structure, which is about 30 m, considering that we do not want to use columns in the floor space in the middle of the store. The distances between the columns over the length of the store are normally 5–6 m.

Loading Bays

A public coldstore is usually provided with a loading bay about 5–7 m; the height is adapted to rail and road traffic, and is about 1–1.2 m above rail or road level as vans and trucks normally have a loading height of 0.6–1.4 m. The problems of differences in level are solved by using movable loading platforms, such as that shown on figure 10.30.

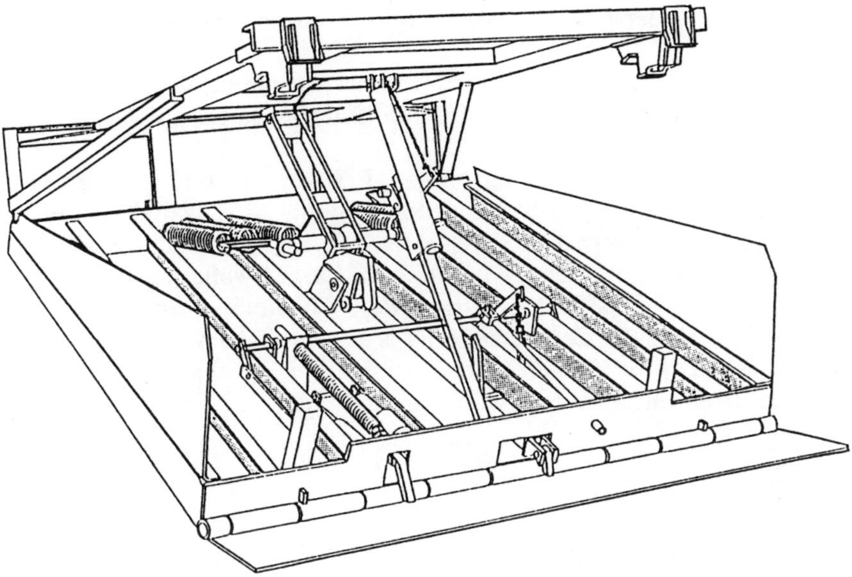

Figure 10.30 *Adjustable loading bay*

It is useful, especially in the case of freezer stores, to build a cool-lock, which can also serve as a covered loading bay and order-picking room. This area is chilled by air coolers with wide fin spacing, and frequent defrosting is arranged in order to cope with the large amount of moisture entering with the outside air.

The openings in the loading bay are provided with special joints, against which the van opening is placed during loading and unloading, in order to save refrigerating energy.

Smaller freezer stores can also be equipped with anterooms, of which the outside door has to be closed before the door of the freezer is opened.

Air curtains are not very effective. The best solution is to place a plastic strip curtain in the freezer door opening.

10.3 Pipework and Vessel Insulation

All previously mentioned materials can be used for pipework and vessel insulation. Prepared segments, mostly of PUR foam, are used or it is injected into a metal sheet casing.

When aluminium foil is used as a vapour barrier, care must be taken to finish the joints with silicone sealant; PVC tape can also be used as a vapour barrier.

For hot-gas defrost lines, glass wool is best used as the insulation material.

Where there are supports in vertical lines, segments of PU of extra high density, 100 kg/m^3, are used. Horizontal lines are supported by putting them in steel sheet sleeves.

Flexible cellular insulating material is often used for pipework, especially for smaller plants using copper tubing. It is available in tubes of 2 m length with internal sections corresponding to standard copper pipe sizes. Material with a density of 90 kg/m^3 has a λ-value of 0.04 W/(m K) and a σ-value of 2.3×10^{-8} g/(m s Pa). The temperature limits range from −75°C to +105°C. The material is also available in slabs. No vapour barrier is necessary. When placed outside it is necessary to protect the material from infra-red rays with a special paint.

Thickness of the Insulation

The thickness of this type of insulation is calculated so that condensation cannot occur on the surface. Henckey's equation states that

$$\delta = \lambda[1/\alpha_2 - (t_2 - t_1)/(t_2 - t_d) - 1/U]$$

where $1/U$ is the heat transfer resistance of the non-insulated tube and equals

$$1/(2\pi r_1\alpha_1) + (r_o - r_i)/(\pi\lambda\{r_o + r_i\}) + 1/2\pi r_o\alpha_2$$

where r_i = inside radius
r_o = outside radius
α_2 = outside surface coefficient
α_1 = inside surface coefficient
t_1 = temperature inside
t_2 = temperature outside
t_d = dew-point temperature.

A more simplified version is

$$1/U = (1/2\pi\lambda)\ln(d_2/d_1)$$

This can be controlled using the equation for surface temperature t_o:

$$t_o = t_2 - 2\lambda(t_1 - t_2)/(\alpha_2 d_2 \ln\{d_2/d_1\} + 2\lambda)$$

but then t_o must be higher than t_d to prevent condensation. The graphs on figure 10.31 can also be used. The values obtained from the tables contain safety factors which will show differences when compared with the theoretical figures obtained by the calculation shown above.

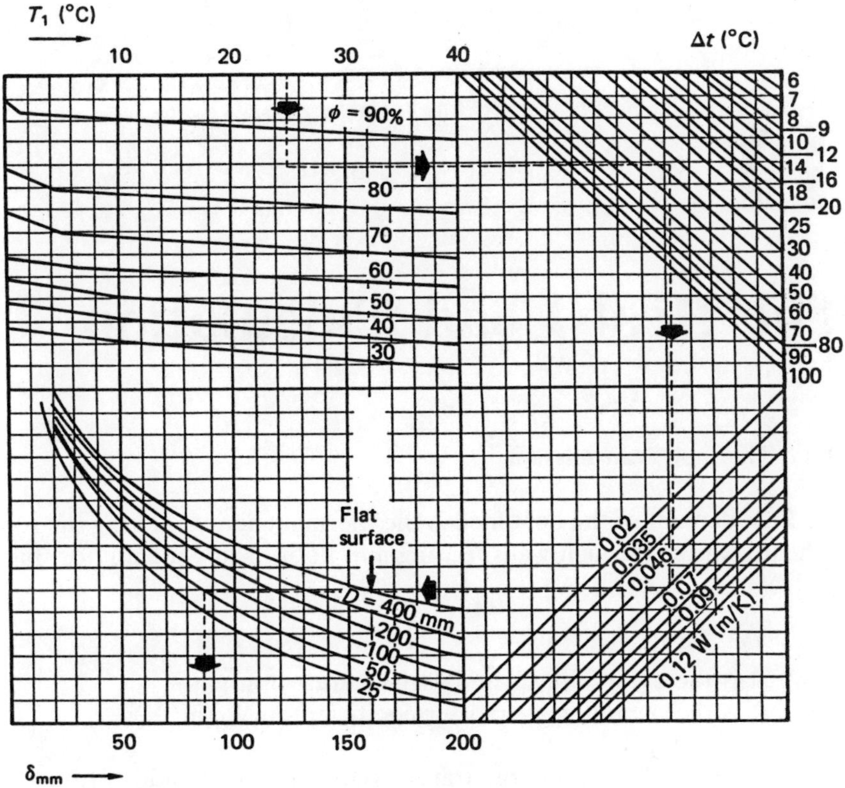

Figure 10.31 *Graphs to determine minimum insulation thickness to prevent condensation*

11

The Heat Load Calculation

11.1 Thermal Transmission

The first element of the heat load is the thermal energy entering the room by heat transfer through walls, ceiling and floor. The equation for this is

$$Q_1 = UAdT \quad (\mathrm{W})$$

where dT is $T_o - T_i$,
A is the external insulated surface area in m^2,
U is the overall heat-transfer coefficient in $W/(m^2\,K)$.

In cases of high solar energy transmission we have to add:

$$Q_1' = U_iAdt' \quad (\mathrm{W})$$

where $dt' = t_i - t_{\text{outside surface}}$.

In northern Europe it is usual to establish the outside temperature at +25°C and the soil temperature at between +10°C and +15°C. A 15% safety factor can be added to Q_1' since the insulation quality decreases after some time.

11.2 Air Infiltration, Q_2

This is a very important element of the total heat load. It is most difficult to calculate the extra heat load that results from fresh air entering from the outside. However, for forced fresh air intake it is a simple matter.

The amount of fresh air intake depends on the product stored and varies from 1 to 5 times the total room volume per 24 hours. In small rooms, where no fan is used, the opening and closing of the doors provide adequate natural air intake. In freezer rooms and gas-tight rooms there is, of course, no forced ventilation.

The outside air can be cooled by a special air cooler, but normally it is done by the normal air coolers. This heat load can be determined using the formula

$$Q_2 = m\mathrm{d}h \quad \text{(W)}$$

where m is the mass flowrate of the entering air,
dh is the difference in enthalpy between outside air and air under coldstore conditions.

It is difficult to calculate exactly how much air is entering during the time that the doors are open. Empirical equations are often used as follows:

$n = 35/\sqrt{}$(room volume) for low temperature rooms
$n = 70/\sqrt{}$(room volume) for chilled rooms

where n is the number of times the complete air volume of the room is exchanged in 24 hours. In cases where intense traffic can be expected, n is multiplied by 2.

Some engineers assume for calculation purposes that Q_2 is 10% of the total heat load. This can be too small a figure, especially in public coldstores or stores with heavy traffic.

There are other factors which affect this load: whether the doors are hand or automatically operated; whether plastic strip curtains or an air-curtain is used; whether a cold-lock or a loading bay, refrigerated or not, is part of the layout. It also depends on the ratio between coldstore volume and total doorway area.

Underestimating Q_2, the infiltration heat load, can lead to big variations in heat load calculations being made by different engineers.

The following theoretical treatment is one attempt at evaluating Q_2. We will first take a close look at the air-stream entering or leaving the coldstore when doors are open as shown in figure 11.1. What is the condition of the air-layer in an open doorway?

The cold air inside has a higher density than the warmer air outside. At floor level there is a certain overpressure on the inside of the door. At the top of the door opening the overpressure is on the outside of the door. At the top warm air enters, while at the bottom cold air leaves the coldstore. Somewhere at a specific level labelled z there is a balance point where there is no air stream at all. At this level:

$$p_{\mathrm{iz}} = p_{\mathrm{oz}} \tag{11.1}$$

The pressure difference is caused by difference in density, which is the result of the difference in the temperature. At z there is also equilibrium between the entering and the leaving air volume:

$$V_{\mathrm{i}} = V_{\mathrm{o}} \tag{11.2}$$

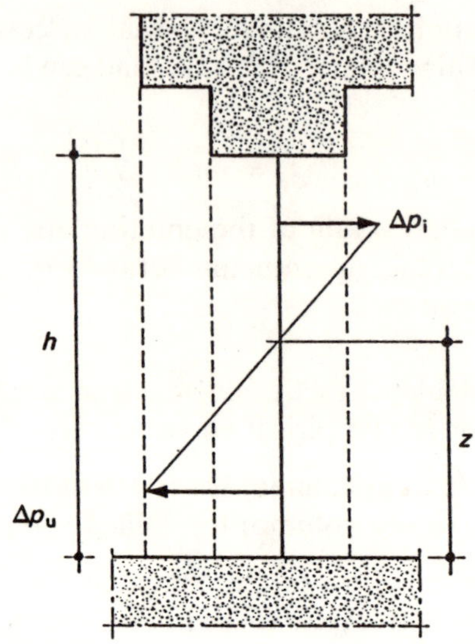

Figure 11.1 *Airstream entering or leaving the coldstore through an open door*

At a point $x < z$, the outside pressure is given by the equation:

$$p_{ox} = 0 - x\rho_o + p_w \tag{11.3}$$

Atmospheric pressure on the bottom is considered to be 0, ρ_0 = the density of outside air and p_w = wind pressure on the doorway.

The inside pressure is given by

$$p_{ix} = p_{ib} - x\rho_i \tag{11.4}$$

where ρ_i = density of the inside air and p_{ib} = inside air pressure at the bottom. Then

$$dp_x = p_{ix} - p_{ox} = p_{ib} - p_w - x(\rho_i - \rho_o) \tag{11.5}$$

When $x = z$, equations (11.1) and (11.5) give:

$$z = (p_{ib} - p_w)/(\rho_i - \rho_o) \tag{11.6}$$

Applying Bernoulli's equation:

$$U_x^2/2 = dp_x/\rho_i$$

where U_x is the velocity of the air at x. Then

$$U_x = \sqrt{(2dp_x/\rho_i)}$$

or

$$U_x = \sqrt{\{(2/\rho_i)[p_{ib} - p_w - x(\rho_i - \rho_o)]\}} \tag{11.7}$$

$$U_x = dV_i/bdx \tag{11.8}$$

where b = door width.
From equations (11.7) and (11.8):

$$dV_i = b\sqrt{\{(2/\rho_i)[p_{ib} - p_w - x(\rho_i - \rho_o)]\}}dx$$

$$v_i = b\sqrt{\{2/\rho_i\}}\int_0^z \sqrt{\{p_{ib} - p_w - x(\rho_i - \rho_o)\}}dx \qquad (11.9)$$

After integration:

$$V_i = -\frac{2b}{3(\rho_i - \rho_o)}\sqrt{\left[\frac{2}{\rho_i}(p_{ib} - p_w)^{3/2} - (p_{ib} - p_w - z(\rho_i - \rho_o)^{3/2}\right]}$$

and as in equation (11.6) the second factor in the denomination becomes 0:

$$V_i = -\frac{2}{3} \times \frac{b}{\rho_i - \rho_o}\sqrt{\left[\frac{2}{\rho_i}(p_{ib} - p_w)^3\right]} \qquad (11.10)$$

For the moment of air V_o entering the top of the doorway, we can calculate it in the same way:

$$V_o = +\frac{2}{3} \times \frac{b}{\rho_i - \rho_o}\sqrt{\left[\frac{2}{\rho_o}\{p_w - p_{ib} + h(\rho_i - \rho_o)\}^3\right]} \qquad (11.11)$$

Where $V_i = V_o$ (see equation 11.2) p_{ib} can be found from equations (11.10) and (11.11):

$$p_{ib} = p_w + \frac{h(\rho_i - \rho_o)}{\{1 + \sqrt[3]{(\rho_o\rho_i)}\}^3} \qquad (11.12)$$

Putting this into equations (11.10) and (11.11), then the entering or leaving air-volume V is:

$$V = \frac{2}{3}bh\sqrt{\left[\frac{2h[1 - (\rho_o/\rho_i)]}{[1 + \sqrt[3]{\rho_o\rho_i})]^3}\right]} \qquad (11.13)$$

If the wind pressure falls, the influence on the entering and leaving air is equal but reversed.
Approximately, one can say that $\sqrt[3]{(\rho_o/\rho_i)} = 1$, and as $b \times h = A$, the equation becomes:

$$\begin{aligned} V = {} & (2/3)A\sqrt{[2h/(1 + 1)3]}\sqrt{[1 - (\rho_o/\rho_i)]} \\ & (2/3)A\sqrt{(h/2^2)}\sqrt{[1-(\rho_o/\rho_i)]} \\ & (2/3)(1/2)A\sqrt{h}\sqrt{[1 - (\rho_o/\rho_i)]} \\ & (1/3)A\sqrt{h}\sqrt{[1 - (\rho_o/\rho_i)]}\ (m^3/s) \end{aligned} \qquad (11.14)$$

This gives, compared with equation (11.13), a deviation of 6–10% at 273–303K.

If V is the entering volume flowrate, then $V\rho_i$ kg/s of air enters the coldstore. This amount of outside air entering needs to be refrigerated. The difference in enthalpy $dh = h_o - h_i$ kJ/kg. Putting this information into equation (11.14) gives Tamm's equation:

$$Q_2 = (1/3)\rho_i A \surd\{h(1 - \rho_o/\rho_i)\}dh \quad (\text{kW/s}) \qquad (11.15)$$

This has been modified by Fritzsche to:

$$Q_2 = t\rho_i A \sqrt{\left[h\left(1 - \frac{T_i}{T_o}\right)\right]}\,\Delta h \ (\text{kW/s})$$

This is a short-term load which applies only when the door is open. The open time can be reduced to 30 seconds by using automatic doors.

In later work, Fritzsche and Lilienblum gave a correction factor k to be applied to the equation:

$$k = 0.48 + 0.004(t_o - t_i)$$

11.3 Moisture Load, Q_3

Moisture enters the coldstore in different ways before it finally reaches the evaporator coils and freezes there, a process which adds an extra load, a latent heat load, Q_3, to the overall heat load. The different entry methods of the moisture are:

(a) Air leakage through the insulation. Little is yet known about this amount. Experiments are in progress at the University of Delft, The Netherlands. Standards will be created, expressing the leakage in a unit called Selo, the specific leakage opening. Not enough information is yet available to take its effect into account in heat load calculations.
(b) Humidity entering as a result of water vapour diffusion through the insulation and vapour barrier. In modern prefabricated construction, this is negligible. Using traditional vapour barriers, calculations can be made as discussed above.
(c) Moisture entering as water vapour content of the infiltration air. This quantity is significant enough to be included in the heat load calculation.
(d) Moisture coming from the stored goods. Though it is an important factor few accurate figures are available. Experiments on the subject are still in progress at the Sprenger Institute in Wageningen, The Netherlands. For some products the values are already available. See figure 8.4 for data applicable to one variety of apple. In those graphs the moisture production is shown at different temperatures and relative humidities. For other products the figures are given according

to the difference in water vapour pressure. For example, the value for carrots is 7.1×10^{-10} kg/(kg Pa s). Figures are as yet unreliable, as there is much variation, according to varieties, dimensions, etc. The driving pressure is calculated as: $dp = p_s - p$, where p_s is the saturated water vapour pressure and $p = \phi p_s$. In the theory of mass transfer we met the coefficient β. Lewis states that there is a relationship between β and α. Under certain circumstances, in cold-stores:

$$\alpha/\beta = \sigma c_p \text{ J/(m}^3\text{ K)} \quad \text{and} \quad m_b = \beta dp/(R_o T)$$

11.4 Product, Q_4

The goods loaded into store bring yet another heat load, the product load. This load may consist of one or more of the following:

(a) Sensible heat, when the goods only have to be lowered in temperature.
(b) Latent heat, when crystallization or freezing takes place.
(c) Respiration heat, when living matter like fruit, plants and vegetables are involved. The reader is referred to the chapter on food products.

Values are available in various standard works such as *Handbuch fur Kalte Techniker* by Pohlman, and the data books and product guides published by ASHRAE.

In order to calculate the thermal energy to be taken out of the products during chilling or even freezing, we need to know:

1. mass m (kg)
2. specific heat above the freezing point c_1
3. specific heat below the freezing point c_2
4. solidification heat, crystallization heat or at least the water content x of the product, so that we can calculate the solidification heat of the water content of the product, knowing that the solidification heat of water is 333 kJ/kg
5. The initial temperature of the goods T_1 when brought into the store
6. The freezing point of the goods T_f.
7. The final temperature of the goods T_2.

Then

$$Q = m\{c_1(T_1 - T_f) + (x \times 333) + c_2(T_2 - T_f)\}$$

is the equation necessary to calculate the heat load for a complete frozen product. See table 11.1 for typical food values to use in the above equation.

Table 11.1 Latent heat and specific heat capacity for typical foods

Food	*Specific heat capacity* {*kJ*/(*kg K*)}		*Latent heat capacity* (*kJ*/*kg*)
	Above freezing	*Below freezing*	
Apples			
Bananas	3.85	1.76	281
Pears	3.35		251
Beer	3.85	1.76	209
Butter	3.77		301
Eggs	2.51–2.68	1.26	196
Strawber-	3.18	1.67	234
ries	3.85	1.97	300
Fish – lean	3.43	1.80	255
– fat	2.85	1.85	209
Beef – lean	3.25	1.76	234
– fat	2.55	1.49	172
Veal – lean	2.95	1.67	209
Pork – fat	2.14	1.34	130–153
Chicken	2.93–3.18	1.67	247
Carrots	3.64	1.88	276
Oranges	3.85	1.84	285

There are also tables that consider the enthalpy of the product at any temperature, so we can determine the difference in enthalpy dh/kg of the goods both before chilling or freezing and after.

11.5 Miscellaneous Heat Loads, Q_5

Finally, we must not overlook the additional thermal energy brought into the coldstore by the following:

(a) people,
(b) electric motors, such as those in the air-cooler fans or secondary fans, or motors in production equipment found in cutting or packing rooms,
(c) electric lights.

Source (a) involves very small quantities of energy since one person gives out only about 350 W.

Source (b) can only be estimated at the start of the calculation since at that point the fans have not yet been in operation. Assume a value based on 0.145 W/m^3 of coldstore volume.

Source (c) depends on the intensity of the lighting chosen.

Generally, one may assume that roughly 2% of the total heat load comes from these three sources.

Normal lamps have a relatively large thermal energy production; fluorescent tube lamps produce less warmth but their efficiency at low temperatures is not very good. With a 1.5 W fluorescent tube lamp the light capacity is 7100 lumens; the surface temperature is +40°C. With a normal lamp, the light capacity is 1380 lumens; the surface temperature is +200°C. The TL lamp has an operating life at least 10 times that of a normal lamp, and gives maximum light after 20 minutes.

A lighting loading of 115 W/50 m^2 normally gives satisfactory visibility in coldstores. Special 125 W sodium lamps give out 5600 lumens.

11.6 Sample Heat Load Calculation

The following data applies to a coldstore for apples.

Dimensions: 30 × 20 × 6 m or $V = 3600$ m^3.
Quantity of goods: 1000 tonne.
Daily input: 100 tonne including packing at +20°C.
Insulation: prefabricated sandwich panels with 10 cm PU foam.
Floor not insulated.
Above the ceiling there is a ventilated space under the roof.
The store is surrounded by other coldstores. In front there is a packing room with a temperature of +20°C, 70% RH.
Running time of the installation is 18 hours.
Door dimensions: 2 m wide by 3 m high.
The door opens for 20 minutes during a 24 h running period.
The coldstore conditions are: +3°C, 95% RH.
Q_5 is about 5%, not including fans.
Q_o for fans is about 3%.
There is no supplementary fresh air intake.

Calculation

$$Q_1 = UA\mathrm{d}T \quad \text{(W)}$$

$1/U = 0.1/0.03$ so $U = 0.3$ W/(m^2 K).
$Q_{\text{front wall}}$: 0.3 × 20 × 6 × 17 = 612 W.
$Q_{\text{back-and side walls}}$: 0.3[2(30 × 6) + (20 × 6)]5 = 720 W.
Q_{ceiling}: 0.3 × 30 × 20 × 27 = 4860 W.
Q_{floor}: 2 × 30 × 20 × 12 = 14 400 W.

$$Q_1 = 20.59 \text{ kW}$$

Q_2 = fresh entering air through the door:

$$Q_2 = (1/3)\rho_i A\sqrt{\{h[1 - (\rho_o/\rho_i)\}}dh$$

$\rho_o = 1.175$ kg/m^3.
$\rho_i = 1.28$ kg/m^3.
$A = 6$ m^2.
$h_o = 59.760$ kJ/kg.
$h_i = 14.135$ kJ/kg.
$dh = 45.625$ kJ/kg.

$$Q_2 = (1/3)1.28 \times 6\sqrt{\{3[1 - (1.175/1.28)]\}}45.625 \text{ kW} = 57.4 \text{ kW}$$

The correction factor to be applied under these conditions is $0.48 + 0.004 \times 22 = 0.568$.

$$Q_2 = 0.568 \times 57.4 = 32.6 \text{ kW}$$

This only occurs during a total of 20 minutes in 24 hours; the effective heat load on a 24-hour basis is given by $32.6(0.333/24) = 0.45$ kW.
Q_3 = the entering moisture load. First we calculate the massflow rate of the entering air:

$$m = (1/3)1.28 \times 6\sqrt{\{3[1 - (1.175/1.28)]\}} = 1.25 \text{ kg/s}$$

Then with the water vapour quantity difference between inside and outside air = 0.010 kg/kg, we can calculate the rate of entry of water vapour as 0.0125 kg/s. Then:

$$Q_3 = 0.0125 \text{ kg/s} \times 333.4 \text{ kJ/kg} = 4.17 \text{ kW}$$

As with the previous load, this only occurs for a period of 20 minutes, so the daily effect is $(0.333/24)4.17$ kW = 0.058 kW. Q_4, respiration of the goods has three elements.
$Q_{4.1}$, respiration heat:

$$900 \times 10 \text{ W/tonne} = 9000 \text{ W} + 100 \times 60 \text{ W/tonne} = 6000 \text{ W}$$

Total respiration heat = 15 kW.
$Q_{4.2}$, moisture production:

$$900 \times 15 \text{ g/h} = 13\,500 \text{ g/h} + 100 \times 30 \text{ g/h} = 3000 \text{ g/h}$$

Total moisture production = 16 500 g/h = 16.5 kg/h = 16.5/3600 kg/s.
Then: $Q_{4.2} = (16.5/3600)$ kg/s $\times$ (333.4 + 2500)kJ/kg = 12.99 kW.
Note that (333.4 + 2500) kJ represents the sublimation heat of 1 kg of water vapour freezing on the air-cooler coils.
$Q_{4.3}$, chilling down the goods entering:

$Q_{4.3} = 100\,000 \times 3.8 \times 17 = 6\,460\,000$ kJ = 77.77 kW when calculated on a 24-hour day

Q_5 is given by $(0.05 + 0.03) \times 131.20 = 10.50$ kW.
This gives Q_{total}, the total heat load, as 137.36 kW.
The necessary plant capacity to handle this, based on a running time of 18 hours per day, pull-down time for each day's input of 24 hours and the air-cooler fans running continuously, is given by:

$$\text{Plant capacity} = 24Q_{total}/\text{Running time (hours)}$$

Therefore our calculated Plant capacity is $24 \times 137.36/18 = 183.1$ kW. When we check this result against empirical values we get the kW per m^3 volume of the store as:

$$183\,100/3600 = 50.9 \text{ W/m}^3$$

Note 1: Normally apple coldstores with an intake of goods of 10% of the total load are calculated on the basis of 40 W/m^3. We see here that this figure corresponds to our theoretical calculation.

Note 2: In the case of ULO coldstores, the heat load for the evaporators is calculated during the storage period and with a 4 h running time. However, the heat load for the compressors is calculated as described in this example but with a 20% product intake.

References

C. G. Franke (Technishe Hogeschool Delft), Labo voor koude techniek, Delft, The Netherlands, *Koeltechniek*, 64 (1981).

W. Tamm (Munich, Germany), Kälteverluste durch kühlraum öffnungen, *Kältetecnik*.

12

Economic Considerations of Refrigeration

This chapter should be of special interest to anyone responsible for comparing quotations for refrigerating plants from different competitors. The problem lies in the fact that several different solutions are possible for the same project. It is true to say that refrigerating plants, even of the same refrigerating capacity, can be built for a wide range of prices. It all depends on the choice of the design concept; however, because of manufacturing standardization and the small number of plant-manufacturers now in business, price differences between 5 and 15% can only be explained by differences in the individual overall concept of the plants – no longer can they be attributed to the quality of components.

In international tendering, the price of labour and its productivity can contribute in part to any price difference. However, it is competition that tends to keep most prices in the same range. One must be most careful when there are unexplained but large price differences. Sometimes for budgetary reasons and for the anticipated limited lifetime of the project, the choice is based solely on the price – without due regard to quality, working life, power consumption, maintenance costs and, most importantly, effects on the quality of the stored, frozen or chilled products.

The importance of the quality of the product depends greatly on the value of the product, its marketing position and its final destination.

Where only a low budget is available, it may be better to increase this budget and balance the subsequent increased interest charges against lower running costs. Note particularly that maintenance costs are not just the direct costs involved in preventive maintenance and repair. The indirect costs which occur during an installation failure may be very high. Breakdowns can also lead to production losses, delivery interruptions and a subsequent loss of clients. A safe and reliable installation, with full maintenance facilities, will involve higher investment costs.

Selection of the lowest price often means in reality purchasing expensive 'credit', with interest on the 'loan' increasing every year, caused by the running costs escalating year by year. From this point of view it is important to invest for the foreseeable future by ordering vital components like vessels and main lines with spare capacity, so that the company will not be burdened later with the high costs of having to replace these components at the first expansion of production capacity.

So what do you do when you receive the various quotations? First make sure that all quotations are based on the same refrigeration capacity and design temperatures. Therefore it is advisable to request all tendering companies to include their heat-load calculations to verify that the basic requirements of your company are being met.

Then comparison can be made by drawing up a table of all important data. Table 12.1 shows a suitable example. If you do not find all the necessary data you need in a quotation, do not hesitate to ask your tenderer for the missing information.

Make sure that all data is given in the same units. When all the necessary information is available, make a running cost calculation.

The data needed is as follows:

A summation of the power consumption of the various electric motors of the installation in kW and running hours, together with the actual cost of 1 kWh and the number of running hours of the installation per year.

For production plants, such as deep freezing plants and tunnels – the production per hour and per year.

The rate of water consumption per hour running time, together with the unit cost of water.

The actual manpower labour costs, where there is a difference in the number and type of operator required for the alternative systems offered.

An estimate of the product weight loss for the different designs. The unit value of product produced per hour and per year.

Maintenance costs.

Depreciation time for the project.

The investments costs – total cost.

The capital costs – loans, credits.

Note 1: The cost of the building, the land, etc. must also be taken into consideration.

Note 2: *Maintenance costs*

It is difficult to give general figures about maintenance costs, as they are critically dependent on the design conception of each

Table 12.1 Important data to be supplied by tenderers

	Supplier				
	1	*2*	*3*	*4*	*5*
A: Compressor					
Number					
Make					
Types					
open/semi-hermetic/hermetic					
piston/screw/other					
in-built volume ratio					
adjustable					
no. of stages					
Refrigerant					
Capacity at design conditions					
Evaporating temperature					
Condensing temperature					
Power consumption at working conditions					
Installed motor-power:					
(a) Is capacity control anticipated?					
(b) What kind of capacity control?					
(c) Does the compressor start unloaded?					
Direct drive/V-belt drive?					
Will a frame be supplied?					
Are gauges supplied?					
Are pressure controls fitted?					
Number of cylinders					
S/D					
Maximum speed of rotation					
Stroke					
In the case of screws:					
• female drive possible					
• variable in-built volume ratio					
• rotor length					
• rotor diameter					
• oil cooler type					
B: Condenser					
1. Air-cooled					
• number of condensers					
• working conditions (temperatures)					
• heat-rejection capacity at working conditions					
• airflow direction					

- make
- type
- d*t* outside air/condensing temperature
- surface area – not always available from the manufacturer
- air volume
- fan motor power
- condensing pressure control? which system?
- corrosion protection
- fin thickness and spacing
- noise level
- weight and dimensions

2. Water cooled

- number of condensers
- heat-rejection capacity
- make
- type
- design horizontal, vertical, double pipe, others
- removable water-covers
- corrosion protection
- tube section and wall-thickness
- water pressure drop
- water temperature in/out
- condensing temperature
- water volume per hour
- water inlet control?
- weight and dimensions

3. Evaporative condenser

- number of condensers
- heat-rejection capacity
- condensing temperature
- wet bulb temperature
- make
- type
- air volume flowrate
- water volume flowrate
- fan motor power
- pump motor power
- frost protection
- type of fans (axial, centrifugal)
- condensing pressure control system
- noise level
- corrosion protection
- is there a dc-superheater?
- weight and dimensions

Table 12.1 *continued*

	Supplier				
	1	*2*	*3*	*4*	*5*
A: Compressor					
4. Cooling tower					
• number of towers					
• heat-load rejection capacity					
• make					
• type					
• design					
• construction materials					
• smallest section for water/air passing in the filling packet (obstructions?)					
• water volume flowrate					
• air volume flowrate					
• wet bulb temperature					
• water in/out temperature					
• fan motor power					
• pump motor power					
• corrosion protection					
• noise level					
• frost protection					
• weight and dimensions					
C. Evaporators					
1. Air coolers					
• number of coolers					
• make					
• type					
• operation (flooded, dry)					
• drain pan design					
• drain pan heating system					
• fan position					
• electric heater for fan ring					
• construction materials					
• surface area (not always available)					
• tube dimensions					
• tube spacing					
• fin thickness					
• fin spacing					
• refrigeration capacity at operating condition					
• evaporation temperature					
• d*t*					
• which kind of d*t*?					
• air temperature in/out					

- air volume
- fan motor power
- throw
- outlet air velocity
- corrosion protection
- defrost system (in the case of negative in-temperatures: is there an air diffuser which can be closed during defrosting?)
- dimensions and weight

2. Liquid chillers
- number of chillers
- make
- type
- operation (flooded, dry)
- number of circuits
- tube dimensions
- separator conception
- removable water covers or a bundle of tubes?
- surface area
- corrosion protection foreseen?
- frost protection foreseen?
- insulation foreseen?
- liquid pressure drop
- refrigerant capacity at working conditions
- evaporating temperature
- temperature of the liquid in/out
- volume of the liquid?
- dimensions and weight

D. Refrigerant separators and intercoolers
- number of separators
- section and length
- design (open/flash)

E. Auxiliary components
- oil separators (primary secondary, ppm efficiency)
- oil washer (ppm efficiency)
- oil-collecting receiver
- heat exchangers (in CFC plants)
- high-pressure receiver (volume)
- if there are de-superheaters in stainless steel, which quality of stainless steel is used?
- are there sufficient stop valves?

Table 12.1 continued

	Supplier				
	1	2	3	4	5
F. Extra controls					
• filters					
• level controls					
• safety valves					
• hand-operated valves to facilitate replacement of defective components					
• temperature control and recording					
• humidity control and recording					
• equalizing valves on coldstores					
• protection heaters (for crank-case, fans, siphon, etc.)					
G. Insulation					
• product					
• type					
• density					
• vapour barrier					
• protection					
• basic environmental conditions					
H. Piping					
• quality					
• steel/copper					
• insulation (product type and density, vapour barrier, protection barrier, environmental conditions)					
• water-piping (stainless steel, galvanized)					
• paint finish					
I. Power supply					
• switch panel included					
• central panel or decentralized					
• voltage					
• frequency (50 or 60 Hertz)					
• mechanical or plc					
• cables					
• power supply cable + overload protection					

J. Miscellaneous • are transport, unloading (if necessary by crane), assembly, start-up, testing, commissioning, primary and secondary refrigerant, oil, site-cleaning, spare parts, maintenance manual, frames, foundations, paint-work, included?					

particular installation, plus the cost of labour. Approximate rules are:

For buildings: 2%

For refrigerating plants: 2–5%, according to design, whether industrial or commercial, centralized or decentralized, ammonia or CFC plant

For stacking and handling systems: 8%

During the first year the plant is still under building warranty and the costs are low; the older the plant the higher the cost of maintenance.

Note 3: *Depreciation time*

For depreciation time, various figures are usually available. Approximate rules are:

Refrigerating equipment	5–10 years
Building	20–24 years
Stacking and handling equipment:	5 years

Note 4: *Lifetime*

Light 'commercial' equipment – 10 years.
Heavy 'industrial' equipment – 20 years.

Example 1

Examination of a coldstore of 8000 m^3 at −25°C with a nominal refrigeration capacity of 100 000 kW at −35°C evaporating temperature for which three different designs are proposed:

(a) Industrial ammonia installation
- 2 open-type two-stage industrial piston compressors
- air coolers with steel tubes and fins, hot-dip galvanized
- automatic hot-gas defrosting
- centralized installation

- flooded system with ammonia pumps
- evaporative condenser: condensation temperature +35°C, wet-bulb temperature +22°C

(b) Industrial R22 installation

Same design features as under (a).

Note that the evaporators here are also the flooded type using a liquid pump. However, usually thermostatic valves are used in CFC designs.

(c) Expanded commercial R502 installation

- semi-hermetic compressor
- single-stage compression
- air-cooled condensers (condensing temperature +45°C)
- air coolers with copper tubes and aluminium fins, electrical defrosting and thermostatic expansion valves
- decentralized installation – in this case packaged roof-top units were used.

Note that it is also possible to use a centralized conception, instead of the separate roof-top units.

Oil equalizing between the compressor bodies is absolutely necessary in this case. If there are only two compressors, the bodies can be connected with a simple equalizing line. However, if there are more compressors on the same circuit, it is better to plan a common oil separator to which all the compressors are connected. This oil separator serves as an oil reservoir as well. Measures have to be taken to ensure that there is always extra oil in this separator.

The vessel is connected to each compressor body by float-valves. The pressure in the vessel is kept at a constant low pressure (constant-pressure valve connected to suction line) in order to avoid big pressure differences over the float-valve, which could result in oil/refrigerant foaming giving lubrication problems.

Maintenance costs will be the highest for the scheme under (c), the lowest for the scheme under (a). The lifetime of small semi-hermetic compressors is shorter than for open compressors by about 50%. Maintenance costs due to breakdown of semi-hermetic compressors are higher in general because they have to be completely replaced; whereas open compressors can be overhauled. When the motor-winding burns out, the semi-hermetic compressor must be replaced entirely and the piping circuit must be cleaned. With open compressors, one can simply replace the electric motor.

In the case of (c) (electric defrosting), maintenance costs for replacement of electric heaters have to be added.

Note especially that the combination of hot-gas defrost systems with thermostatic expansion valves can cause maintenance problems as a result of temperature and pressure shocks caused by contact of the hot gas with the valves. Great care must be taken when using this particular design.

For the purpose of this example only, the energy costs are compared based on an average running time of 12 h a day over a 1-year period and assuming a price per kW of $0.066.

(a) Power consumption: 77 kW
Defrosting energy: nil
Cost of electricity per year: 365 × 12 × 77 × 0.066 = $22 259

(b) Power consumption: 80 kW
Defrosting energy: nil
Cost of electricity per year: 365 × 12 × 80 × 0.066 = $23 126

(c) Power consumption 132 kW
Defrosting energy 50 kW
Total power consumption 182 kW
Cost of electricity per year: 365 × 12 × 182 × 0.066 = $52 613

Conclusions are as follows:

The investment costs in case (a) and (b) are about $30 000 – higher than in case (c).
However, the energy costs in case (c) are higher than in case (a) and (b) by about $30 000 a year.
So, when choosing solution (a) or (b), the extra investment costs are recuperated in one year.

Example 2

A deep freezing plant handling about 1500 kg soft fruit per hour: compare (a) blast freezer (IQF tunnel) and (b) liquid nitrogen tunnel.

Running time:	225 days a year, 8 h per day
Cost of electricity:	0.066 $/kWh
Depreciation:	10 years
Capital costs:	10% (annual interest charge)
Maintenance costs:	blast tunnel installation, 2.5%
	nitrogen tunnel installation, 1%
Cost of 1 kg nitrogen:	$0.10

For every kg of product, 1 kg of nitrogen is needed
Investment for IQF tunnel plant: $200,000
Investment for liquid nitrogen tunnel plant: $52,000
Power consumption for an IQF tunnel plant: 150 kW
Power consumption nitrogen tunnel: 7.5 kW, negligible.

(a) IQF tunnel

Energy costs per year: 225 × 8 × 150 × 0.066	= $17 820
Capital cost including interest and depreciation	= $32 000
Maintenance costs	= $ 5 000
Yearly running costs	$54 820

Cost per kg product: 54 820/225 × 8 × 1500 = $0.02.

(b) Energy costs of nitrogen: $0.1 per kg
1. Capital costs: $7750 per year
2. Maintenance costs: $500 per year

Total of 1 and 2 is $8250 per year or 8250/225 × 8 × 1500 = $0.003 per kg of product. So the total cost per kg is $0.1 + 0.003 = $0.103.

Since we are talking about unpacked product, there will be 1.5% less weight loss when using liquid nitrogen. If the cost of the product is $2 per kg, there is a gain of $ 0.03 per kg. The difference in cost per kg product then becomes: 0.103 − 0.02 − 0.03 = 0.053.

In this case, nitrogen freezing is about 3–4 times more expensive than in a traditional freezing plant.

This example is based on a year-round 8-h production schedule per day. A plant where soft fruit is being processed normally operates only during harvest time, but because operators work 2 or 3 shifts a day, which requires the plant to work practically around the clock, the cost calculation on a yearly base stays the same.

It is the same if the plant works only 8 h a day but other products are frozen with the same equipment outside the fruit harvesting season.

It should not be concluded from this example that there is no field of application for liquid nitrogen or other liquid gases such as CO_2 in freezing processes. It has already been stressed that limiting the weight loss for unpacked goods can change the cost picture, particularly when the value of the product is high, as with lobster or prawns.

When the plant is used only for short periods during the year, low investment may be an advantage from a depreciation point of view.

Example 3

A firm has a constant production rate of 2000 kg/h of fishfingers for 225 days a year. Fishfingers are a relatively cheap product. The depreciation of investment costs is divided over 225 × 8 = 1800 hours a year.

The same firm also has a production rate of 2000 kg/h of an expensive kind of fish for only 60 days of 8 working hours each year. The depreciation is divided over 480 hours.

If the whole plant design were based on a production capacity of 400 kg/h, the plant would only achieve the optimum production capacity for 60 days of the year. For the other 165 working days, the plant would be under-used.

The best solution in this case is to build a traditional plant with a capacity of 2000 kg/h for the fishfingers and a cheap nitrogen or carbon dioxide plant for the 2000 kg/h for the fishfingers and a cheap nitrogen or carbon dioxide plant for the 2000 kg/h of expensive product.

We have the same situation where we expect over-production of about 1000 kg of fishsticks during a very short period every year. Another application for this concept would be in an industrial kitchen for fast-food or TV dinners. Production is often small – during short periods per day, and only a few days per week. A traditional plant with compressors and a tunnel might be too expensive compared with a simple N_2 or CO_2 unit. In those cases tunnels are not used, but simple freezing cupboard-type compartments.

13

Plant Maintenance

We refer the reader to the safety standards summarized at the end of this chapter for instructions regarding the start-up, testing, commissioning and maintenance of industrial refrigeration installations.

Many of the simple maintenance operations in the refrigeration field can be handled by a conscientious worker without special skills. However much maintenance work requires skill and knowledge of refrigeration techniques. European standards and laws, as well as laws in individual European countries, allow only qualified refrigeration engineers to work on refrigeration systems.

13.1 Regular Maintenance Plans

Regular plant maintenance can be divided into the following groups:
Daily inspections of items such as:

- compressor motor current
- compressor suction, discharge and oil pressures
- oil level of the compressors
- evidence of oil leaks on compressor driveshaft or pipe joints
- liquid level in the refrigerant receivers
- liquid stream in the sight glass (bubbles)
- condensing temperatures
- cooling water temperature and flow
- outside air temperature
- air temperature at condenser inlet
- room temperatures and humidity
- temperature of machine room
- condition of the defrost and drain pipe heaters
- operation of the cooler fans

- abnormal vibration
- condition of the stored product.

Monthly inspections of items such as:

- temperature of the liquid refrigerant upstream and downstream of the thermostatic expansion valve
- inspection of all the fuses in the electrical system
- operation of the safety controls
- verification of the filter/dryers in the system
- acidity of the oil and, if necessary, replacement of the oil (refer to the instruction manual for the compressors)
- leak testing with leak detector
- testing the density of the brine if used in the system
- operation of all heaters
- checking the instruction books of the manufacturers or contractors to ensure maintenance work is in agreement with their instructions
- operation of air purge.

Periodical inspections of items such as:

- operation of defrosting system
- operation of general fans and motors
- security of connectors and couplings of the electrical installation
- operation of signal lamps and alarms
- operation of contacts and relays
- cleanliness of electrical panels (use compressed air)
- V-belt tension of compressor motor drives
- drive alignment of direct coupled compressor motors
- cleanliness of the condensers
- cleanliness and integrity of filters
- operation of non-return valves
- operation of float valves
- integrity of insulation and sealing of doors, and operation of heater
- cleanliness of switch panels
- cleanliness of contact-points
- temperature in switch panel
- alarm system operation
- condition of the stored goods
- air-circulation in coldstores.

Annual inspection of items such as:

- cleanliness of machine rooms
- compressor assemblies
- plant corrosion

- integrity of all insulation, vapour barriers and protective coverings
- maintenance of plant log-book – noting daily temperatures, pressures, energy consumption, abnormalities, repairs and maintenance work carried out.

In addition to the above there are plenty of small checks that can be carried out by all the operators. Items such as oil traces on the floor, on connections, valves, etc. – all can be a sign that there is a leak.

Abnormal vibrations of compressors or components of the installation can be detected by hand. Abnormal noise of electrical motors etc. can be detected simply by ear.

13.2 Oil and Refrigerant Charges

13.2.1 Oil Level

During normal operation the oil level should be in the middle of the oil sight glass or, in the case of two glasses, in the middle of the upper glass. The two fault conditions that the maintenance staff should be on the lookout for are:
(a) oil level too low, and
(b) oil level too high.

In the case of (a), care must be taken before the diagnosis of an oil undercharge is arrived at. If, at the end of a long standstill, the oil level is low, then proceed with caution. If it is safe to start the compressor, then do so and carefully monitor the oil level. Should the level remain low after the operating conditions of the system have settled down, then correct the shortage of oil. In a situation where an operating compressor shows a low level, repeat the recommendations above. It is very easy to overcharge a system with oil, so producing a fault which is difficult to diagnose.

In the case of (b), the high oil level causes excessive churning, overheating and even breakdown of the oil which could eventually cause breakdown of the compressor. It is difficult to detect a high oil level while the compressor is running owing to the churning effect; however, if the level is obviously high immediately after shutdown then there is definitely an oil overcharge. The surplus oil should be removed from the crankcase and the compressor restarted. If after a reasonable running period there is still a high level at shut down, the process should be repeated until the correct level is achieved. It may help to remove sufficient oil to take the level below halfway, so as to eliminate any churning and make obvious the surplus in the plant finding its way to the compressor.

It is therefore very important to be on the lookout for variations in the oil level. However, the compressor should always be allowed to run and the operating conditions to stabilize before taking any steps.

At the end of a long standstill period the oil level in a CFC system may be higher owing to the oil absorbed by the refrigerant. Immediately after starting a compressor the oil level will usually fall rapidly. However, it should reach near normal level in the course of only a few minutes. At other times the oil level may fall if the compressor is working with an unusually low suction pressure, such as during pump-down or while the capacity control system is operating.

The oil level should always be visible in the sight glass, whether the compressor is working or is idle.

13.2.2 Refrigerant in the Lubricating Oil

Owing to the difference in vapour pressure between refrigerant and oil, the refrigerating machine oil has a very considerable ability to absorb refrigerant vapour, especially at low temperatures or high pressures. A prolonged period of standstill increases the amount of refrigerant absorbed by the oil. When starting a compressor under circumstances where the oil temperature is close to the saturation temperature on the suction side of the plant, the greater part of the absorbed refrigerant will be freed by the lowering of the pressure in the crankcase and this will cause violent foaming, and in a short time the oil can completely vacate the crankcase. In a refrigeration compressor this results in:

1. Loss of oil – subsequent risk of draining the crankcase.
2. Problems with the oil pump – the pump cannot draw the boiling mixture of refrigerant/oil, with the result that sufficient oil pressure cannot be built up.
3. Owing to the mixing with refrigerant, the viscosity of the oil is reduced to such a degree that its lubrication ability is greatly reduced. Coupled with insufficient oil pressure, this causes increased wear and risk of seizure.
4. Oil hammer in the cylinders, causing risk of destruction of valve plates, valve springs and the unloading mechanism.

Foaming in the oil sump can be reduced as follows:

1. Stabilization in the operation of the refrigeration compressor should be aimed for. It is typical for a piston compressor that the compressor oil has a tendency to leave the crankcase immediately after start-up. It is therefore necessary that the start-up is followed by a sufficiently long

operation period to re-establish the oil level. Short cycling caused by the safety automatic controls – such as the low pressure control – deserves immediate attention, as there is a risk that the crankcase of the compressor might remain short of oil for a dangerously long period. Leaks – for example in a solenoid valve in the piping system may cause unstable operation, with the consequences mentioned above.

2. At start-up special precautionary measures should be taken to avoid overflow coming from the evaporator.
   (a) In refrigeration plants with forced air circulation over the evaporators, the fans should always be prevented by some delay device from starting before the refrigeration compressor.
   (b) At the slightest sign of liquid knocks, the refrigeration compressor should be stopped immediately. Reduce the compressor capacity, at restart, until the knocking sounds have ceased. It might be advisable to throttle the suction stop valve.
3. Expansion valves should be correctly adjusted, according to their individual specifications.
4. During standstill periods, the compressor is best isolated from the main system by closing the suction and discharge stop valves.
5. Rapid pressure drops in the crankcase must be avoided while the oil sump is cold. If necessary, start the compressor at reduced capacity.
6. A cold return line from the oil separator indicates the presence of refrigerant in the oil separator. The stop valve in the return line ahead of the crankcase should be kept closed until the oil separator is at its normal operating temperature.
7. Many compressors are fitted with a crankcase heater; this normally comes into operation during the compressor off-cycle. In the event of the compressor being out of use for a long period, then it is wise to energize this heater some 4–8 hours before starting the compressor. Heating the oil to about 50°C reduces the concentration of refrigerant in the oil by about 80% compared with that when the oil is fully saturated at 20°C.

13.2.3 Checking the Refrigerant Charge of the Plant

Leaks in the refrigeration system, whether they be in the pipework, at joints or within a major component, must be located as soon as possible and repaired. When the leak is on the high-pressure side of a system (or the low-pressure side of a system operating above atmospheric pressure), the refrigerant charge will be lost at a rate depending on the internal pressure and the severity of the leak. Insufficient refrigerant charge reduces the capacity of the refrigeration plant and increases the risk of the system absorbing air and moisture. Leak detection on CFC plants must be carried out on a preventative basis by an approved method. As a back-up, the refrigerant charge of the plant should be checked at regular intervals, as the leak could be very small and a loss of charge would indicate the existence of an undetected leak in the system.

Ammonia leaks are nearly always detectable by the smell. When the leak is on the low-pressure side of a system operating below atmospheric pressure then there will be no loss of charge, but either an increase in operation of the air purging equipment or an increase in the condensing pressure caused by the air migrating to the condenser. In either case, the plant will operate inefficiently.

Bubbles in the Sight Glass

A sight glass is installed in the liquid line to allow examination of the liquid stream leaving the condenser. When bubbles are visible in the sight glass it is usually a sign that the system is short of refrigerant. However these bubbles can also indicate other abnormal conditions, for example, failure of subcooling of the refrigerant liquid in the condenser or too great a pressure drop in the pipe line. A potential pressure drop, for example due to a plugged strainer, a partially obstructed or choked stop valve or the narrowing of a tube, can often be located by the choked point being too cold and, even immediately after start-up and during cooling down, bubbles are often visible in the sight glass.

Leak Detection

Leaks are located either by brushing a soap solution over the joints and watching for tell-tale bubbles, or by using detectors that react to leaking refrigerant from the system. Additives in the CFC make detection of leaks in the system easier. Often an electronic leak searcher is used; it is possible to manage with a leak-detector torch – such torches are very sensitive. In the case of serious leaks it is necessary to ventilate the space, otherwise the torch cannot locate the leak with sufficient accuracy owing to this great sensitivity.

Note that flanges and joints may sag during the initial operation of the plant; therefore, check and retighten them.

13.2.4 Conversion from R12 and R502 to R22

In the coming years, many plants containing R12 or R502 will have to be converted to use the HCFC R22. The following parts of the system will need adaptation:

The compressor Deep-freezing plants using R502 will need converting from single-stage to two-stage owing to the different pressure/temperature relationship. For medium–low evaporating temperatures down to −30°C, single-stage compression with liquid injection can be used.

When replacing R12 with R22 it must be remembered that:

- The compressor capacity using R22 is higher
- At this higher capacity, the absorbed and hence the motor power will be greater
- Not all R12 compressors can operate with R22; this also applies to some R502 compressors.

Whereas the capacity of a belt-driven open motor-compressor can be adjusted, this is however not possible with hermetic and semi-hermetic motor-compressor units.

In a large multi-compressor system, the capacity reduction is achieved by taking out one or more compressors – a method which reduces the flexibility of the system. In certain cases there is no solution other than to fit a smaller compressor.

Condenser As we have seen above, the capacity of a condenser varies depending on whether it is using R12, R502 or R22. This needs to be checked. In the case of condensing units some manufacturers quote equal capacity when using R22, however in many cases the condenser will be too small when using R22.

Expansion valves These will need changing and the use of electronic valves with smaller dt values is preferred.

Filter/driers These will require changing.

Oil The oil needs changing with a grade meeting the specification for R22 and its higher discharge temperatures.

Oil return system The oil-carrying velocity of vertical rising lines will need checking to ensure oil return to the compressor. Some systems may require the installation of an oil separator.

Suction/liquid line heat exchangers R12 systems regularly use suction/liquid line heat exchangers which produce high superheat in the suction line vapour reaching the compressor. This condition results in high discharge temperatures which cannot be tolerated in R22 compressors. The solution is to fit a liquid trap if there is any danger of liquid slugging.

Distributor When the evaporator is fitted with a distributor, the manufacturer should be consulted in case the distributor needs replacing.

Insulation The suction line may need insulating in order to minimize superheating of the suction gases.

Test pressures The higher working pressures of R22 make it essential that the safe working pressures of all the components of the system are satisfactory.

13.3 Fault-finding and Correction

13.3.1 Compressors

Trouble observed	*Cause of defects*	*Action to be taken*
Compressor fails to start	Power failure because of: • master switch, not made • fuses blown • failure in motor starter • bad electric connections.	Disconnect the power, and change possible blown fuses. Determine the reason for the interruption.
	The magnetic coil in the motor starter is not activated owing to open control circuit at: • high/low pressure control – reset • oil pressure control – reset • thermal overload release – reset • open circuit at possible open auxiliary relays.	Locate open control circuit, and rectify the cause for the interruption. Auxiliary relays are often arranged in connection with fans and pumps that start before the compressor. The adjustment of the overload release should be 1.05 times the full-load current of the motor.
	Oil pressurestat cut-out.	Compressor will start at reset. Check oil level.
	Low-pressure pressurestat cut-out.	The compressor does not start before the suction pressure has risen over the set point of the pressurestat. Reset and test for leaks.
	High-pressure pressurestat cut-out.	Reset the pressurestat and find the reason for the high condenser pressure.

Cont'd

Trouble observed	*Cause of defects*	*Action to be taken*
	Compressor or motor failures.	Overhaul compressor and motor.
Compressor starts and stops frequently	The high-pressure pressurestat cuts out.	High condenser pressure. Check condenser cooling and adjust the pressurestat to correct cut-out pressure. Replace defective pressurestat.
	The low-pressure pressurestat cuts out at too low suction pressure.	Low suction pressure. If the low-pressure pressurestat is set too high, adjust it.
	Low-pressure pressurestat differential set too close.	Check the operating conditions and reduce, possibly, the capacity.
	Compressor discharge valves leak.	At stop, the pressure between suction and discharge side is equalized comparatively quickly. Repair or replace the discharge valves.
	Strainer in suction line clogged.	Check the suction strainers of the compressor.
	Leaky liquid line solenoid valve.	Check flow direction. Replace defect valve.
Compressor starts, but stops immediately	Insufficient oil charge.	Add an adequate amount of oil and determine the cause for the lack of oil.
	Oil pressure fails owing to formation of foam in the oil.	Reduce the capacity.
	The low-pressure pressurestat cuts out.	Open possibly closed suction stop valve.

Cont'd

Trouble observed	*Cause of defects*	*Action to be taken*
	The motor starter cuts out.	Determine the cause of overloading. At star–delta start, adjust the starting time to minimum.
Compressor runs continuously	Thermostat or low-pressure pressurestat does not cut-out at too low temperature/ pressure.	Adjust operation setting.
	Limited supply of refrigerant to evaporator. Compressor works at too low suction pressure.	Clean strainers, and check the function of the expansion devices.
Abnormally noisy compressor	Refrigerant liquid in suction line.	Liquid hammer. Risk of broken valves and crankshaft. Adjust expansion valves or float valves.
	Safety valve opens.	Check condenser cooling. Adjust opening pressure of the safety valve.
	Compressor capacity too high at start.	Reduce the capacity. Refrigerant in the oil.
	Too low oil pressure.	See chapter 4 on too low oil pressure.
	Too much oil circulating through the plant.	Oil hammer. Check oil level. Solenoid valve, filter or nozzle in oil return system may be clogged. Leaky suction valve plates, piston rings, and worn cylinder may also cause oil consumption.

Cont'd

Trouble observed	*Cause of defects*	*Action to be taken*
	Loose foundation bolts.	Tighten.
	Misalignment of motor and compressor.	Check alignment according to special instruction.
	Coupling bolts loose.	Tighten (use torque wrench!)
	Worn or defective bearings.	Overhaul, or replace.
	Capacity regulator hunting, owing to failing oil pressure.	Low oil pressure.
System short of capacity	Too low oil pressure.	Oil pressure. Adjust the oil pressure valve as the oil pressure is not sufficient to activate the unloading mechanism. See table of oil pressures.
	Defective oil pump.	The oil pump does not give sufficient pressure to activate the unloading mechanism. Repair.
	Defective capacity regulator.	Repair. The reason is often refrigerant in the oil.
	Insufficient oil charge.	Low oil pressure. Recharge.
	Compressor working at too low suction pressure.	See section on low suction pressure.
	Limitation of refrigerant supply to evaporator, owing to: • too small expansion valve • too high superheat	Locate fault.

Cont'd

Trouble observed	*Cause of defects*	*Action to be taken*
	• dirt in filters • too low condenser pressure.	
	Iced-up evaporator.	Defrost evaporator – adjust the defrosting time.
Liquid knock in compressor during start-up	Absorption of refrigerant in the oil. A sudden reduction of the pressure over the oil sump (the suction pressure) causes foaming.	Heating element in crankcase should be switched on 6–8 hours before start-up to get the dissolved refrigerant boiled out before start-up.
	The suction line has free fall to the compressor.	Start with throttled suction stop valve – stop at knocking sound. Liquid trap should be installed.
	Overflow of refrigerant from evaporator, owing to: • expansion valve too large • not enough superheat.	See adjustment of expansion valves: • change to smaller orifice • adjust superheat, • change valve orifice.
	Liquid regulator has too high capacity.	Check the valve types if necessary. Change valve.
	Wrong placement of expansion valve sensor.	See section on thermostatic expansion valve.
	Superheat of expansion valve set too low.	Adjust the superheat – normally it should be between 5 and 8°C.
	Refrigerant gas in the liquid line.	The expansion valve is hunting.

Cont'd

Trouble observed	*Cause of defects*	*Action to be taken*
Condensing pressure too high	Inadequate condenser cooling, for example fan has stopped or impurities are present.	Regulate the air supply, or possibly reduce the compressor capacity.
	Presence of non-condensable gases (especially air) in the condenser.	Purge the air from the condenser.
	Overcharge of refrigerant.	The refrigerant is backing-up in the condenser, thus reducing the effective area of the condenser. Remove refrigerant.
Condensing pressure too low	The compressor is short of capacity.	Check that the compressor capacity corresponds to the loading of the plant. Reduce condenser cooling.
	Defective or leaky discharge valves.	See compressor instructions. Check valve plates and piston rings.
	Too high condenser capacity.	Regulate the condenser cooling.
	By-pass between high-pressure side and suction side of compressor.	Check the system for possibly by-pass – can be detected as a warm spot.
	Defective piston rings or worn cylinders.	Replace worn parts. See compressor instruction manual.
Suction pressure too high	The compressor is short of capacity.	Regulate the compressor capacity. Check that all cylinders are working. Check function of capacity regulator.

Cont'd

Trouble observed	*Cause of defects*	*Action to be taken*
	Leaky suction valves.	See compressor instruction. Remove cylinder covers, check valve plates and piston rings, renew them if necessary.
	Open by-pass between suction side and high-pressure side of compressor. Leaky safety valve, or the safety valve opens too early.	Check the system for possible by-pass. Can be detected as a warm spot. Adjust valves – repair leaky valves.
	Wrong adjustment of the liquid-regulating valve.	Liquefied refrigerant in the suction line. Adjust the regulating (expansion) valve – repair or replace.
Suction pressure too low	Compressor capacity too high.	Reduce the compressor capacity. Check system for capacity regulation.
	Lack of refrigerant. Bubbles in sight glass and possibly, warm liquid line.	Check the refrigerant charge. Recharge if necessary. Locate the leak and repair.
	Oil in evaporator.	Drain the oil.
	Clogged filter in liquid line.	Investigate and clean the filter.
	Too high superheat of the suction gas.	Regulate the expansion valves so more liquid can pass.
	Frosted expansion device.	Thaw the expansion device by applying hot wet cloth, and let the liquid from the receiver pass through dehydrator. Change the desiccant.

Cont'd

Trouble observed	*Cause of defects*	*Action to be taken*
	Expansion valve has lost charge.	The valve does not open – replace the valve.
	Solenoid valve in liquid line does not open.	Coil may be burned out – replace coil.
	Expansion valve too small.	Change to correct size.
Discharge temperature too high	Suspend temperature too high owing to reduced refrigerant supply to evaporator (high superheat): • low refrigerant charge • clogged filters • incorrectly adjusted liquid regulating valves.	Check the charge. Clean the filters. Check the thermal expansion valves.
	Leaky discharge valves.	Leak in the discharge valves causes heat generation. Replace defect valves.
	The compressor is working at a too high compression ratio.	Check condenser cooling and adjust the suction pressure to pre-determine operation conditions. If necessary, mount system for cooling of compressor.
	Open by-pass between high and low pressure side of compressor, for example, leaky safety valve.	Locate the by-pass and remedy possible leak.
	Bad insulation of suction pipe.	Check superheat at suction.

Cont'd

Trouble observed	*Cause of defects*	*Action to be taken*
Discharge temperature too low	Low suction temperature owing to overflow of refrigerant liquid from the evaporator.	Adjust the liquid-regulating valve. Increase the superheat.

13.3.2 Oil

Trouble observed	*Cause of fault*	*Action to be taken*
Oil level in crankcase falling	Clogged filter in solenoid valve or nozzle in oil return line.	The oil return line should be warm during operation. Clean the filter.
	Defective solenoid valve in oil return line.	Coil in solenoid valve burned out – replace coil.
	At the first start-up some oil will find its way into the system.	Especially in plants working with fluorinated refrigerants, part of the oil will be circulating in the system. When balance is achieved in the plant, recharge with new oil if necessary.
	Leaky piston rings or worn cylinder.	Replace the piston rings, and perhaps replace pistons and cylinder liner.
	Liquid in suction line.	Check if the liquid supply to the evaporator is correct.
	Leaky suction valves.	Replace valve plates.
Oil in crankcase foaming heavily	Adsorption of refrigerant in the oil.	

Cont'd

Trouble observed	*Cause of defects*	*Action to be taken*
	Capacity of compressor too large during the cooling-down period.	If liquid is passing through the suction line into the crankcase, this may cause foaming of the oil. Reduce the compressor capacity.
Oil pressure too low	Refrigerant in the oil.	
	Defective oil pump.	The oil pump does not give sufficient oil to create the necessary pressure to activate the unloading pistons.
	Oil pump pressure set too low.	The oil pressure does not rise high enough to move unloading pistons. Adjust for correct oil pressure. See table for adjustment pressures.
	The compressor has lost oil.	Lack of oil in the compressor may cause too low an oil pressure for moving the unloading pistons. Add the necessary amount of oil. Check the reason for the fault.
Sweating or frosted crankcase	Liquid in suction line.	Adjust the expansion valves. See entry on suction in section 13.3.3 for details of thermal expansion valves.
	Capacity too large during the cooling-down period.	Too large capacity during cooling-down may cause liquid to

Cont'd

Trouble observed	*Cause of defects*	*Action to be taken*
		enter the suction line to the compressor. Run the machine with reduced capacity to prevent liquid entering the suction line.
	Liquid-regulating valve or float valve giving too much liquid.	Reduce the superheat at thermal expansion valve.
	Expansion valve bulb not correctly located.	See section 13.3.3 on expansion valves.
Shortage of refrigerant charge	Abnormally warm liquid lines: when the liquid level in condenser/receiver falls below the outlet for the liquid line, this will be filled with a mixture of liquid and warm gas.	Each of these symptoms may be due to other circumstances; if they occur at the same time, however, the system should be checked for leaks and charged with refrigerant as necessary.
	Bubbles in sight glass in liquid line: the gas/liquid mixture is visible in the sight glass. Heavy bubbling indicates need of adding refrigerant. The gas mixture in the liquid pipe will often result in a hissing noise in the expansion valve. Low suction pressure: the temperature in the cooling system at cooling surface and in	

Cont'd

Trouble observed	*Cause of defects*	*Action to be taken*
	suction line is comparatively high in relation to the suction pressure observed, because of reduced supply of refrigerant to the evaporator. Reduced capacity of the plant. Low condenser pressure – reduce the condenser cooling. If the condenser pressure does not rise rapidly, this indicates shortage of refrigerant.	

13.3.3 Expansion Valve

Trouble observed	*Cause of fault*	*Action to be taken*
Low suction pressure – high superheat	The pressure-drop over the expansion valve may be too low as a consequence of: • low condensing temperature • exceptionally high vertical lift from receiver to evaporator • undersized liquid line.	Low condenser pressure.
	The valve capacity is too low because of refrigerant vapour in the liquid line as a consequence of:	Clean the filters. Adjust condenser cooling. Check refrigerant charge.

Cont'd

Trouble observed	*Cause of defects*	*Action to be taken*
	• pressure loss in the liquid line – filter plugged • failing subcooling of the refrigerant liquid • insufficient refrigerant charge.	
	The thermal expansion valve does not function, because of faulty external pressure equalization.	Check the external pressure-equalization line concerned.
	Ice, wax, oil or other impurities clogging the expansion valve. Ice formation or paraffin at the regulating cone will often be characterized by a rapid increase of the suction pressure, after a stop period during which the system was heated suddenly.	Heat the thermal expansion valve by means of warm, wet cloth and pass the liquid from the receiver through a drying filter. **Warning**: Never add methanol to the system to avoid freezing, as it may cause conditions for corrosion and chemical attack in compressors, etc.
	Leaky expansion valve capillary tube or element, or the charge is lost.	Where charge is lost, the valve closes. Check the element as follows: (a) stop the compressor, (b) place the bulb in ice and water, (c) start the compressor, (d) then heat the bulb.

Cont'd

Trouble observed	*Cause of defects*	*Action to be taken*
		Check at the same time if a rapid change of the temperature in the suction line takes place, indicating overflow of refrigerant liquid. If this is confirmed, the expansion valve is able to function.
	Wrong bulb position, or the superheat for the expansion valve is set too high.	The bulb should be positioned correctly. Adjust the superheat.
Low suction pressure – low superheat	Poor distribution of refrigerant in evaporator. Liquid is allowed to affect the expansion valve before the evaporator receives sufficient refrigerant – which means poor utilization of the evaporator capacity.	If possible, mount a liquid distributor. Check that the liquid distributor is mounted vertically.
	Poor air distribution over the cooling surface causes unequal liquid distribution.	The air distribution should be uniform all over the cross-section of the cooler.
	Evaporator filled with refrigeration machine oil.	After drainage of the oil, oil separator and device for oil return should be checked.
High suction pressure – high superheat	The compressor is too small in relation to the capacity under the actual conditions. The evaporator is too big.	Check and adjust the operating conditions.

Cont'd

Trouble observed	*Cause of defects*	*Action to be taken*
High suction	Compressor undersized.	Increase the capacity.
	Expansion valve superheat setting too low.	Adjust the superheat setting.
	Vapour in liquid line and oversized expansion valve.	Shortage of refrigerant.
	Expansion valve nozzle leaks or stays in open position.	Change valve nozzle after valve.
	Exterior equalization not connected or badly connected.	Connect equalization.
Fluctuating suction pressure	The evaporator is hunting.	Check the superheat.
	The bulb inadequately placed.	Check bulb location and equalizing pipe.
	Liquefied refrigerant passing into the suction line, that is, as a consequence of inadequate liquid distribution in the evaporator.	

References: P. C. Koelet, *Code Van Goede Praktijk/Code de bonne practique*, UBF/BVF, Belgium: Sabroe Refrigeration A/S Product Division, Høbjerg, Denmark.

13.4 Standards

Whenever a refrigeration plant is being designed, built or maintained it is essential to observe the strictest safety standards. In view of the potential impact of refrigeration plants, especially the CFCs, upon the environment, governments will increasingly impose stricter laws to ensure compliance with the necessary standards. In many countries standards are already in existence, but soon there will be two international standards – the ISO standard and the CEN standard.

The ISO standard is in the draft stage and is titled *Draft international standard ISO/DIS 5149 Mechanical refrigeration systems used for cooling and heating. Safety standards.*

The European standard is being prepared by the European Communities for standardisation CEN and is titled *EN 378. Refrigeration systems, safety and environment requirements.* This will be ready in 1992.

At the time of writing many European countries, such as the United Kingdom, France, Germany, Sweden, Norway, Belgium and The Netherlands all have their own standards. Once the CEN standard comes into force, all the members of the EC will be obliged to adopt the CEN standard or bring their own standards into line.

It is relevant to append here a list of some of the standards in present use in the USA.

American National Standards
ASME Code for Pressure Piping B31

ANSI/ASME B31.5–1983, and addenda of 1985 and 1986, Refrigeration piping
ANSI/ASME B31.5–1987 Refrigeration piping

ASHRAE Standards
ANSI/ASHRAE 41.4–1984 Standard method for measurement of proportion of oil in liquid refrigerant

ANSI/ASHRAE – 83–1985 Capacity measurement of field-erected compression type refrigeration and air conditioning systems

ANSI/ASHRAE 15–1989 Safety code for mechanical refrigeration. New draft March 1991

ASHRAE Guideline 3–1990 Reduced emission of fully halogenated chlorofluorocarbon (CFC) refrigerants in refrigeration and air-conditioning equipment and applications

ARI Standards
ANSI/ARI 495–79 Refrigerant liquid receivers

International Institute of Ammonia Refrigeration Standards
ANSI/IIAR – 2–1984 Equipment, design and installation of ammonia refrigeration systems

IIAR: Ammonia mechanical
(a) Refrigeration systems

Safety practice
First aid

(b) Start-up, inspection and maintenance of ammonia mechanical refrigeration systems (Bulletin no. 110–1983)

IIAR: A guide to good practice for the operation of an ammonia refrigeration system 1983.

Epilogue

This last chapter contains only superficial information on the basics of refrigeration maintenance. The reader is reminded how important it is that regular preventive maintenance is carried out by skilled and licensed technicians. This includes the opening of compressors for inspection. For reasons of both safety and the protection of the environment, plants should only be worked on by appropriately qualified and approved personnel when regular general tests, and those concerning leak-tightness and mechanical strength, are being carried out. In many countries, either laws covering this field apply or at least good codes of practice are in use.

It must never be forgotten that a refrigeration installation contains a chemical substance under a relatively high pressure, a situation that can be a potential threat to people and the environment, whether the refrigerant is ammonia or a CFC.

The designer must always be aware of the importance of the refrigeration system's reliability, whether it be a link in a production chain, storing food or conserving precious materials.

For a final word, we quote L. Mattarolo of Padua University, Italy, who said at the IIoR Congress in Dresden in September 1990: "In order to ensure the nutrition of the growing world population in the future the following factors must be considered:

1. increase of food production
2. measures to reduce spoilage
3. improvement of uneven distribution of food among different regions.

Refrigeration and food processing show their value not only by avoiding or reducing losses but also by employing systems and structures to preserve foods and maintain higher nutritive values. It is well-known that losses in industrialized countries represent up to 30% of the harvest. In developing countries this figure often reaches 50%.

Only some 20% of all perishable food is presently being treated by refrigeration. This leaves 80% of the 1.7×10^9 tonnes still to be tackled. A bright future and a noble task for everyone engaged in refrigeration."

References

List 1 contains references to works referred to in the text. List 2 contains all firms, books and papers from which information was taken (with the permission of the authors and the companies) for use in this book, insofar as they have not already been quoted in the references given at the end of individual chapters.

Reference List 1

Verein Deutsche Ingenieure, *VDI Wärm Atlas*, 5th edition, 1988, and the editions of 1973 and 1974, Düsseldorf, Germany.
W. Frost, *Heat Transfer at Low Temperatures*, Plenum Press, New York, 1975.
Data books on Heat Transfer and Fluid Flow, General Electric Corporate Research and Development, Schenectady, NY (USA).
Robert H. Perry and Don Green, Perry's Chemical Engineers Handbook, 6th edition, McGraw-Hill, New York.
John van de Vechte, *Feed-back Control Systems*, 2nd edition, Prentice-Hall, Englewood Cliffs, NJ (USA).

Reference List 2

Alfa Laval N. V., Brussels, Belgium (Plate evaporator).
Baltimore Air Coil, Heyst o/d Berg, Belgium (Evaporative condensors).
Broderne Gramm, Vojens, Denmark (Ice-cream freezer).
Carrier Corporation Manual, New York, USA (Psychrometric chart, 1975).
Centrale School voor Tuinbouw Techniek en Technologie, C. M. Iedema, Ede, The Netherlands, 'koeltechniek 1981'.
Code van goede praktijk/code de bonne practique, P. C. Koelet, UBF/BVF, Belgium.
Danfoss, Nordborg, Denmark (Controls).
Danish Meat Research Institute, Roskilde, Denmark.
Deutsche Kälte Verein (DKV) Arbeitsblätter, Verlag C. F. Müller, Karlsruhe, Germany.

Delair, Etten Leur, The Netherlands (Psychrometric graph, humid air).
Du Pont de Nemours International S. A., Freon Product Division, Genève, Switzerland (Pipeline graphs).
Fermod, Paris, France (Equalizing valves).
Guide Pratique de l'Isolation Frigorifique, G. Ballot, Pyc Edition, Paris.
Handboek voor koudetechniek, The Netherlands, Uitgeverÿ P. C. Noordervliet, Zeist, The Netherlands.
Het grote Handboek Isolatie, Uitgeverij L. A. van Beek BV, The Netherlands.
IARR, Chicago, USA.
Industrie Micro-Ondes International, Epone, France.
International Institute of Refrigeration IIF/IIR, 177 bl. Malesherbes F 75017, Paris (h/log p graphs).
Koeltechniek en Klimaat Regeling, Uitgeverÿ P. C. Noordervliet Zeist, The Netherlands (Leergang ontwerpen van koelinstallaties, 1977).
Schroef compressoren in de koeltechniek, P. C. Koelet, 1962–1969.
Kältebehandlung Schnell Verderblicher Lebensmittel, Dr. Ing. E. Emblik, Brücke Verlag, Germany.
Polyko Eeymoutiers, France (Equalizing valves).
Rijks Universiteit Gent, Belgium, Labo voor Levensmiddeltechnologie, Prof. M. Feys and Prof. P. T. Tobback.
Sabroe Refrigeration, A/S Product Division, Høbjerg, Denmark.
Solvay, Central Laboratory, Brussels, Belgium.
Sprenger Instituut, Wageningen, The Netherlands.
Storflex, Aaesch, Switzerland (Mobile racks).
VDI (Verein Deutsche Ingenieure), *Wärme Atlas*, 1973, 1974 and 5th edition, 1988, Düsseldorf, Germany.
Ziegra, 3004 Isernhagen, Germany (Flake ice machines).

Bibliography

ASHRAE 1990 Refrigeration Handbook, American Society of Heating, Refrigerating and Air-conditioning Engineers, 1791 Tulli Circle, Atlanta, GA 30329, USA.

Bulletin of the International Refrigeration Institute, 177 Boulevard Malherbes, F75017, Paris, France.

CIBSE Guide 1986, Delta House, 222 Balham High Road, London SW12 9BS.

Institute of Refrigeration, Kelvin House, 76 Mill Lane, Carshalton, Surrey SM5 2JS.

International Institute of Ammonia Refrigeration (IIAR) publications, One Illinois Center, 111 East Wacken Drive, Chicago, Illinois 60601, USA.

Glossary

Air-cooled condenser A condenser that includes means for forcing air circulation over the external surface of the condenser coil for the heat removal necessary to liquefy refrigerant vapour on the inside of the tubes.
Air-cooled desuperheater Part of the system designed to cool the ammonia refrigerant vapour after it is discharged from the compressor and before it enters the condenser. It is provided with a means of forcing air circulation over the external surface of the desuperheater coil for the heat removal necessary to cool the refrigerant vapour on the inside of the tubes. It does not include desuperheaters that are integral components of condensers.
Brine Any liquid used for the transmission of heat without a change in its state.
Check valve A control device which permits fluid flow through the device in one direction, but prevents return of the fluid in the opposite direction.
Coil or **grid** A part of the refrigeration system constructed from bent or straight pipes or tubes suitably connected and serving as a heat exchanger (evaporator or condenser).
Compressor A specific machine, with or without accessories, for compressing refrigerant vapour. A booster compressor is a compressor, with or without accessories, for compressing refrigerant vapour and discharging to the suction system of a higher stage compressor.
Condenser That part of the system designed to liquefy refrigerant vapour by removal of heat.
Container A cylinder for the transportation of ammonia refrigerant.
Evaporative condenser A condenser that obtains cooling effect through evaporation of water in an air stream on the external surface of the tubes for the heat removal necessary to liquefy refrigerant vapour on the inside of the tubes.
Evaporator That part of the system designed to vaporize liquid refrigerant to produce refrigeration effect.
Evaporator coil That part of an evaporator constructed of pipe or tubing other than a shell and tube type.
Evaporator pressure regulator A controlling device which regulates the flow of primarily gaseous refrigerants from an evaporator section of the

system into a lower pressure section, and which is actuated toward open by a pressure above set-point ahead of the valve.

Gauge pressure The difference between the absolute pressure in the system and the atmospheric pressure at the site. Note that atmospheric pressure is generally assumed to be 1.013 bar.

Header A pipe or tube component of a refrigerating system to which several pipes or tubes are connected.

Heat pump A refrigerating system used for the purpose of heating.

Heat-transferring liquid Any liquid used for the transmission of heat without any change in its liquid state.

High side Those parts of a refrigerating system subjected to condenser pressure.

High-side float valve A controlling device which regulates the flow of volatile liquid refrigerant from a higher pressure section of the system into a lower pressure section, and which is actuated toward open by a rising liquid level upstream of the valve.

Hot-gas by-pass regulator A controlling device which regulates the flow of a refrigerant hot gas through the device from a higher pressure section of the system to a lower pressure section of the system, and which is actuated toward open by a pressure falling below regulator set-point downstream of the regulator orifice.

Isolating valves Valves arranged in a way to isolate sections of systems for the purposes of assembly or maintenance while other sections remain charged with refrigerant.

Liquid receiver A pressure vessel permanently connected to a refrigerating system by inlet and outlet pipes for storage of liquid refrigerant.

Low side: Those parts of a refrigerating system subjected to evaporator pressure.

Low-side float valve A controlling device which regulates the flow of volatile liquid refrigerant into an evaporator pressure section of the system from a higher pressure section, and which is actuated toward closed by a rising liquid level downstream of the valve.

Mechanical refrigeration system A combination of interconnected refrigerant-containing parts constituting one closed refrigerant circuit in which a refrigerant is circulated for the purpose of extracting heat, and in which a compressor is used for compressing refrigerant vapour.

Oil drain An arrangement in a refrigerating system whereby lubricating oil carried from the compressor may be drained from the system in the machinery room or elsewhere.

Pilot-operated valve The valve which regulates the flow in response to a signal from a pilot.

Piping The pipe or tube mains for interconnecting the various parts of a refrigerating system. Piping includes pipe, flanges, bolting, gaskets, valves, fittings, the pressure-containing parts of other components such as expansion joints, strainers, and devices which serve such purposes as mixing, separating, snubbing, distributing, metering or controlling flow, pipe-supporting fixtures and structural attachments.

Pressure relief valve A pressure-actuated valve held closed by a spring or other means and designed automatically to relieve pressure in excess of its setting; also called a *safety valve*.

Refrigerant A substance used to produce refrigeration by its expansion or vaporization.

Refrigerant pressure-activated condenser water regulator A device which regulates the flow of cooling water through the device to or from a water-cooled condenser and which is actuated toward open by refrigerant high-side pressure rising above the regulator set-point.

Refrigerant pump A mechanical device for moving liquid refrigerant within a closed circuit mechanical refrigeration system.

Refrigerating system A combination of interconnected, refrigerant-containing parts constituting one closed refrigerant circuit in which a refrigerant is circulated for the purpose of extracting and rejecting heat.

Saturation pressure of a refrigerant The pressure at which there is stable co-existence of the vapour and liquid or the vapour and solid phase.

Shell and tube condenser A type of condenser where several tubes in a bundle with each end secured in a tube sheet are enclosed in a shell which contains a refrigerant. A shell and tube condenser with refrigerant in the shell is a *pressure vessel*.

Shell and tube evaporator A type of evaporator where tubes or coils are enclosed in a shell. The flooded type is one with the evaporation refrigerant in the shell. The direct expansion type is one with the evaporating refrigerant in the tubes or coils.

Solenoid valve A valve which is opened or closed by the magnetic action of an electrically energized coil. The opposite action is accomplished by gravity, pressure or spring action.

Stop valve A device to shut-off the flow.

Surge drum or **liquid separator** A vessel containing liquid refrigerant at low pressure and temperature and connected by liquid feed and vapour return pipes to an evaporator.

Thermostatic expansion valve A controlling device which regulates the flow of the volatile refrigerant into an evaporator of a refrigeration system, and which is actuated by changes in evaporator pressure and superheat of the refrigerant gas leaving the evaporator. The basic aim is to superheat.

Index